W9-CSO-817

Exploration Geophysics
of the
Shallow Subsurface

Macintosh is a registered trademark of Apple Computer, Inc.

Finder and MultiFinder are trademarks of Apple Computer, Inc.

Microsoft is a registered trademark of Microsoft Corporation.

Warranty Information

EXPLORATION GEOPHYSICS OF THE SHALLOW SUBSURFACE

H. Robert Burger
Smith College

Accompanying Macintosh Computer Software
by
Douglas C. Burger
and
H. Robert Burger

Prentice Hall P T R
Englewood Ciffs, New Jersey 07632

Library of Congress Cataloging-in-Publication Data

BURGER, HENRY ROBERT (date)
    Exploration geophysics of the shallow subsurface / by H. Robert Burger;
with accompanying Macintosh computer software by Douglas C. Burger
and H. Robert Burger.
      p.   cm.
    Includes bibliographical references and index.
    ISBN 0-13-296773-1
    1. Prospecting—Geophysical methods.   2. Prospecting—Geophysical
methods—Computer programs.   I. Burger, Douglas C.   II. Title.
TN269.B86 1992
622′. 15 —dc20   91-30335
                                             CIP

Editorial/production supervisor and
   interior design: *Kathryn Gollin Marshak*
Cover design: *Joe DiDomenico*
Manufacturing buyer: *Susan Brunke*
Prepress buyer: *Mary Elizabeth McCartney*
Acquisitions editor: *Michael Hays*

© 1992 by Prentice Hall P T R
Prentice-Hall, Inc.
A Simon & Schuster Company
Englewood Cliffs, New Jersey 07632

Printed in the United States of America

10 9 8 7 6

ISBN   0-13-296773-1

Prentice-Hall International (UK) Limited, *London*
Prentice-Hall of Australia Pty. Limited, *Sydney*
Prentice-Hall Canada Inc., *Toronto*
Prentice-Hall Hispanoamerica, S.A., *Mexico*
Prentice-Hall of India Private Limited, *New Delhi*
Prentice-Hall of Japan, Inc., *Tokyo*
Simon & Schuster Asia Pte. Ltd., *Singapore*
Editora Prentice-Hall do Brasil, Ltda., *Rio de Janeiro*

To my parents, Emily and Henry, who sacrificed so that I might have the opportunity;
to my professor, Judson Mead, who gave me the tools and the philosophy;
to my wife, Ann, who convinced me I could do it and supported me during the process;
and to my sons, Doug and Bob, who by being great kids gave me the pride
and the peace of mind so necessary to see this project through to completion.

Contents

Contents

4 *SEISMIC EXPLORATION: THE REFLECTION METHOD*   *134*

5   *ELECTRICAL RESISTIVITY*   *241*

Preface

This book is based on an exploration geophysics course that I have taught for 15 years. The level is directed at advanced undergraduates who typically are majoring in geology. However, this text also is appropriate for geology graduate students and students majoring in other disciplines (hydrology, engineering, archaeology) who require some knowledge of geophysical methods.

Because of the increasing interest in the shallow subsurface, I focus on methods and approaches that are appropriate for the shallow target. Examples, illustrations, applications, and problem sets all emphasize shallow exploration, as contrasted with the more typical geophysics text which contains material much more directed toward targets at depth.

The organization of this book follows, in a general way, the relative utility of various geophysical methods for exploring the shallow subsurface. Seismic methods are discussed first. After a brief chapter devoted to some basics (propagation of elastic waves, instrumentation, etc.), refraction seismology is presented, as this still is the most common method employed for shallow work. The longest chapter in the book concentrates on reflection seismology. This technique is just beginning to find applications in shallow work, holds great potential, and its shallow applications are not covered in existing texts. Electrical resistivity methods follow seismic methods, as these often are used in conjunction with seismic work for investigations at shallow depths. The next to last chapter in the text is concerned with gravity. Although not as commonly used for shallow surveys, gravity, especially microgravity investigations, has quite a lot to offer in obtaining much useful information, and applications to shallow as well as deep surveys are elucidated, compared, and contrasted. Magnetic surveying is discussed last, as this is the least common method currently used for shallow investigations other than for metals exploration and locating buried iron and steel objects.

Each geophysical method is introduced by presenting the basic theory and considerations necessary for understanding its application in exploring a given target. Although the fundamental equations for each method are always developed in detail, mathematical sophistication on the part of the student is not required. Calculus is used sparingly but is included where necessary. Each chapter includes a brief discussion of the instruments used in field surveys. This discussion is intended to convey the most fundamental principles of design in order to support discussions of field procedures and data and interpretation limitations. Data collection and reduction procedures are discussed in sufficient detail to enable the student to plan well-conceived field surveys. Approaches and techniques for analyzing field data follow the section on field operations. This discussion typically covers the classic methods but also attempts to pay sufficient attention to the inherent uncertainties in a given method and some common pitfalls. The fact that uncertainties in interpretation are frequently encountered is considered by discussing the goals of the field survey, the accuracy of results likely with a given method, and the time available in the field to gather data. The final material presented for each exploration method focuses on applications and illustrates data and interpretations. Problem sets accompany each chapter.

Because geophysics exploration depends heavily on computer processing and analysis, students must be introduced to hands-on experience as soon as possible. Accompanying the written text is a series of computer programs, written for the Macintosh®, that perform direct and inverse modeling and support all the major exploration methods covered in the book. These programs are true Macintosh applications, conform to the Macintosh user-interface guidelines, and are as similar to one another in operation as possible. This means that they are easy to learn to use and that data input is quick and flexible. Maximum use is made of the graphics capability of the Macintosh and of the mouse and menu items for data input and modification. Students can access these programs to explore concepts discussed in the text (exercises occasionally are suggested in box inserts) as well as to work on many of the problems in the problem sets. A total of nine programs are available: three for reflection seismology (one for modeling, two for interpretation); two for refraction seismology (one for modeling, one for interpretation); two for electrical resistivity (one for modeling, one for interpretation); one for gravity, and one for magnetics.

The programs are standard routines that typically are used in geophysics courses. They are not designed as research tools for a geophysicist and should not be assumed to be so. However, their capabilities are more than adequate for the persons using this text and will support the field surveys that commonly accompany a course of this sort. The strength of the programs is their ease of use and uniformity of design.

Because of the utility and graphics capability of these computer programs, fundamental concepts are often illustrated and supported with examples created by them. Therefore, students can reproduce the same diagrams used in the text to elucidate concepts and can test variations in starting assumptions as their curiosity dictates. For instance, discussion of the hidden layer in refraction surveying or of the optimum offset in reflection surveying are accompanied by suggestions for "discovering" these relationships through experimentation with the software. Problem sets also emphasize program use and include samples of real field data for reduction and analysis. It is important to realize, however, that the text was written to stand alone and does not depend on using these programs. Rather, the programs add an important added dimension that is missing from most existing texts.

Almost all the tables in the book are produced from spreadsheet templates and are referred to as *dynamic tables*. These Microsoft® Excel templates are also on the disks accompanying the text. As a student works through text material, she or he can access the template, change values in certain cells, and study the effect. Once again, the tables stand on their own, as the data they contain help to support concepts or relationships discussed in the text. Most table data also are presented in diagram form as graphs. However, these tables support additional learning activities if an Excel program and a Macintosh computer are available to the student.

My major objective in designing my own course and this book was to cover the most fundamental principles of the common methods of exploration geophysics while making it possible for a student to interact with and explore these ideas. Such hands-on experience has proved to be successful both in learning and retention of basic concepts, but, even more important, it has fostered an atmosphere in which students not only ask questions but often are able to seek their own answers by utilizing the computer templates and programs. Creating an environment for such independent investigation is my major goal, and I hope this text will facilitate independent inquiry.

H. Robert Burger

Preface
to Computer Materials

The Macintosh applications contained on the disks included with this book were developed to allow a student to perform direct and inverse modeling for each geophysical method covered in the text. Most are familiar, as they are standard programs used in many geophysics courses. These programs are designed for student use in order to aid comprehension of the topics presented in the text and to work with field data collected during field exercises typically accompanying a course in exploration geophysics. In order to keep the programs easy to use (and to keep development time within reasonable bounds) these programs are not intended to substitute for the full-featured programs necesssary for research applications. Instructions for using these applications are in Appendices A–E.

The programs included execute on any Macintosh computer equivalent or superior to the Macintosh Plus. The programs have been tested on a Macintosh Plus, Macintosh SE and SE30, and Macintosh II computers and are MultiFinder compatible (but see the following). Although the programs require at least System 6.0 to run, they were tested with System 6.0.4 and Finder 6.1.4. To be safe, do not use older versions of the operating system. The programs will run under MultiFinder but occasionally quit if several applications are open and several switching operations take place. The programs do not check for a "disk full" error when saving data or model files. Because the disks supplied with the text are almost full, you should copy some folders to additional disks to create free space. At any rate be sure to make copies of these disks before using them. To be absolutely safe either save data and models frequently if you will be switching to other applications or do not have other applications open when running these programs. These programs have been thoroughly tested and are believed to be essentially bug-free. Calculations agree with those from other programs using the same data and, therefore, are believed to be accurate. However, these programs are included with this text as supplementary material and should be treated as such. Bugs and errors undoubtedly still exist. We hope these will be reported directly to us so we can correct them in future versions.

The Excel templates that also are included on disks were created with Microsoft Excel Version 2.2, which is required to access the templates. Templates are included for almost all tables in the text. Cell items in boldface in the tables are meant to be changed by the user. Results of calculations should normally be graphed using the charting capabilities of Excel or by transferring these values to a special graphing program.

We hope both the applications and templates will prove valuable in your exploration geophysics course. They constitute a rich teaching and learning resource that should be utilized if at all possible. We wish to thank the W. M. Keck Foundation of Los Angeles for providing the support that made development of the Macintosh applications possible.

H. Robert Burger
Douglas C. Burger

1

Approaching
the Subsurface

What Are the Options?

Presumably you are reading this book because you want to learn about exploring the subsurface. Your interest may be an active one in that you actually intend to go into the field and collect information, or you may be interested only in learning how subsurface information is gathered so that you will be better equipped to judge recommendations for action based on field surveys. In either case you probably intend to lend your talents to improving humanity's lot by identifying sources for much-needed resources, protecting water supplies, providing safe building sites, locating areas for safe disposal of certain wastes, or extending knowledge of our forebears by finding places where they lived and worked. We hope this book will provide you with some of the tools to accomplish your objectives and that you will be stimulated to continue with your quest.

This first chapter is not presented as a survey of what is to come. Rather, we mention a few important points about the practice of exploration geophysics that we hope you continually will keep in mind as you work through the pages ahead. If you continue to refer to these points frequently as you cover new topics, you likely will emerge as a more-reasoned observer and/or practitioner of the discipline.

Various approaches are available to gather information about the subsurface. The best is direct observation of the sediments and rocks themselves. Of course, this is rarely possible to the extent that we would like. It's not often that numerous, deep valleys cut through an area of interest to the necessary depths so that all we need to do is correlate geologic sections from one position to another. It also is relatively unusual for an investigation site to contain numerous wells that were logged by a professional geologist and are uniformly distributed. More commonly when subsurface information is necessary, we acquire physical measurements on the surface and use these to deduce subsurface geology. The value of geophysics is its ability to acquire information about the subsurface over a substantial area in a reasonable time frame and in a cost-effective manner.

Geophysics essentially is the measurement of contrasts in the physical properties of materials beneath the surface of the earth and the attempt to deduce the nature and distribution of the materials responsible for these observations. Variations in elastic moduli and density cause seismic waves to travel at different speeds through different materials. By timing the arrivals of these waves at surface observation points, we can deduce a great deal about the nature and distribution of subsurface bodies. Variations in the electrical conductivity of rocks and sediments produce varying values of apparent resistivities as the distance between measuring probes is increased or as the position of the probes is changed on the surface. Density variations in the subsurface lead to variations in gravitational acceleration at surface instrument stations, and variations in magnetic susceptibilities produce measurable differences in the magnetic field at field observation sites. These variations in physical properties must be sufficiently large so that their effects can be determined by our instruments. A frequent problem is that insufficient contrasts exist to detect the subsurface target of interest. Other times the presence of nearby bodies of great contrasts create effects that mask those created by our target.

To the uninformed it seems almost magical that these contrasts in physical properties permit the explorationist to map in detail bedrock topography that is present at depths of perhaps 150 m. Equally impressive is our ability to extract the form and distribution of alluvial sand and gravel deposits or to trace the course of a buried river channel. Perhaps you will find the stone walls of an ancient city or determine the distribution of contaminated waters flowing from a waste disposal site. Maybe you will find a placer deposit or develop a new technique to interpret electrical resistivity data. Or your contribution may be in defining guidelines for groundwater use based on the results of geophysical surveys. Whatever awaits you in the future, if you continue to deal with matters depending upon or affected by subsurface geology, we hope this text played a small part in helping you to make informed decisions.

Some Fundamental Considerations

Throughout the chapters that follow we tend to treat the earth and the subsurface as an ideal subject for study. Subsurface geology often is visualized as constituted by bodies of constant thickness with perfectly planar contacts. Dipping beds are only infrequently pictured, and even then dips rarely depart from those of constant inclination. Lateral variations in physical properties are mentioned as something to be aware of but are rarely treated in as much detail as homogeneous bodies. Lateral variations also always seem to be pictured as abrupt and always with vertical boundaries—gradational, inclined contacts are not the rule in geophysics texts. Every geologic unit pictured always has a sufficient physical-property contrast to enable it to be differentiated from all adjacent units. The earth's surface is almost always horizontal, and differences in materials at the surface are nonexistent. Finally, vibrations from wind and traffic or currents induced by electrical transmission lines are never illustrated.

Of course, the real world departs significantly from the ideal of the next several chapters. We must make certain assumptions in order to deduce the effects produced by the variations in physical properties that we have mentioned. We also take pains to point out how the nonideal subsurface affects our assumptions and results. On occasion definite techniques are developed to deal with these departures from uniformity. However, there is no question that the parade of diagrams depicting planar contacts and homogeneous strata creates a subtle mental trap. If you are not careful, you will tend to focus on the world of the text and will try to force your observations into its orderly geology. Try to avoid this trap. Always try to determine what information about nonuniformity you can extract from your observations rather than trying to decide how far you can stretch the assumption of homogeneity to enable interpretation of less-than-ideal data.

Defining Objectives

As you begin your work in geophysics, your natural excitement and curiosity will quite naturally make you impatient to go into the field and begin to accumulate observations for analysis. Without sufficient exposure to proper exploration techniques, it is only too easy to go somewhere close by, lay out some seismic cable, and begin taking measurements. If this approach is followed too often, it will develop an undesirable operational mode. Always have a specific survey objective in mind and define it as explicitly as possible. This should be true even as you are learning a specific method such as electrical resistivity profiling. Even the most standard elements of survey procedure will vary somewhat depending upon what subsurface information is of primary importance. For example, suppose you elect to use the seismic reflection method to map buried bedrock topography. To do this as well as possible, you may have to sacrifice some details from within the sediments above the bedrock surface. You therefore need to be sure that this degree of detail is truly warranted in view of what you will lose.

In the real world there always will exist constraints which necessitate that survey objectives be defined well and in detail. Funding always will be tight, and time is money. You will not enjoy the luxury of laying out a few more seismic lines. Transportation costs, crew salaries, prices for expendables (such as explosives), and deadlines will impose severe restrictions. Even if you are working alone on a summer research project, you may have only 2 months and $500 to define the geometry of a local aquifer.

The first step, of course, is to understand the fundamentals of the various exploration methods so that you know what each can accomplish and what it cannot. Ask what is most important and what constitutes an added dividend but is not essential. Before planning the data-acquisition stage, be sure to determine what information already exists. It's not particularly exciting to search through town archives for drilling logs, but in some localities that's where you will find them, and even one or two good well logs could be invaluable for the interpretative phase of your project. After survey objectives are carefully defined and existing information researched, then plan your field design, carefully and in detail. Think about a lot of "What if?" questions and prepare contingency plans for unforeseen but not unlikely difficulties during data acquisition (e.g., a farmer plowing a field close to your most impor-tant seismic line).

If you adopt this approach now, essentially at the beginning, it will become second nature, and you will become one of those individuals others always envy because of your successes.

Limitations

Although geophysical methods are extremely useful and in the majority of cases provide valuable information about the subsurface, it is important to recognize their limitations. A common limitation is the lack of sufficient contrast in physical properties. Even though seismic wave velocities vary for saturated sediments of different compositions, they typical-ly are similar enough in value that layers of different compositions cannot be differentiated by the seismic refraction method.

A second and very common limitation is the nonuniqueness of many interpretations. In geophysics we tend to engage in two types of modeling: direct and inverse. In *direct modeling* we develop an equation to describe the effect of the specific physical-property variation we are studying. This equation then tells us what effects this variation produces. In most cases such an equation can be derived and leads to unambiguous results. However, we most often are engaged in the reverse process of observing effects and then modeling the cause. This is referred to as *inverse modeling* and typically is more difficult and more ambiguous. The computer programs included with this book fall into one of these two cate-gories. Those which perform direct modeling contain the suffix "Model." Those which per-form inverse modeling contain the suffix "Solve."

Ambiguity tends to be most obvious in gravity work, when it often is difficult to differentiate the effect of a small body near the surface from that of a larger body at depth. However, many apparent resistivity curves can be due to a variety of subsurface lay-ering configurations, and many magnetic curves do not lead to only one possible explana-tion for the observed magnetic anomaly. A good geological background and knowledge of the area being studied are essential if you are to arrive at the correct interpretation. You will perform better as an exploration geophysicist if you are a good geologist.

Another limitation is resolution. All methods are saddled with this restriction. As seismic wavelengths typically are many meters, there is a lower limit to stratigraphic thick-nesses that can be resolved by reflection surveys. Thin beds also may not contribute their signature to an apparent resistivity curve unless they create an especially high resistivity contrast with overlying or underlying strata. Resolution limits often can be reduced but nor-mally at increased cost.

Finally, we must mention *noise*. Almost all geophysical data contains some undesired signal (or noise) to a greater or lesser extent. Today most noise is not due to equipment design. It is relatively straightforward to reduce noise due to operator error to a very minor amount. Noise due to human activities and natural phenomena (e.g., water pumps and wind) normally can be eliminated by careful survey planning. The noise that remains car-ries important information about ways in which the subsurface varies from our idealized picture. The continuing challenge is to develop ways to utilize such information. Maybe someday you will make a contribution of this sort.

The Advantage of Multiple Methods

Often specific survey objectives cannot be met by applying only one geophysical method. In seismic reflection surveys, seismic refraction lines provide important information about velocities of near-surface materials which make it possible to apply needed corrections to reflection travel times. As electrical resistivity surveys depend on different physical proper ties than seismic surveys, the two taken together may provide important cross checks. For example, consider a survey where an accurate bedrock depth is essential. Drilling is out of the question. Seismic refraction results are perfect, but the nagging question of the "hidden" layer and its effect on computed depths remains. However, resistivity results are not affected by the hidden layer and could be used to confirm the seismic thickness value.

Assume that we are mapping the confined aquifer diagrammed in Figure 1-1. A thick sequence of saturated clay overlies a saturated sand and gravel layer which overlies a heavily fractured bedrock surface. Our goal is to map the thickness of the sand and gravel. The clay-sand contact does not have a sufficient velocity contrast to be detected by seismic refraction, but the resistivity contrast is sufficient for mapping by the electrical resistivity method. Conversely, the fractured bedrock surface is filled with water and possesses a resistivity very similar to the saturated sand and gravel. However, a sufficient velocity contrast is present here for bedrock depth to be determined by the refraction method.

As we now proceed to learn about the basic methods of exploration geophysics, keep in mind some of the suggestions in this brief chapter. Learn the strengths and weaknesses of each method and determine how and when you will benefit by employing more than one method. Remember to define field survey objectives carefully beginning with your first excursion. Recognize the limitations of geophysics, but don't become negative about the ambiguities inherent in the various methods. Of course, after you complete analyzing your field observations, your conclusions must deal with this lack of certainty. But learn to seek clues and to make maximum use of preexisting data in order to keep the degree of uncertainty to a minimum. Don't be misled by illustrations in this and other texts, and never for

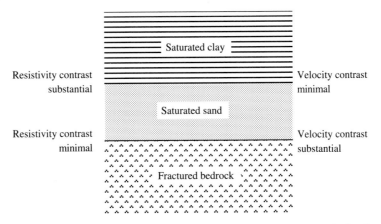

Figure 1-1   Determining the thickness of a confined aquifer by using both seismic refraction and electrical resistivity.

get that the real world is messy and that your data reflect this. Rather, strive to solve the sticky problems. The easy ones aren't as much fun. Geophysics plays a more and more important role in our lives with each passing year as populations increase and resources dwindle. Your journey is therefore an important one. We hope we have provided the means for you to enjoy it.

2

Seismic Exploration: Fundamental Considerations

SEISMIC WAVES AND WAVE PROPAGATION

The physical basis upon which the science of exploration seismology rests is remarkably straightforward. A sudden disturbance, or displacement, occurs at a small localized region in a mass of rock. This displacement propagates outward from its origin as a spherical wavefront and eventually arrives at distant points where it can be recognized by the motion it induces in the materials present. You surely are familiar with two very different phenomena, each of which illustrates just such an event: the impact of a pebble in a standing body of water and an earthquake.

When a pebble is thrown into a lake, the impact momentarily displaces a volume of water. This disturbance then radiates outward in the form of circular wavefronts or ripples. If the toss occurs on a day when the lake surface is quiet, the regularity of the ripple pattern is quite striking. The arrival of these ripples at distant points on the lake is recognized by the motion the water particles induce as the wavefronts pass by. Many inexperienced young boys and girls, fishing for the first time, almost assuredly have linked the up and down motion of their bob to causes other than the motion of the water particles induced by the passage of ripples. Although it doesn't occur to most people admiring the almost hypnotic appeal of orderly trains of ripples spreading across a lake or pond, these could be used to determine some useful information. It would be fairly simple to calculate the speed with which the ripples are moving. Then, with the help of a companion, one could use this new bit of information to calculate various distances of interest.

Similarly, the sudden and violent displacement of two masses of rock across a fault plane is a disturbance that sends seismic waves propagating through the earth. Anyone who has taken an introductory geology course knows that these waves are used to elucidate the details of Earth structure and, in conjunction with other measurements of Earth properties, provide insights into the nature of materials at depth. In a very general sense, seismologists are attacking a much more sophisticated version of the lake-distance problem. Determining the time of the disturbance (earthquake) and measuring the arrival of the waves is not terribly difficult once equipment is available that is needed to measure the small ground motions present at significant distances from the earthquake. Because many of the seismic waves that are received at recording stations far from the earthquake have passed through the earth, it seems evident that in order to make deductions about internal structures and compositions, we must understand the types of waves with which we are dealing, what factors govern their transmission, and the speeds with which they travel.

In this book our immediate focus is on the materials just beneath the surface. Usually, we will be concerned with those at the relatively shallow depths of 0 to 500 m, even though most of the principles we establish are applicable to exploration efforts at considerably greater depths. For the moment, let us consider a very simplified model for our discussion of the generation and transmission of seismic waves (Fig. 2-1). In this case we assume an isotropic and homogeneous body of granite which is separated from the atmosphere by a perfectly planar surface. A disturbance is produced at a point on the surface by a blow with a heavy hammer. This disturbance propagates outward as a series of wavefronts quite analogous to the ripples on the lake. The passage of the wavefronts by each point on the surface is marked by the motion of that surface, which can be measured and recorded by sufficiently sensitive detecting and recording instruments. As in the case of our lake discussion, we can calculate the distance of the recording instrument from the hammer blow if we know the time interval between the impact and the arrival of the first wavefront at the instrument and if we can calculate the speed of the wave. Such an exercise obviously is trivial, since at the scale of our investigation we could easily measure this distance directly. Determining the speed of the wave is not quite so trivial, since we cannot directly observe its passage as we could in the case of ripples on a lake surface. However, given the distance and assuming the presence of an instrument that is able to measure the time interval between impact and wave arrival, the value for the velocity of the wave is easy to obtain.

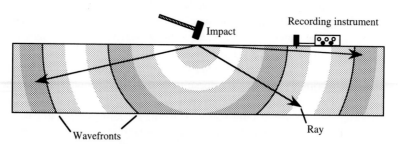

Figure 2-1   A hammer blow creates a disturbance that propagates through a homogeneous granite as seismic waves. Wavefronts and rays are indicated.

At this point, as a curious scientist, you probably have many questions about this procedure and analogy. What kinds of wave speeds are we likely to encounter? Would these speeds be different if the surface material was sand and gravel instead of granite? In what way does the ground move when a wavefront arrives at our sensing instrument? How does the design of the instrument take advantage of this motion? Is only one type of wave generated by the hammer blow? What if the material beneath the surface is not homogeneous? Can we detect this? Could we possibly identify the composition of material beneath the surface if the seismic waves arriving at the surface passed through this material? Let's begin to seek some answers to these questions.

Wave Terminology

Before pursuing some of these questions in greater detail, we need to be sure that we are familiar with the terminology used to describe various aspects of wave form and motion. For the purposes of the following discussion, we will assume that we are dealing with periodic waves of sinusoidal shape (Fig. 2-2).

The *wavelength* λ is the distance between two adjacent points on the wave that have the same phase or similar displacements. The distance between successive ripple crests is one wavelength. The *amplitude A* of the wave is the maximum displacement associated with the particle motions that occur as the wave passes through the material. In the case of our ripple crests this would be the height of a ripple crest above the level of the undisturbed lake surface. In the case of granite the amplitude would be the maximum particle displacement from the undisturbed position.

Wavelength and amplitude provide information concerning the shape of the waveform, but we also require information concerning wave movement through a material. Figure 2-3 illustrates how a given water particle in a lake behaves as ripples pass through its location. This diagram equally well could represent the behavior of a rock particle in granite as seismic waves pass through a region. Imagine that the wave train is moving and that the arrow marked "observation point" is your point of reference. The *period T* is the time it takes for two successive wave crests to pass this reference point and, therefore, for the motion to complete one cycle. When dealing with seismic waves, we often are interested in how often such repetitions or cycles occur in a given unit of time. In other words, What is the repetition rate? The numbers of repetitions per unit of time is defined as *frequency f*. Consider a seismic wave that has a period $T = 0.1$ s. Wave crests must pass a reference

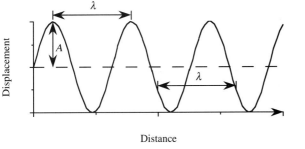

Figure 2-2 Pattern of particle displacements during passage of wave created by hammer blow illustrated in Figure 2-1. λ = wavelength; A is amplitude of maximum particle displacement from normal position.

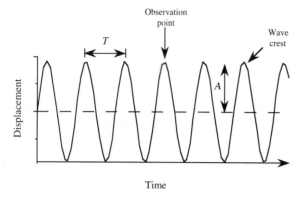

Figure 2-3 Motion of a single water particle in lake or a particle in granite as waves pass by. *T* (wave period) is the time for two successive wave crests to pass the observation point.

point every tenth of a second. The frequency, or repetitions per second, is then 10. It should be clear that frequency and period are related in the form

$$f = \frac{1}{T} \tag{2-1}$$

The unit of frequency is the *hertz* (Hz) which is the number of repetitions per second.

   If we refer to lake ripples once again, it is evident that the speed with which the ripple crest or wavefront is moving can be determined by noting the time it takes the wavefront to traverse a known distance

$$V = \frac{distance}{time} \quad (\text{ms}^{-1}) \tag{2-2}$$

Similarly, speed can be determined if we know the wavelength and the frequency

$$V = f\lambda \quad (\text{ms}^{-1}) \tag{2-3}$$

If this relationship is not immediately clear, consider two separate waveforms, each with the same cycle of repetition or frequency but with vastly different wavelengths. The wave with the greater wavelength must also have a greater velocity in order to have the same frequency.

   Even though wavelength, amplitude, and frequency are straightforward terms, you should constantly be solving simple mental problems using these relationships as you read through the material in this and subsequent chapters. Calculate the frequency or the wavelength of the waves with which you are working whenever possible. What kinds of amplitudes can you expect? What wavelength does a given speed and frequency imply? What are characteristic frequencies for the waves with which you are working? Once these questions and their answers become second-nature, they will be invaluable in planning, evaluating, and interpreting field studies using the seismic methods to be outlined in subsequent chapters.

   As the ripples spread across the lake, we could draw a surface through a given ripple crest. Along this surface all water particles are in the same phase of motion. This surface is referred to as a *wavefront*. It moves along as the disturbance that initiated it spreads outward. If the medium through which the wave is moving is isotropic and homogeneous, as for instance in the granite in Figure 2-1, the direction of propagation is always perpendicular to the wavefront. Thus, we could draw lines perpendicular to the wavefront, and these lines would indicate the direction in which the wave is moving through the material. Such lines are called *rays* and are used extensively in deriving the relationships for the seismic refraction and reflection methods. Rays will appear in most diagrams in this book rather than wavefronts. Nevertheless, they are a simplification, and you will be well served to constantly imagine the wavefront to which these rays are perpendicular.

Elastic Coefficients

If we refer to Figure 2-1 and consider the waves spreading out from the impact point of the hammer blow, we might wonder what determines the speed with which the waves travel. Considering the physical state of the granite before the hammer blow and after the waves have passed through the rock, it seems logical to propose that the waves are propagated by displacements of rock particles in an analogous fashion to water waves being propagated by displacements of water particles. Following the passage of the disturbance the particles return to their original state (except perhaps at the exact point of impact of the hammer blow). In other words, the rock has been subjected to a stress (force/area), there has been a shape change (strain), and the rock returns to its original state (shape). Such behavior is termed *elastic*. Perhaps the manner and speed with which seismic waves travel through materials is controlled by their elastic properties.

In order to determine if this proposition makes intuitive sense, it is a good idea to examine several basic relationships between the application of stress (a given force divided by the area over which it is applied) to an object and its resultant change in shape (strain). When determining these basic elastic constants, we assume Hookean behavior. Materials exhibiting Hookean behavior are elastic and exhibit an instantaneous linear relation between stress and strain (Means, 1976, p. 240). In other words, when a stress is applied to such materials, the resulting strain is immediate. When the stress is removed, the material returns to its original shape.

If a Hookean material is subjected to a uniaxial compression or tension, the linear relationship between the applied stress σ and resulting strain ε is given by

$$\sigma = E\varepsilon \qquad (2\text{-}4)$$

where the constant of proportionality E is termed *Young's modulus*. As this constant directly relates resultant strain to a given stress, it seems probable that rocks with different values for E may well have different velocities.

Elongation ε is the change in length of a line in its final or deformed state divided by its original length

$$\varepsilon = \frac{l_f - l_o}{l_o} = \frac{\Delta l}{l_o} \qquad (2\text{-}5)$$

If, as is illustrated in Figure 2-4, a Hookean solid is subjected to a uniaxial compression, it

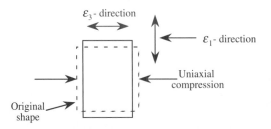

Figure 2-4 A uniaxial compression pro-
duces a positive elongation $+\varepsilon_1$ and a negative elongation $-\varepsilon_3$. The ratio of these two
strains is referred to as Poisson's ratio μ.

will shorten in the direction of the applied stress but at the same time will lengthen in a direction at right angles to the compression. Elongations can be measured for each direction, and their ratio is referred to as *Poisson's ratio* μ:

$$\mu = -\frac{\varepsilon_1}{\varepsilon_3} \qquad \text{where} \quad \mu \leq 0.5 \qquad (2\text{-}6)$$

Two additional elastic coefficients also are important. Given an isotropic material, subject it to a general pressure change (Fig. 2-5). A volume change will occur, and this change from original volume V_o to final volume V_f, when compared to the pressure change that induced it, is termed the *bulk modulus K*:

$$K = \frac{\Delta P}{\Delta} \qquad \text{where} \quad \Delta = \frac{V_f - V_o}{V_o} \qquad (2\text{-}7)$$

The bulk modulus is a measure of the incompressibility of the material.

Finally, it is possible to deform a solid by simple shear (Fig. 2-6). In this case a shear

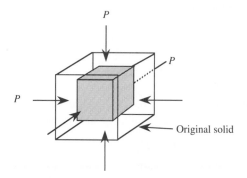

Figure 2-5 A change in volume (reduction) produced by a change in pressure
(increase). The ratio of the pressure change
to the volume change is a measure of the
incompressibility of the material and is
known as the bulk modulus K.

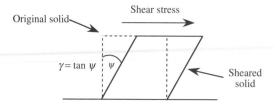

Figure 2-6 A measure of the shearing
resistance of a material is the ratio of the
shear stress to the shear strain, which is
termed the rigidity modulus G. The angular
shear is ψ.

strain γ is induced by applying a shear stress σ_s. The ratio of these quantities is the *rigidity modulus G*:

$$G = \frac{\sigma_s}{\gamma} \tag{2-8}$$

If we pause for a moment to consider Hookean materials, it seems reasonable to propose that the quantities E, μ, K, and G reasonably thoroughly define elastic behavior and, therefore, must be important factors in governing the speeds with which disturbances travel through Earth materials. Fortunately, values for all of these quantities can be determined in the laboratory. Some representative values for common rock types are given in Table 2-1. As you might guess, these four quantities are not mutually independent. Specifically, G and K can be defined in terms of E and μ:

$$G = \frac{E}{2(1 + \mu)} \tag{2-9}$$

$$K = \frac{E}{3(1 - 2\mu)} \tag{2-10}$$

TABLE 2-1 Elastic Coefficients and Seismic Velocities for Selected Common Rocks

Rock Type	Density ρ	Young's Modulus E	Poisson's Ratio μ	Vp (m/s)	Vs (m/s)	Vp/Vs	Vs as %Vp
Shale (AZ)	2.67	0.120	0.040	2124	1470	1.44	69.22%
Siltstone (CO)	2.50	0.130	0.120	2319	1524	1.52	65.71%
Limestone (PA)	2.71	0.337	0.156	3633	2319	1.57	63.84%
Limestone (AZ)	2.44	0.170	0.180	2750	1718	1.60	62.47%
Quartzite (MT)	2.66	0.636	0.115	4965	3274	1.52	65.96%
Sandstone (WY)	2.28	0.140	0.060	2488	1702	1.46	68.42%
Slate (MA)	2.67	0.487	0.115	4336	2860	1.52	65.96%
Schist (MA)	2.70	0.544	0.181	4680	2921	1.60	62.41%
Schist (CO)	2.70	0.680	0.200	5290	3239	1.63	61.24%
Gneiss (MA)	2.64	0.255	0.146	3189	2053	1.55	64.38%
Marble (MD)	2.87	0.717	0.270	5587	3136	1.78	56.13%
Marble (VT)	2.71	0.343	0.141	3643	2355	1.55	64.65%
Granite (MA)	2.66	0.416	0.055	3967	2722	1.46	68.62%
Granite (MA)	2.65	0.354	0.096	3693	2469	1.50	66.85%
Gabbro (PA)	3.05	0.727	0.162	5043	3203	1.57	63.51%
Diabase (ME)	2.96	1.020	0.271	6569	3682	1.78	56.05%
Basalt (OR)	2.74	0.630	0.220	5124	3070	1.67	59.91%
Andesite (ID)	2.57	0.540	0.180	4776	2984	1.60	62.47%
Tuff (OR)	1.45	0.014	0.110	996	659	1.51	66.20%

Units for Young's modulus are $(N/m^2) \times 10^{11}$. Velocities computed from p, E, and μ. Values selected from Press (1966, pp. 97–173).

Seismic Waves

Using advanced analysis, we can derive equations of motion which demonstrate that for an isotropic homogeneous body of infinite extent deforming elastically, it is possible for two types of waves to be transmitted. One type of wave is transmitted by particle movements back and forth along the direction of propagation of the wave and is referred to as a *longitudinal wave*. The other wave type is termed a *transverse wave*, because particle motions are transverse to the direction of movement of the wavefront, or perpendicular to the ray. It is, of course, somewhat satisfying that these two wave types are observed in laboratory experiments and in the field.

The particle motions that transmit longitudinal waves consist of a series of compressions and dilations, which can be envisioned as the centers of rock particles being moved closer together than normal and then moved farther apart than normal (Fig. 2-7). The familiar analogy is a coiled spring under tension in which an oscillation is created and which then is propagated along the spring by the coils moving together and pulling apart. Longitudinal waves usually are termed *P-waves* (primary waves) for, as we shortly demonstrate, they have the greatest speed and therefore appear first on traces illustrating wave arrivals at a specific location. P-waves also may be referred to as compressional waves due to the particle compressions during their transport.

As a transverse wave is moving through material, particles are subjected to shearing stresses as adjacent points move in a plane at right angles to the direction of propagation of the wave (Fig. 2-8). These *S-waves* have speeds less than P-waves, appear on seismograms after P-waves, and may be referred to as secondary waves or shear waves. The familiar analogy for this type of wave is a rope anchored at one end and then set into oscillation by a flick of the hand holding the rope. The wave travels along the rope, but the rope particles move in a direction at right angles to the propagation direction. The rope analogy can be

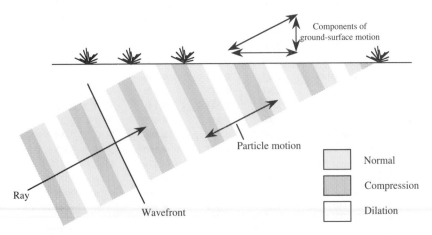

Figure 2-7 P-wave incident on surface. Particle motion is parallel to direction of travel. Since the wavefront plane is canted to the horizontal plane, ground-surface motion will have horizontal and vertical components.

taken a step further in that particle motion can be generated in any direction in a plane per-pendicular to the direction of propagation of the wave depending on which direction the hand holding the rope displaces it. Similarly, for transverse waves traveling through the earth, the particle motion is not constrained to a specific direction as in the case of longitu-dinal waves and may be in any direction in the plane perpendicular to propagation direction. Often, especially in advanced analysis, S-wave motion is resolved into two components (Fig. 2-8)—a component parallel to the ground surface (SH-component) and a component lying in a vertical plane containing the incident ray (SV-component). Both components must be in the plane perpendicular to the ray.

As stated at the outset of this discussion of seismic wave types, the material in which the disturbance occurs is assumed to be isotropic, homogeneous, and infinite in extent. Since most of our efforts in utilizing the seismic method involve generating energy at or close to the surface, this assumption clearly is not valid. The presence of this surface per-

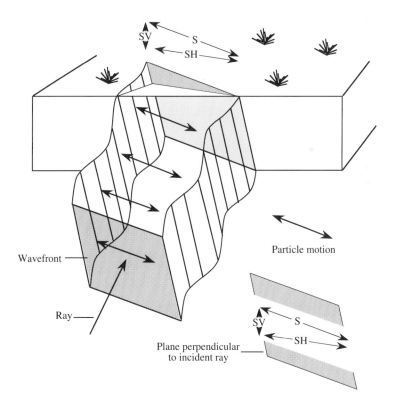

Figure 2-8   S-wave incident on ground surface. Particle motion is at right angles to the direction of travel. Since the wavefront plane is canted to the horizontal plane, ground-sur-face motion will have horizontal and vertical components. Normally, shear wave motion S is resolved into two components which are contained in the plane perpendicular to the inci-dent ray: (1) an SH-component which is parallel to the ground surface (horizontal) and (2) an SV-component in a vertical plane that contains the incident ray.

mits other types of waves to be generated. These are referred to as *surface waves*, since they are propagated within and confined to the region of the bounding surface. In contrast, P- and S-waves often are referred to as *body waves*. There are two types of surface waves that are observed: *Rayleigh* waves and *Love* waves. They are named after the scientists (Lord J. W. S. Rayleigh (1842–1919) and A. E. H. Love (1863–1940) who proved theoretically that these waves could exist.

Rayleigh waves propagate by particle motion that is confined to a vertical plane, is retrograde elliptical (Fig. 2-9(a)), and is in the direction of wave travel. Particle displacements are greatest at the surface and decrease exponentially downward (Richter, 1958, p. 241). Because of this substantial vertical displacement at the surface, Rayleigh waves produce a pronounced signature on exploration seismic records. For this reason, surface

Figure 2-9   (a) Rayleigh waves. All particle motion is contained in a vertical plane, and disturbance decreases downward. (b) Ground surface motion associated with the passage of Love wave. All particle motion is in the horizontal plane and decreases downward.

waves often are referred to as *ground roll* by those involved in field studies. Rayleigh waves illustrate dispersion, which means that velocity is not constant but varies with wavelength. Love waves travel by a transverse motion of particles that is parallel to the ground surface (Fig. 2-9(b)) and in this sense are similar to S-waves.

You might ask, and rightly so, why we are so concerned with particle motions connected with the various wave types. Remember that our observations usually are restricted to the surface, and our sources of information concerning the subsurface consist predominantly of the times at which waves energies arrive at recording instruments and the direction and magnitude of ground-surface motions. Clearly, those displacements depend on the angles at which the waves arrive at the surface, their vibration directions, and their relative energies. These factors all will be discussed or reviewed again in the sections that follow, but our knowledge of wave types and behavior will be critical in interpreting field data.

Seismic Wave Velocities

In this book we will follow the convention of Dobrin and Savit (1988, p. 32) and treat velocity synonymously with speed. Recall, however, that velocity possesses direction as well as magnitude (speed) and is, therefore, actually a vector quantity. A result of the mechanical analysis that demonstrates the existence of P- and S-waves is equations that give velocities in terms of the density ρ and elastic coefficients of a material. The velocities of P- and S-waves are

$$V_P = \sqrt{\frac{K + 4/3\,G}{\rho}} = \sqrt{\frac{E}{\rho}\frac{(1 - \mu)}{(1 - 2\mu)(1 + \mu)}} \tag{2-11}$$

and

$$V_S = \sqrt{\frac{G}{\rho}} = \sqrt{\frac{E}{\rho}\frac{1}{2(1 + \mu)}} \tag{2-12}$$

If we note that the bulk modulus K and the rigidity modulus G are always positive and recall that Poisson's ratio μ is less than or equal to 0.5, then it is evident that the velocity of P-waves always must be greater than that of S-waves by a considerable factor. Establishing the actual ratio consists simply of applying the preceding equations and simplifying

$$\frac{V_P}{V_S} = \sqrt{\frac{1 - \mu}{1/2 - \mu}} \tag{2-13}$$

Because G is equal to zero for liquids, the velocity of S-waves in liquids goes to zero. In other words shear waves cannot be propagated by liquids. It is this fact and the observation that direct S-waves are not recorded in the earth's "shadow zone" that constitute a primary line of evidence for a liquid outer core.

Refer once again to Table 2-1. P- and S-wave velocities calculated from the appropriate values of elastic coefficients and densities are listed, as are the velocity ratios.

> Table 2-1 is a *dynamic table*, which means that it is a spread-sheet template on the computer disk that comes with this book. You can insert values in any cell that is in boldface, and values will be recalculated. In order to gain an appreciation for how density, Young's modulus, and Poisson's ratio affect velocities, insert a range of values similar to those in Table 2-1 and note the effect on P- and S-wave velocities.

Because compressional waves are used almost exclusively in seismic exploration, especially at shallow depths, we primarily are concerned with P-wave velocities in rocks and sediments normally encountered in our work. Velocities are determined either by laboratory measurements, velocity logging in wells, or through seismic field methods. All have advantages and disadvantages. However, for much exploration work only approximate values are needed for planning the actual survey, as more refined values are determined during the survey itself.

Table 2-2 lists velocity ranges for a range of common materials. Many ranges overlap, so there are no unique values for rocks or sediments. However, a few general rules are suggested by these values.

- Unsaturated sediments have lower values than saturated sediments.
- Unconsolidated sediments have lower values than consolidated sediments.

TABLE 2-2 Representative P-wave Velocities

Unconsolidated Materials		Consolidated Materials		Other	
Weathered layer	300–900	Granite	5000–6000	Water	1400–1600
Soil	250–600	Basalt	5400–6400	Air	331.5
Alluvium	500–2000	Metamorphic rocks	3500–7000		
Clay	1100–2500	Sandstone and shale	2000–4500		
Sand		Limestone	2000–6000		
Unsaturated	200–1000				
Saturated	800–2200				
Sand and gravel					
Unsaturated	400–500				
Saturated	500–1500				
Glacial till					
Unsaturated	400–1000				
Saturated	1700				
Compacted	1200–2100				

Velocities in m/s

Velocity ranges compiled from Press (1966, p. 195–218).

- Velocities are very similar in saturated, unconsolidated sediments.
- Weathered rocks have lower values than similar rocks that are unweathered.
- Fractured rocks have lower values than similar rocks that are unfractured.

In general, on the basis of velocity alone, you should have no trouble in distinguishing dry sediments from wet sediments, sediments from rocks, and sedimentary rocks from igneous and metamorphic rocks. Of course, it likely would be impossible to distinguish a wet, compacted glacial till from a weathered sandstone or poorly consolidated sandstone. Although we hesitate to suggest specific values when in the first stages of field survey design, it is not unreasonable to use 500 m/s for dry unconsolidated materials, 1500 m/s for wet unconsolidated materials, 4000 m/s for sedimentary rocks, and 6000 m/s for unweathered igneous and metamorphic rocks.

The velocity of the air wave also is listed in Table 2-2. When hammer blows or explosives located near the surface are used as energy sources, a compressional wave is formed which travels through the air to the seismometers or geophones. This is referred to as the *air wave* and has a velocity that can be slightly greater than some common surficial materials. We can expect it to appear in much of our work and always should look for it on our records.

In some field studies we also will want to estimate velocities for S-waves and Rayleigh waves. Love waves typically are not generated during exploration work, so we have little interest in estimating their velocities. General rules of thumb for such estimates are

$$
\begin{aligned}
V_S &= 0.6\,V_P \qquad &&\text{for crystalline rocks} \\
V_S &= 0.5\,V_P \qquad &&\text{for sedimentary rocks} \\
V_S &= 0.4\,V_P \qquad &&\text{for soils and unconsolidated materials} \\
V_R &= 0.9\,V_S
\end{aligned}
\tag{2-14}
$$

Earlier in this discussion we noted the relation between wavelength, velocity, and frequency. Energy sources typically used to generate seismic waves for exploration purposes produce waves in the 10- to 100-Hz range. Now that we have some representative P-wave velocities we can calculate what types of wavelengths might be expected. As we will discover in our discussions of reflection seismology, wavelength will affect both our horizontal and vertical resolution. If we take our "typical" value for saturated unconsolidated materials to be 1500 m/s, then a wave with a frequency of 10 Hz will have a wavelength of 150 m, and a 100-Hz wave will have a wavelength of 15 m. These certainly are substantial wavelengths, and often in our field surveys, other factors being equal, we will try to work with values in the lower range.

RAY PATHS IN LAYERED MATERIALS

In the preceding several pages our discussion was based on very simple assumptions for the generation and propagation of seismic waves. For the most part we assumed an infinite, homogeneous, and isotropic medium. Our greatest complication was to allow for a horizontal, planar interface between rock (or water) and the atmosphere. Of course, these assumptions are unbelievably simplistic when we consider the actual subsurface, but they are necessary in order to manage the physics of the phenomena. Once again, based on your own past experiences, you should be curious about what happens to these waves when they encounter "obstructions." You may well have observed ripples striking a planar surface and reflecting and probably have heard or read about the bending, or refraction, of light waves when they pass from air into water. Even if you have taken relatively little physics or geology, it probably is clear that you expect seismic waves to reflect and refract as they travel through sediment or rock and encounter surfaces bounding materials with different elastic coefficients and densities.

Two important insights are very useful in developing the relationships for refracting and reflecting waves that we will use throughout our development of seismic exploration methods. These are Huygens' principle and Fermat's principle.

Huygens' Principle

Christian Huygens (1629–1695) was a Dutch mathematician, physicist, and astronomer who formulated a conceptually simple but remarkably powerful principle as part of his development of the wave theory of light. His principle states that all points on a wavefront can be considered as point sources for the generation of spherical secondary wavelets; after a time t the new position of the wavefront is the surface of tangency to these wavelets. If we apply this principle to the wavefront at time t_1 in Figure 2-10, we can construct the wavefront at time t_2. For simplicity in the example we assume constant velocity throughout the medium. We next select a few point sources on the initial wavefront, calculate the radius of the secondary wavelets based on the velocity of the medium and the elapsed time, Δt, construct these wavelets, and draw a surface tangent to them. This produces the wavefront at time t_2.

Fermat's Principle

The Frenchman Pierre de Fermat (1601–1665) was an outstanding mathematician who developed what has come to be known as the *principle of least time*. This principle states that in the propagation of waves, the wave path between any two fixed points is that one along which the time of travel is the least of all possible paths.

In our development of reflection and refraction, we assume that velocity in a given medium remains constant. By applying Fermat's principle we see that the wave path or ray must be a straight line in a medium of constant velocity. Because a straight line is the shortest distance between two points and as velocity is constant, a straight line between two points also is the one for which the time of transit is a minimum.

In previous diagrams wavefronts often were illustrated as spherical surfaces, but in many of the derivations in following sections the wavefront is drawn as a plane. As a wave increases its distance from its center of origin, the radius of curvature increases to such an extent that we can treat the wavefront as a plane near any particular point of interest.

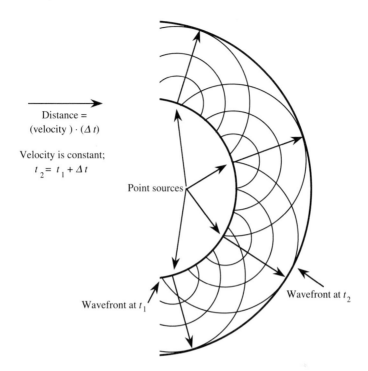

Distance =
(velocity) · (Δt)

Velocity is constant;
$t_2 = t_1 + \Delta t$

Point sources

Wavefront at t_2

Wavefront at t_1

Figure 2-10   Applying Huygens' principle to determine the position of the wavefront at t_2
after an interval of time Δt. Given the position of a wavefront at time t_1 and applying
Huygens' principle, the position of the wavefront at time t_2 is determined.

Reflection

As everyone who ever purchased a pair of sunglasses should know, a wavefront is reflected
when it encounters a surface bounding two materials with differing velocities of transmission of the wave. For our discussion we assume a planar wavefront that strikes a planar,
horizontal surface separating materials with seismic velocities V_1 and V_2. The angle between
a ray and a normal to the horizontal surface is termed the *angle of incidence* and is labeled
θ_1 in all diagrams. If we refer to Figure 2-11 or 2-12, we can ask, What is the relationship of
angle θ_2 to θ_1? In other words, What path will the reflected ray take?

It is possible to answer this question by applying either Huygens' principle or
Fermat's principle, and since it is instructive to utilize both of these principles, we will take
a look at both in developing the basic geometries of reflection and refraction of waves.

First, let us examine reflection through an application of Huygens' principle.
Consider a wavefront with velocity V_1 approaching a horizontal interface (Fig. 2-11). When
ray 1 strikes the interface, it creates a disturbance that spreads outward both into the upper
and lower materials (for the moment we are concerned only with the propagation of the
reflected wave into the upper material). At some later time ray 2 strikes the interface and,
still later, ray 3 arrives at point Y. The secondary wavefront generated at A must travel outward a certain distance during the time it takes ray 3 to travel from B to Y. As velocity is

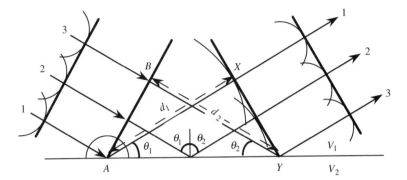

Figure 2-11   Using Huygens' principle to demonstrate that the angle of incidence equals the angle of reflection ($\theta_1 = \theta_2$). Analysis detailed in the text.

constant, this distance must be equal to BY. Similar reasoning can be applied to ray 2. Therefore, we can construct the new wavefront by drawing an arc with an origin at A and radius $d_1 = d_2$. In like fashion an arc can be constructed for ray 2. We then construct a line through Y that also is tangent to these arcs and locates the position of the wavefront at the moment ray 3 is at point B. Because the two right triangles ABY and AXY are congruent, θ_1 and θ_2 are equal, and the angle of incidence equals the angle of reflection.

In utilizing the principle of least time to discover the relationship between θ_1 and θ_2, we want to discover which path is the shortest for a ray passing through A, reflecting from the interface, and emerging through B. Figure 2-12(a) illustrates the problem. Does travel along path ARB take the least time or is some other path, such as APB, the one along which time of travel is a minimum?

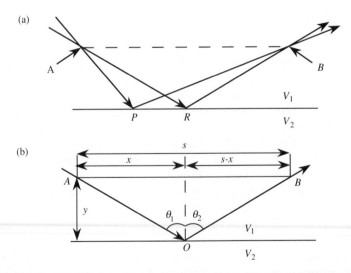

Figure 2-12   Using Fermat's principle to demonstrate that the angle of incidence equals the angle of reflection ($\theta_1 = \theta_2$). (a) Which path takes the least time? (b) Symbols used for analysis in text.

By referring to Figure 2-12(b), we can state that the time (path distance/velocity) for a ray to travel from A through some point O to B is

$$t = \frac{(x^2 + y^2)^{1/2}}{V_1} + \frac{((s - x)^2 + y^2)^{1/2}}{V_1} \qquad (2\text{-}15)$$

In order to determine the minimum value of time t, we take the first derivative of the function and set it equal to zero:

$$\frac{dt}{dx} = \frac{x}{V_1(x^2 + y^2)^{1/2}} - \frac{(s - x)}{V_1((s - x)^2 + y^2)^{1/2}} = 0 \qquad (2\text{-}16)$$

Using the relationships

$$\sin\theta_1 = \frac{x}{(x^2 + y^2)^{1/2}} \quad \text{and} \quad \sin\theta_2 = \frac{(s - x)}{((s - x)^2 + y^2)^{1/2}} \qquad (2\text{-}17)$$

we see that

$$\frac{\sin\theta_1}{V_1} - \frac{\sin\theta_2}{V_1} = 0 \quad \text{and, therefore,} \quad \theta_1 = \theta_2 \qquad (2\text{-}18)$$

Thus, the path for which the time of travel is least is the one for which the angle of incidence equals the angle of reflection.

Refraction

The approach in developing the relationship between incident and refracted rays is almost identical to that utilized for reflected rays. Let us apply Huygens' principle first and then employ Fermat's principle.

The construction in Figure 2-13 is essentially the same as in Figure 2-11, except here we deal with refracted rays. When ray 1 arrives at point A, it creates a disturbance in the material with velocity V_2. The disturbance spreads outward in this layer and will travel a distance d_2 during the time t_1 it takes ray 3 to travel from point X to point Y (a distance d_1). We construct the position of the new wavefront at the moment ray 3 arrives at Y by drawing a line connecting Y and B (which is the tangent to the wavelet with radius d_2 that was generated at A). Because

$$\sin\theta_2 = \frac{d_2}{AY} \quad \text{and} \quad \sin\theta_1 = \frac{d_1}{AY} \qquad (2\text{-}19)$$

$$d_1 = t_1 V_1 \quad \text{and} \quad d_2 = t_1 V_2 \qquad (2\text{-}20)$$

it holds that

$$AY = \frac{t_1 V_1}{\sin\theta_1} = \frac{t_1 V_2}{\sin\theta_2} \quad \text{and} \quad \frac{\sin\theta_1}{\sin\theta_2} = \frac{V_1}{V_2} \qquad (2\text{-}21)$$

Thus, approaching the geometry of refraction using Huygens' principle, we learn that the

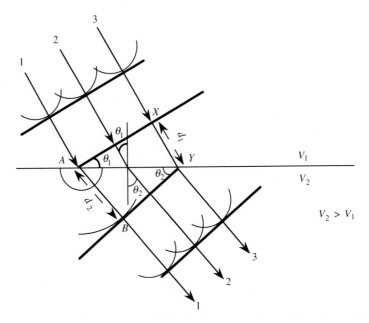

Figure 2-13 Using Huygens' principle to demonstrate the relationship between the angles of incidence and refraction. Analysis detailed in the text.

ratio of the sines of the angle of incidence and angle of refraction is equal to the ratio of the velocities of the two materials.

When applying Fermat's principle, we can ask the same sort of questions that we did for the reflection problem. When examining Figure 2-14(a), we would like to know what distinguishes the path of least time of a ray traveling from A to C as it passes through an interface separating materials of differing velocities. Once again we write an equation expressing the time it takes for a ray to travel from A through B to C:

$$t = \frac{(x^2 + y^2)^{1/2}}{V_1} + \frac{((s - x)^2 + z^2)^{1/2}}{V_2} \tag{2-22}$$

Continuing, we take the derivative and arrive at the expression:

$$\frac{dt}{dx} = \frac{x}{V_1(x^2 + y^2)^{1/2}} - \frac{(s - x)}{V_2((s - x)^2 + z^2)^{1/2}} = 0 \tag{2-23}$$

In taking the derivative, note that z and y are constants as they retain the same value regardless of where B is placed along the interface. As before, we can use identities to arrive at the final form relating angles and velocities:

$$\sin \theta_1 = \frac{x}{(x^2 + y^2)^{1/2}} \quad \text{and} \quad \sin \theta_2 = \frac{(s - x)}{((s - x)^2 + z^2)^{1/2}} \tag{2-24}$$

$$\frac{\sin \theta_1}{V_1} - \frac{\sin \theta_2}{V_2} = 0 \quad \text{and, therefore,} \quad \frac{\sin \theta_1}{\sin \theta_2} = \frac{V_1}{V_2} \tag{2-25}$$

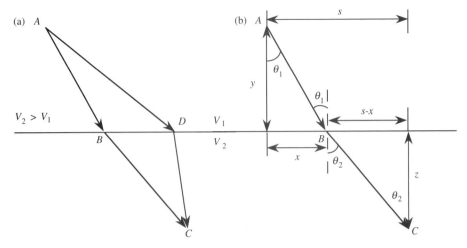

Figure 2-14 Using Fermat's principle to demonstrate the relationship between the angles of incidence and refraction. (a) What are the distinguishing features of the path of least time (*ABC*) as opposed to others (*ADC*)? (b) Symbols for analysis used in text.

As you probably expected, this is the same relationship derived using Huygens' principle and, if V_1 is substituted for V_2, is the same as deduced for reflected rays. Those interested in a more detailed treatment of the material discussed in this and the preceding section should consult Slotnick (1959).

Snell's Law

The relationship $\sin \theta_1/\sin \theta_2 = V_1/V_2$ is commonly referred to as *Snell's law* and is fundamental in deriving other expressions to utilize the reflection and refraction of seismic waves in deducing subsurface relationships. Snell's law often is written as $\sin \theta_i/\sin \theta_r = V_i/V_r$, where *i* represents incidence and *r* represents refraction. Because ray paths are confined to V_i in reflection, Snell's law reduces to $\sin \theta_i = \sin \theta_{\text{reflection.}}$

    Up to this point we did not identify the incident and reflected or refracted waves as P- or S-waves but assumed that the incident and refracted/reflected waves were of the same type. The actual situation is somewhat more complex. A P-wave incident upon a surface separating materials with different velocities (and thus different elastic coefficients and densities) creates a disturbance that gives rise to a reflected P-wave and S-wave (Fig. 2-15(b)) and a refracted P-wave and S-wave (Fig. 2-15(d)). Snell's law still applies to the angles of refraction and reflection of the S-waves. The only adjustment required is to substitute the appropriate S-wave velocity. For example, the angle of refraction for the S-wave (θ_{rfr_S} in Figure 2-15(d)) is determined by the relationship

$$\frac{\sin \theta_{i_P}}{\sin \theta_{\text{rfr}_S}} = \frac{V_{1_P}}{V_{2_S}} \tag{2-26}$$

Similarly, an incident S-wave gives rise to reflected and refracted P-waves as well as

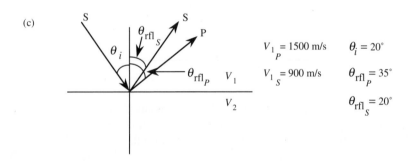

Figure 2-15   (a) General terminology for subsequent diagrams: θ_i = angle of incidence, θ_{rfl} = angle of reflection, and θ_{rfr} = angle of refraction. V_1 is the velocity of a given wave in the material above the interface; V_2 is the velocity below the interface. (b and d) An incident compressional wave produces a reflected P-wave, a reflected S-wave, a refracted P-wave, and a refracted S-wave. The angles of reflection and refraction of these waves depend on the angle of incidence and the velocities of the waves in the upper and lower materials. (c and e) An incident shear wave produces both reflected and refracted P- and S-waves. (f) This summary diagram is intended to reinforce the idea that an incident P-wave or an incident S-wave produces both reflected and refracted compressional and shear waves. Note, however, by studying (b) through (e), that the ray paths of reflected and refracted P-waves due to an incident P-wave are not the same as the ray paths of reflected and refracted P-waves due to an incident S-wave. This also is true for reflected and refracted S-waves produced by incident P- and S-waves.

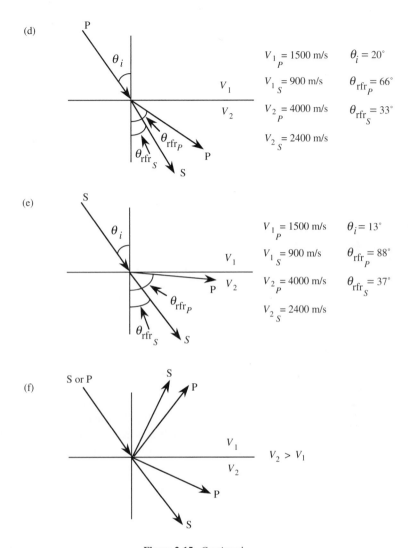

Figure 2-15 *Continued.*

reflected and refracted S-waves (Fig. 2-15(c) and (e)). Specific examples with values are illustrated in Figure 2-15(b), (c), (d), and (e). You might want to apply Snell's law to these situations as practice in calculating the appropriate angles. Table 2-3 also lists angles for incident P- and S-waves, their velocities, and the calculated angles of reflection and refraction.

The only case for which an incident S-wave does not give rise to reflected and refracted P-waves is when the S-wave is entirely of the SH type. In this case particle motions are confined to the horizontal plane and are, therefore, parallel to the velocity discontinuity.

Table 2-3 also is a dynamic table. Experiment by inserting
various velocities in place of those present or by altering
the value of the *first* cell in the column labeled "Angle of
incidence."

Critical Refraction

You may have noticed when examining Table 2-3 that in three of the columns for refracted
rays angles increase to a maximum of 87° or 88° and then are replaced by asterisks (*). If
you experimented with the dynamic table, you noticed these cells are labeled *#NUM!*. This
message appears in a cell when an unacceptable argument is used for a function. If you
changed values for the angles of incidence or velocities, the *#NUM!* message may have dis-
appeared or perhaps appeared in even more cells. What is the significance of these relationships?

Because Snell's law was used to calculate the angles of refraction, we should examine
the quantities in the equation with more care. First, for simplicity, let us restrict our discussion to compressional waves. As is most common in shallow exploration, the velocity of the
layer below the interface is greater than the velocity above the interface ($V_2 > V_1$). As the
angle of incidence increases, the value of the sine of the angle of refraction also must
increase in order to maintain the equality of the ratios in Snell's law. At some point the
angle of incidence will be such that $\sin \theta_i = V_1/V_2$, which requires that the sine of the angle
of refraction = 1.0 (and the angle = 90°). If θ_i is increased beyond this value, an error ensues
because the sine of an angle cannot take on values greater than 1, and, therefore, the equality cannot be maintained. In physical terms this means that the angle of refraction increases
as the angle of incidence increases until rays are refracted parallel to the interface between
the two materials. If the angle of incidence increases beyond this special value, then no
refraction occurs, and the ray is *totally reflected*.

The refraction seismic method is based on this special case, and so we speak of the
critical angle and *critical refraction*. If θ_{ic} denotes the critical angle, its value can be determined by

$$\theta_{ic} = \sin^{-1}\left(\frac{V_1}{V_2}\right) \qquad (2\text{-}27)$$

As before, it is trivial to prove that this general form also holds for all cases of refraction
such as refracted shear waves produced by incident compressional waves.

We can visualize this critically refracted ray as traveling parallel to the interface, but
in the V_2 medium, and producing a disturbance of the interface. This disturbance generates
secondary wavelets as the refracted wave passes which return to the surface through the V_1
medium. We once again could utilize Huygens' principle to construct the wavefront travel-

TABLE 2.3 Angles of Reflection and Refraction for P- and S-waves

Angle of incidence	For incident P-waves				For incident S-waves			
	Reflected P	Reflected S	Refracted P	Refracted S	Reflected S	Reflected P	Refracted S	Refracted P
10	10	6	28	16	10	17	28	51
11	11	7	31	18	11	19	31	58
12	12	7	34	19	12	20	34	68
13	13	8	37	21	13	22	37	88
14	14	8	40	23	14	24	40	*
15	15	9	44	24	15	26	44	*
16	16	10	47	26	16	27	47	*
17	17	10	51	28	17	29	51	*
18	18	11	55	30	18	31	55	*
19	19	11	60	31	19	33	60	*
20	20	12	66	33	20	35	66	*
21	21	12	73	35	21	37	73	*
22	22	13	87	37	22	39	87	*
23	23	14	*	39	23	41	*	*
24	24	14	*	41	24	43	*	*
25	25	15	*	43	25	45	*	*
26	26	15	*	45	26	47	*	*
27	27	16	*	47	27	49	*	*
28	28	16	*	49	28	51	*	*
29	29	17	*	51	29	54	*	*
30	30	17	*	53	30	56	*	*

Velocity 1-P (m/s)	**1500**
Velocity 1-S (m/s)	900
Velocity 2-P (m/s)	**4000**
Velocity 2-S (m/s)	2400

Wavefront

Head wave

θ_{ic}

θ_{ic}

Critical refraction

Figure 2-16 A wave that strikes an interface at its critical angle θ_{ic} is refracted parallel to the interface producing what is commonly referred to as a head wave.

ing upward at velocity V_1. This wave generally is referred to as the *head wave*, and any of its rays also are at the critical angle (Fig. 2-16).

The phenomena of refraction and reflection, critical refraction, and total reflection are summarized in Figure 2-17.

Diffraction

For the most part, only continuous planar surfaces and other simple geometries will be considered in our investigations of the seismic reflection and refraction methods, even though the subsurface is rarely so ideal. However, one important exception to this concerns the phenomenon of *diffraction*. If a disturbance is generated at the surface and encounters a sudden change of curvature at a velocity discontinuity, additional waves are produced that are not predicted by drawing rays for reflection or refraction. Our analysis fails when the dimensions of an object encountered by seismic waves are no longer significantly greater than the wavelengths of these waves.

Once again, studying the behavior of water waves helps in visualizing particular wave phenomena. If water waves with a planar wavefront encounter a barrier containing a slit with a dimension similar to the wavelength of the waves (Fig. 2-18), diffraction occurs and the water waves spread outward from the slit in a semicircular pattern. Actually, if we consider the small portion of the wave that passes through the slit and apply Huygens' principle (every point on the wavefront can be considered as the source of a secondary wavelet), the reason for the diffraction pattern is clear.

Although the process that occurs at a sharp discontinuity such as diagrammed in Figure 2-19, is somewhat more complex, the general pattern is explained quite well by considering the sharp change in curvature as a source of secondary wavelets and thereby reconstructing the wavefront by Huygens' method. Under what conditions are diffractions likely

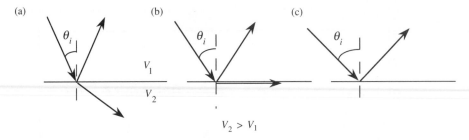

(a)   (b)   (c)

θ_i θ_i θ_i

V_1

V_2

$V_2 > V_1$

Figure 2-17 (a) Incident, reflected, and refracted rays. (b) Increasing θ_i results in a critically refracted ray. (c) Increasing θ_i still further produces total reflection.

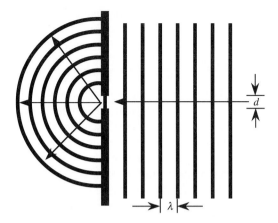

Figure 2-18 Plane wavefronts approaching an opening in a barrier where the width of the opening d is equal to the wavelength λ of the approaching wavefronts. Huygens' principle predicts the diffraction effect observed.

to occur in the subsurface? Refer again to Figure 2-18. As the size of the slit in the barrier is increased, the diffraction effect is decreased and is barely perceptible when the slit dimension is several times the wavelength. This suggests that diffraction will be most important when sharp changes in curvature have radii that are similar in dimension to seismic wavelengths. As seismic wavelengths typically are several tens of meters in length, diffracted waves certainly will not be uncommon. Because of the very different approaches in refraction and reflection seismology, considerations of diffraction phenomena play a much greater role in the reflection technique. This does not mean, however, that diffraction effects are not present on refraction records.

Wave Arrivals at the Surface

In the next section we discuss how wave energy is diminished as the wave propagates through materials and is reflected and refracted by velocity discontinuities. Such a discussion is undertaken primarily in order to establish the magnitude of amplitudes and ground

Figure 2-19 Diffracted waves are produced when wavefronts encounter a sudden change in curvature of an interface.

motion expected at the surface for different types of events. Before undertaking that discussion, however, it seems appropriate to consider the important aspects of the wave events already discussed and to question, in a general way, what a surface observer might expect.

Because compressional waves have the greatest velocities they arrive first at a destination whether it is a seismograph station on the other side of the world or a geophone located a few meters from a hammer blow. For a variety of reasons shear wave arrivals are extremely difficult to detect at short distances from an energy source. Primarily because of this difficulty, seismic exploration methods concentrate almost exclusively on P-waves. S-waves do provide additional important information (such as enabling the determination of elastic coefficient values for engineering purposes), and much research currently is taking place on S-wave applications. Because of their relatively minor role in common and routine seismic methods, S-waves will be mentioned only occasionally in the remainder of this text. Depending on field conditions, geometries, and equipment, however, S-waves will contribute their signature on seismic records, so we cannot ignore them entirely.

What general major arrivals might we expect to see in a configuration such as illustrated in Figure 2-20? If energy is imparted at or near the surface, a P-wave (the ground-coupled *air wave*) will activate sensors (geophones) placed at intervals along the surface. Other compressional waves will follow a direct path through the uppermost layer from the energy source to the sensors (the *direct wave*). P-wave reflections and the critical refraction also will arrive and produce ground motion. Finally, due to the significant ground motion they induce, Rayleigh waves (*ground roll*) certainly will impart their distinctive signature.

An easy way to gain some initial insight into the distribution in time and space of the arrivals of these waves at the surface is to construct a spreadsheet form. One such possibility is provided in Table 2-4. Assume that geophones are placed at even intervals from a shot point (or hammer blow). Decide on this interval and also select compressional wave velocities for the layers above and below a planar interface as well as the depth of the interface. If we calculate the path distances for each wave and the velocities over that path, we can determine the time of arrival at each surface geophone. Although these equations will be developed in subsequent chapters, for present purposes they simply are included in the

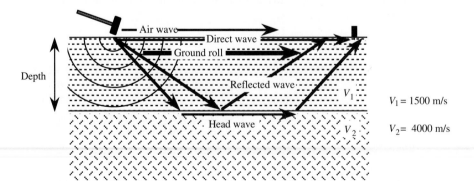

Figure 2-20 Generalized diagram illustrating ray paths for waves included in Table 2-4. Included are the direct wave, reflection, critical refraction, the air wave, and ground roll. All waves are compressional with the exception of ground roll.

TABLE 2-4 Time-Distance Values for Selected Waves

Distance from Shot (m)	Arrival Times (ms)				
	Direct Wave	Reflected Wave	Head Wave	Air Wave	Ground Roll
0	0.00	20.00	18.54	0.00	0.00
3	2.00	20.10	19.29	8.96	4.44
6	4.00	20.40	20.04	17.91	8.89
9	6.00	20.88	20.79	26.87	13.33
12	8.00	21.54	21.54	35.82	17.78
15	10.00	22.36	22.29	44.78	22.22
18	12.00	23.32	23.04	53.73	26.67
21	14.00	24.41	23.79	62.69	31.11
24	16.00	25.61	24.54	71.64	35.56
27	18.00	26.91	25.29	80.60	40.00
30	20.00	28.28	26.04	89.55	44.44
33	22.00	29.73	26.79	98.51	48.89
36	24.00	31.24	27.54	107.46	53.33
39	26.00	32.80	28.29	116.42	57.78
42	28.00	34.41	29.04	125.37	62.22
45	30.00	36.06	29.79	134.33	66.67
48	32.00	37.74	30.54	143.28	71.11
51	34.00	39.45	31.29	152.24	75.56
54	36.00	41.18	32.04	161.19	80.00
57	38.00	42.94	32.79	170.15	84.44
60	40.00	44.72	33.54	179.10	88.89

Velocity 1-P (m/s)	**1500**	Shot offset (m)	**0**	
Velocity 2-P (m/s)	**4000**	Geophone interval (m)	**3**	
Depth (m)	**15**	No head wave before	12.14	

spreadsheet. As in previous instances, Table 2-4 is included as a template on disk, so that you may experiment with various geophone intervals, velocities, and depths if you wish.

The pattern of the wave arrivals at points on the surface is much easier to analyze if we construct a graph of arrival time versus distance from the shot point for each wave type. Figure 2-21 was produced by taking the values in Table 2-4 and placing them into a computer graphing program. Such a graph, usually referred to as a *travel-time curve*, provides a clear sense of arrival relationships at various positions along the surface. Although it is somewhat premature, you still should study this graph and begin to analyze the arrival patterns especially for the air wave, direct wave, and ground roll. Figure 2-22 represents a further step in our introduction to wave arrival patterns. This computer-produced diagram simulates the wave arrivals in Table 2-4 and Figure 2-20 at each geophone position.

Even in this simulated seismogram many of the relationships so evident in Figure 2-21 are much more difficult to discern although myriad factors that would further obscure a real field seismogram are not included. Note especially that it is virtually impossible to deduce individual events at geophones 1–10 in the time range of 0 to 50 ms. An important

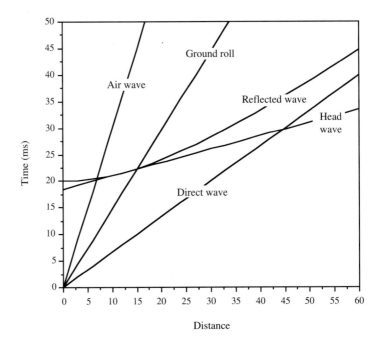

Figure 2-21 A plot of time against distance for the various waves in Table 2-4. Only compressional waves are considered, with the exception of the ground roll (Rayleigh waves). Although projected back to 0 m, no head waves arrive prior to 12.14 m.

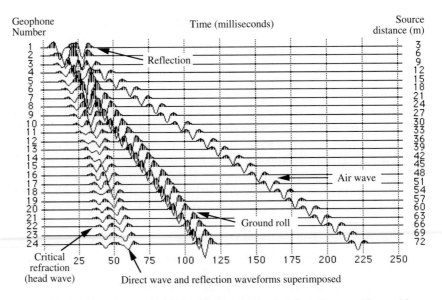

Figure 2-22 A computer-generated diagram that simulates how the waves illustrated in Figure 2-20 would appear on a field record.

task will be to formulate methods that make it possible to extract such information for interpretative purposes.

It is quite helpful to develop a feel for the arrival patterns and relationships calculated in Table 2-4 and graphed in Figure 2-21. Because Table 2-4 is a dynamic table you can change the depth of the interface and the velocities of the layers above and below the interface. You can vary the distance from the shot point or hammer blow to the first geophone and can change the geophone interval. Also, as this spreadsheet program has graphing capabilities, you can plot the results of any changes you make for any waves you want to investigate. Thus, this table is a powerful tool to help develop insights into the arrivals of these waves, and you should spend some time experimenting with the values in boldface if the appropriate computer software is available to you.

WAVE ATTENUATION AND AMPLITUDE

The major goal in seismic exploration is the interpretation of wave arrival patterns at the surface. It already is clear that this will not be a trivial task, even when Earth materials are homogeneous and only one interface is present. The variety of seismic wave types, variations in densities and elastic coefficients, reflection and refraction at interfaces, and the conversion of waves at interfaces all contribute to a complex arrival pattern at any surface point. An aid in deciphering this puzzle is an understanding of what factors are likely to alter wave amplitudes at various detecting stations along the surface.

Spherical Spreading

The first factor influencing wave amplitudes is easy to visualize. Picture wave energy spreading outward from a disturbance as a spherical wavefront. For each meter the wave travels the energy must be distributed over a much larger area; therefore, at any one localized region the amount of energy must decrease. Consider the amount of energy distributed over a small area of the spherical wavefront at one instant and project that area onto the wavefront at a later instant in time. Since the surface area of a sphere is $4\pi r^2$, where r is the radius, the ratios of the areas must equal the ratios of the squares of the radii of the spherical wavefronts. Thus, the energy distributed over the new area of the wavefront must decrease as $1/r^2$. Because wave amplitude is proportional to the square root of wave energy, amplitude decreases as $1/r$. Table 2-5 illustrates the losses in energy for distances from an energy source that are typical for many field surveys where the target is relatively shallow

TABLE 2-5 Energy Losses Due to Spherical Spreading

Distance from shot (m)	3	6	9	12	12	24	36	48	48	96	144	192	240
Spherical spreading loss (db)	9.54	15.56	19.08	21.58	21.58	27.60	31.13	33.62	33.62	39.65	43.17	45.67	47.60
Loss from previous point (db)		6.02	3.52	2.50		6.02	3.52	2.50		6.02	3.52	2.50	1.94

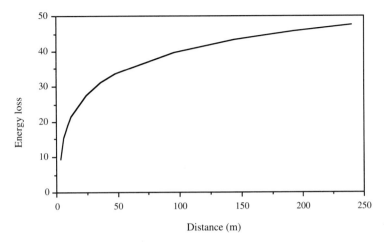

Figure 2-23   Energy losses in decibels due to spherical spreading at various distances from an energy source.

(less than 100 m). Figure 2-23 is a graph of the relationships between rows 1 and 2 of the table.

Absorption

As seismic waves propagate, they do so by distorting (straining) the materials through which they pass. Heat energy thus generated reduces the total energy of a wave. This reduction in the elastic energy of the wave is termed *absorption*. The equation for this process is

$$I = I_0 e^{-qr} \tag{2-28}$$

where I is defined as the energy intensity (the amount of energy passing through a unit area in a unit time), q is the absorption coefficient, and r is distance. Given the energy intensity at a point I_0, I represents the energy intensity at a point a distance r from I_0. The absorption coefficient q is expressed in decibels/wavelength (db/λ). Table 2-6 illustrates energy losses for a situation where q is given as 0.55 db/λ. This value for q is in the midrange of published values for various Earth materials, but the important aspect of the table is not the absolute values but the comparisons at various frequencies and distances.

These figures vividly illustrate that energy losses due to absorption are much greater at higher frequencies than at lower frequencies at a given distance from the energy source. Note the marked difference between the 10-Hz wave and the 100-Hz wave at 240 m. Because many natural seismic pulses contain a broad array of frequency components, the higher-frequency components will be progressively reduced with distance, so that the form of the seismic pulse changes to a smoother and broader shape. This is the reason that the Earth is often referred to as a low-pass filter. Lower frequencies are transmitted with less energy loss, and higher frequencies are progressively attenuated. If a survey is interested in

TABLE 2-6 Energy Losses (in decibels) Due to Absorption

Frequency (Hz)	Compressional Wave				Shear Wave			
				Distance (m)				
	10	60	120	240	10	60	120	240
5	0.0293	0.1755	0.351	0.702	0.0585	0.351	0.702	1.404
10	0.0585	0.351	0.702	1.404	0.117	0.702	1.404	2.808
30	0.1755	1.053	2.106	4.212	0.351	2.106	4.212	8.424
100	0.585	3.51	7.02	14.04	1.17	7.02	14.04	28.08
200	1.17	7.02	14.04	28.08	2.34	14.04	28.08	56.16
300	1.755	10.53	21.06	42.12	3.51	21.06	42.12	84.24

P-wave velocity (m/s)	**4000**
S-wave velocity (m/s)	**2000**
Absorption coefficient	0.55 db/wavelength

high-frequency energy (as it will be when we attempt shallow reflection work), we need to find ways to increase that component when generating energy and to be particularly sensitive to detecting those higher frequencies rather than the stronger lower ones.

All other factors being equal, waves traveling at lower velocities will lose energy more quickly from absorption than waves traveling at higher velocities. On this basis alone, S-waves will lose energy more quickly than P-waves. Because wave amplitude is proportional to the square root of wave energy, amplitudes of S-waves will diminish much more quickly than amplitudes of P-waves. This is just another reason that S-waves tend not to be utilized in exploration seismology.

> You can experiment with the values in boldface in both Table 2-5 and Table 2-6. It is especially interesting to vary the velocities in Table 2-6 and to note the amount of energy loss due to absorption as you increase or decrease velocity.

Energy Partitioning

You already realize that P- and S-waves striking a velocity discontinuity not only may be reflected and refracted but can generate other P- and S-waves as well. Because the total amount of energy must remain constant between the incident and reflected and refracted waves, amplitudes of refracted and reflected waves will be diminished relative to the incident wave.

Given the amplitude of an incident compressional wave, it is possible to calculate the amplitudes of reflected and refracted P- and S-waves using equations developed by Zoeppritz (1919). If the velocities and densities of materials above and below an interface are known, these equations provide solutions for all angles of incidence. Essentially, we are dealing with four equations with four unknowns (the amplitudes), and although the equations are not that complicated, we will not list them here as they are available in several advanced texts such as Telford et al. (1990, p. 155–156).

However, if P-wave incidence is normal to the interface, the equations reduce to a very simple form. No S-waves are generated under normal incidence. The ratios of the amplitudes A are

$$\frac{A_{rfl}}{A_i} = \frac{\rho_2 V_2 - \rho_1 V_1}{\rho_2 V_2 + \rho_1 V_1} \qquad \frac{A_{rfr}}{A_i} = \frac{2\rho_1 V_1}{\rho_2 V_2 + \rho_1 V_1} \qquad (2\text{-}29)$$

where the quantities are as diagramed in Figure 2-24. These equations often are shortened to the form

$$\frac{A_{rfl}}{A_i} = \frac{Z_2 - Z_1}{Z_2 + Z_1} \qquad \frac{A_{rfr}}{A_i} = \frac{2Z_1}{Z_2 + Z_1} \qquad (2\text{-}30)$$

where $Z_1 = \rho_1 V_1$ and $Z_2 = \rho_2 V_2$

Values for these amplitude ratios are given in Table 2-7 for one set of density-velocity relationships in the cells labeled "Reflection coefficient" and "Refraction coefficient." These terms are most commonly used to denote the amplitude ratios. Some geophysicists also use the same terminology to refer to the energy ratios of the refracted and reflected waves. Recall the general relationship between energy and amplitude. As you might expect then, the equations for the energy ratios are quite similar in appearance to those for amplitudes:

$$\frac{I_{rfl}}{I_i} = \left(\frac{\rho_2 V_2 - \rho_1 V_1}{\rho_2 V_2 + \rho_1 V_1} \right)^2 \qquad \frac{I_{rfr}}{I_i} = \frac{4\rho_1 V_1 \rho_2 V_2}{(\rho_2 V_2 + \rho_1 V_1)^2} \qquad (2\text{-}31)$$

or

$$\frac{I_{rfl}}{I_i} = \left(\frac{Z_2 - Z_1}{Z_2 + Z_1} \right)^2 \qquad \frac{I_{rfr}}{I_i} = \frac{4Z_1 Z_2}{(Z_2 + Z_1)^2} \qquad (2\text{-}32)$$

Solutions to these equations also are presented in Table 2-7.

Of course, not all cases of interest have normal incidence, and Zoeppritz' equations must be solved in their original form. For our purposes, however, it is sufficient to illustrate the results of a solution for one set of velocities and densities although the form of the curves can become very different depending on the values used. Figure 2-25 is based on graphs in Tooley et al. (1965). The most useful observations that can be drawn from these curves are: (1) At low angles of incidence (rays nearly normal to the interface) most of the energy is in the refracted P-wave; (2) at high angles of incidence most of the energy is in the reflected P-wave; (3) both the refracted P-wave and the refracted S-wave disappear at the critical angle for each; and (4) the energy in the "partitioned" S-waves is relatively low compared to the P-waves, so these S-waves are likely to be difficult to identify on records.

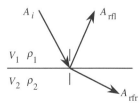

Figure 2-24 Illustration of quantities used in simple form of Zoeppritz' equations.

TABLE 2-7 Reflection Coefficients for P-wave at Normal Incidence

Density (g/cm3)—layer 1	**2.00**	Z1	3000
Density (g/cm3)—layer 2	**2.60**	Z2	11700
P-wave velocity (m/s)—layer 1	**1500**		
P-wave velocity (m/s)—layer 2	**4500**		

Reflection coefficient	0.59	Energy fraction reflected	0.35
Refraction coefficient	0.41	Energy fraction refracted	0.65

The densities and velocities in Table 2-7 are representative for a layer of saturated sediment overlying bedrock. In this case a large fraction of incident energy is reflected, which, of course, diminishes the energy transmitted into layer 2. It is quite instructive to insert various velocity and density values to determine if this relationship is maintained. You should discover that as the velocities become quite similar (as in a sequence of sedimentary rocks), the energy fraction reflected becomes quite small. If a negative sign appears, this designates that the waveform has undergone a 180° phase reversal. Also, insert values such that $\rho_1 V_1 = \rho_2 V_2$. As long as these two products are equal, all the energy is transmitted. The greater the difference between the products, the greater the percentage of the energy that is reflected.

Additional Factors

A number of other factors will affect the amplitudes of waves arriving at a given detecting point in addition to spherical spreading, absorption, and energy partitioning. Although a number of these will be discussed in the following section and in the next two chapters, they are included here to develop a complete picture.

As will become evident in the next section and is self-evident at any rate, the amount of energy imparted to create the seismic waves in the first place will directly affect the amplitudes of these waves at any distance from the energy source. However, it is not only the amount of energy used but how well it is coupled to the ground that is important. An explosion on the ground surface will produce far less impact from the explorationist's point of view than an equivalent explosion buried a meter or two below the surface in a tightly packed hole.

Similarly, good coupling is important for the detecting device (geophone). In the best possible situation, the geophone becomes a part of the ground itself, so that it precisely mimics ground motion. The sensitivity of the instrument to this ground motion also is critical. And, of course, it doesn't matter how well coupled and sensitive the geophone is if the recording instrument is incapable of amplifying and capturing as much information from the geophone response as possible.

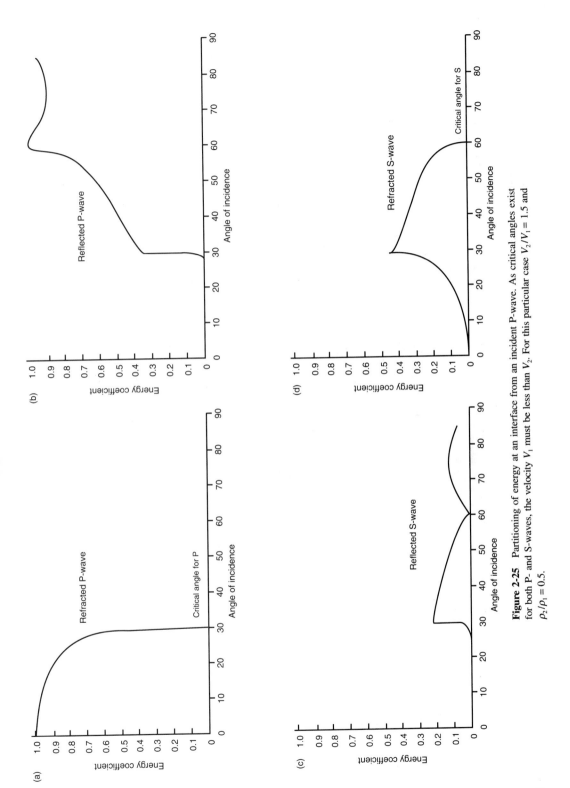

Figure 2-25 Partitioning of energy at an interface from an incident P-wave. As critical angles exist for both P- and S-waves, the velocity V_1 must be less than V_2. For this particular case $V_2/V_1 = 1.5$ and $\rho_2/\rho_1 = 0.5$.

Noise is produced by wind, traffic, airplanes, pumps, microseisms, and myriad other factors. It is present at every exploration site, although the intensity clearly will vary. These ground vibrations may destructively interfere with the various waves passing through the subsurface and may render their arrival at a detecting point impossible to detect.

Although many of these factors are obvious, all play as important, if not more important, a role in a site investigation as the specific details of spreading, absorption, and partitioning. Careful attention to site conditions and exploration design are just as critical for good results as a full theoretical knowledge of wave behavior.

ENERGY SOURCES

A wide variety of energy sources have been and still are used in seismic exploration. Because our emphasis is on the shallow subsurface, we will concentrate on the most appropriate sources for elucidating targets at shallow depths. What are the characteristics of the ideal energy source? First and foremost, it must impart sufficient energy to provide reflections and refractions from the contacts between different materials whether they are rock units or sediments. The amplitudes must be sufficient to be detected and displayed reasonably well for interpretative purposes. Thus, not only the total amount of energy imparted is important but also its frequency content and the shape of the waveform at the source. For most purposes we do not want to generate a broad waveform of complex shape.

Source Types

In land-based work most energy sources fit into one of three categories: weight drops, explosive, or vibratory. Of these, weight drops and explosives are most commonly used in shallow work. The simplest example of the weight drop is the hammer source. A sledge hammer, usually 5.4 or 7.3 kg, is swung against a metal plate in order to more efficiently couple the energy transfer. Obviously, the energy imparted is not great, and in the past a hammer source was used only for very shallow investigations. The hammer blow became a more useful source with the advent of relatively inexpensive instruments that could sum ground motion for a number of impacts and thus enhance the waveforms on records. Under good conditions a hammer source routinely can detect the overburden-bedrock interface at depths of 50 m or slightly more. The majority of other weight-drop systems depend on a truck or trailer to transport the weight, lift it to a height of 2 m or more, and release it. The weights vary from steel plates to leather bags filled with birdshot and can weigh as much as 250 kg or more.

Explosive sources run the gamut from dynamite to 50-caliber rifles. What we tend to consider as "true" explosives most frequently are items similar to primacord (an explosive in the form of a rope from which lengths can be cut and detonated) and containers of ammonium nitrate. These materials can be rigged in the field for a wide range of energy inputs, but all require placement in drilled holes typically on the order of several meters in depth. In a great many cases the target depths are sufficiently large or ground conditions are sufficiently poor that only large explosive charges will produce interpretable results.

However, with the advent of the enhancement seismograph, the shotgun source has become widely used in shallow work (Pullan and MacAulay, 1987). It clearly is superior to the hammer source and rivals or exceeds the smaller weight-drop sources (Fig. 2-26). Note that the vertical scale in Figure 2-26 is logarithmic, so that the energy contrast between the 7.3-kg hammer and the 8-gauge shotgun is substantial.

Vibratory sources are used mainly for deep exploration targets in the range of several hundreds of meters to tens of kilometers. They typically require a large investment in field equipment and computer processing capabilities. Although this type of source is not of direct interest for shallow work, its approach is so different than weight drops or explosives that a few words of description seem warranted. Weight drops and explosions apply a momentary force at the source. The typical vibratory system, Vibroseis, is oscillatory in character and lasts for a few seconds. Essentially, a large mass is placed in contact with the ground surface and vibrated according to a specific pattern. During the several seconds' duration of signal generation, the signal frequency is continuously varied. This long variable signal results in a final record of reflections and other events that cannot be interpreted directly but requires computer processing to separate the events of interest. Because the exact form of the source signal is known, such extraction is possible although complicated.

In marine conditions hammer blows are not very effective, and other sources must be used. The most common of these are air guns and sparkers. Both methods create a shock wave surrounding an air bubble in the water that is similar to one created by an explosive source but with somewhat more control and much less hazard. It is this shock wave produced by rapidly compressing the water molecules in the vicinity of the source that spreads out and travels through the sediment and rock underlying the water column. Sparkers create an air bubble by an electric discharge which vaporizes the water in the vicinity of the

Figure 2-26 A comparison of relative energies and frequency content for shotgun and weight-drop seismic sources.

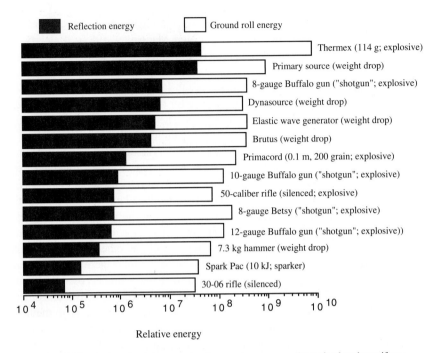

Figure 2-27   A plot of relative energies from various sources determined under uniform conditions at an experimental site. Based on information in Miller et al. (1986, p. 2067–2092).

discharge. Air guns are designed to suddenly release a volume of air under high pressure into the water, thereby creating the bubble. Variations of these methods have been applied to land-based surveys, but for a variety of reasons are not in common use.

Figure 2-27 provides a comparison of the energy output for a variety of sources measured under identical conditions. Be sure to note that the scale is logarithmic. In addition to the energy output of the source the frequency content of the signal (see Fig. 2-26) and the amount of ground roll generated also are important. However, a number of other factors must be taken into account before a source is selected for a particular study.

Source Considerations

Since almost all seismic exploration programs work within specific budget constraints, cost becomes an important factor. In typical surveys a large number of impacts or shots take place. One must therefore take into consideration repeat costs as well as initial cost. A source with high initial cost but with low cost per event may be preferable to one with a low initial cost but high repeat cost.

Convenience and efficiency also enter into an energy source decision. What is the portability of a given source? How easy is it to get the source to the exploration site? Does it take a long time to set up and break down and move? Can energy events be repeated at the same position under virtually similar conditions? Holes must be drilled for explosive sources, which require additional equipment, time, and expense. Access roads for large weight-drop sources may not be present.

The characteristics of the site itself often dictate the energy source. Some materials, such as dry sands, soak up energy like a sponge. If a thick sequence of dry sands is present, hammer and shotgun sources are virtually useless.

Safety and environmental concerns also must enter into the energy equation. Explosives usually may not be used in an urban setting, whereas large weight drops present little hazard and do minimal environmental damage.

SEISMIC EQUIPMENT

Although the sophistication of seismic equipment varies dramatically depending upon the nature of the exploration effort and the resources of the party undertaking the exploration, it is possible to reduce the equipment configuration to three elements: the detection of ground motion, the conditioning of the weak signal from the detectors, and the recording of the conditioned signal.

Signal Detection

As you already are aware, the devices that detect the ground motion from arriving wave energy commonly are called *geophones* or *seismometers*. These are simple but elegant in design (Fig. 2-28), small in size (Fig. 2-29), and quite rugged. Geophone design essentially consists of a cylindrical coil of fine wire suspended in a cylindrical cavity in a magnet. As

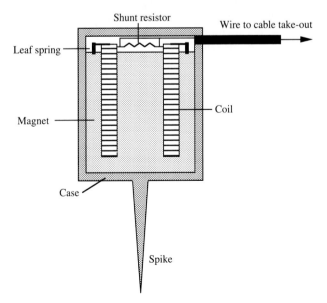

Figure 2-28   Schematic cross section of typical geophone elements.

Figure 2-29   Photograph of typical geophones used for shallow exploration work. From left to right: 14-Hz phone, 30-Hz phone, and 50-Hz phone with marsh case. Each geophone has a spike for insertion into ground and two clips for connecting to seismic cable.

illustrated in Figure 2-28, the coil is held in place by leaf springs. The coil and magnet are mounted within a protecting case, usually plastic, which also has a long metal spike for anchoring the geophone to the ground. When ground motion occurs, the geophone moves, inducing relative motion between the magnet and the coil due to the inertia of the coil. This motion generates a voltage that is proportional to the amount of ground displacement. It is, of course, this voltage that is returned to the equipment and constitutes the "raw" signal before electronic processing.

   Most of the details of geophone design are not required in order to understand and apply the basics of seismic exploration techniques, but there are a few additional important considerations. First, it is critical to be sure the geophone is coupled as well as possible to the ground. Thus, the spike should be inserted as far as possible into the ground but should not be rocked from side to side during insertion as this will loosen the contact between the spike and ground. Also, in the standard geophone the relative motion between coil and magnet is constrained to the vertical, so that horizontal ground motion cannot be measured. Geophones designed to measure shear waves are readily available, however, and function essentially the same as already described except that motion is constrained parallel to the horizontal. In marine surveying, unless geophones are placed on the lake bed or sea floor, conventional geophones obviously cannot be used. In this situation hydrophones are the detector of choice. These contain piezoelectric materials that produce a potential difference when subjected to pressure. Pressure differences occur in the water due to arriving reflected and refracted waves.

One of the most important aspects of a given geophone is its natural frequency. This is the frequency for which the output of the geophone has the greatest value, as illustrated in Figure 2-30. Once such an oscillatory system is activated, it will continue to oscillate even when ground motion has ceased. If such a freely oscillating system were exposed to an arriving seismic pulse composed of many frequency components, the high response at the geophone's natural frequency and its continuing oscillations would make it impossible to faithfully record the details of the ground motion and therefore the arriving waveform. For this reason geophones are damped in order to control the oscillations. Damping reduces the sensitivity of the system, so that the amount of damping is a compromise between controlling oscillations and maintaining sensitivity.

A natural damping already occurs due to the interaction between the magnetic field of the magnet and the magnetic field produced by current flowing in the coil. Additional damping is achieved by connecting a shunt resistor across the coil of the geophone (see Fig. 2-28). Resistors with higher values produce less damping than resistors with lower values. If a geophone is damped to a point where a gentle tap fails to produce an oscillation, the geophone is said to be critically damped. The damping factor usually chosen during manufacturing is 0.7 of the critical damping value. This value produces what is termed a *flat response*, which means that the geophone responds with the same sensitivity to all frequencies above the natural frequency (Fig. 2-30).

Depending upon the seismic method and objectives, geophone selection is important. Recalling that we have characterized the earth as a low-pass filter, you should remember

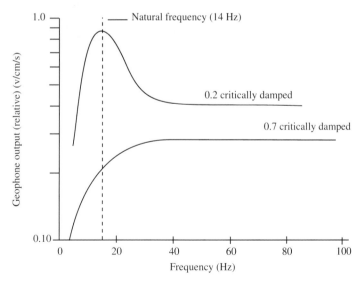

Figure 2-30 Response curves for 14-Hz geophone. 0.2 critical damping still produces a much greater response at the geophone's natural frequency than at other frequencies. 0.7 critical damping produces a more or less equal response to all frequencies above the natural frequency.

(a)

(b)

Figure 2-31   (a) Relative amplitude of seismic waves transmitted by Earth materials plotted against frequency. (b) Contrast in parts of frequency spectrum sampled by 100-Hz geophone (shaded) and 14-Hz geophone (shaded and unshaded areas beneath curve).

that if we plot relative transmitted amplitude of seismic waves against frequency (Fig. 2-31(a)), maximum amplitudes will be at low frequencies. For reasons we soon will discover, refraction surveying design usually seeks the largest possible signal for waves that arrive first at the geophone. Thus, for refraction work we should select geophones with a natural frequency in this same low-frequency range. A typical refraction geophone sold for shallow work has a natural frequency of 14 Hz and can sample the frequency spectrum illustrated in Figure 2-31(b).

Figure 2-32 Diagrammatic representation of basic components of seismic exploration equipment.

In our earlier discussion of seismic wavelengths we learned that the shorter wavelengths have higher frequencies and therefore higher resolution. In most shallow reflection work we are interested in as much resolution as possible and therefore want to isolate the higher-frequency component of the seismic pulse. For detailed shallow work we might choose geophones with natural frequencies of 50 or 100 Hz. These would reduce the low-frequency signals with high amplitudes (Fig. 2-31(b)), emphasize the high-frequency pulse, and increase resolution.

In a typical low-budget shallow-target seismic survey, geophones are used in groups of 12, 24, or, less frequently, 48. The output of each geophone is transferred to amplifying and recording instruments by means of a multiconductor cable. Each geophone connects to the cable with two clips at *take-outs* which are spaced at uniform intervals along the cable to facilitate placing the geophones at equal spacings along the ground (Fig. 2-32).

Signal Conditioning

The term *signal conditioning* is used in several contexts in exploration geophysics. Here we use the term to include signal amplification and filtering. The amplifier is an electronic device that receives and amplifies the very small signals transmitted by the geophones. Most amplification systems in modern seismographs treat the incoming signal so as to maximize the information in the signal. The main objective is to suppress somewhat the early strong signals, especially at geophones close to the energy source, while increasing the signal in the latter part of the record or at geophones at substantial distances from the energy source. Many engineering seismographs have gain controls which permit amplifier gain to be set at low levels for geophones near the energy source and to be systematically increased for geophones farther along the cable (Fig. 2-32). Some systems offer AGC (Automatic Gain Control) which attempts to keep amplified output at a more or less constant level. If you are interested in a more detailed discussion of seismic amplifiers, Telford et al. (1990, p. 197–202) contains a concise and readable discussion of both analog and digital amplification strategies.

Another important procedure in electronic treatment of the geophone signal is filtering. Simply put, electronic filters cut or filter out certain frequencies and pass along others for recording. This capability is useful for two major reasons: (1) the ability to remove unwanted noise and (2) the ability to restrict the seismic pulse frequencies recorded and eventually displayed for interpretation. At some field sites unwanted ground vibrations (noise) generated by one or more of many possible sources (e.g., pumps or wind) are sufficiently great to mask the desired seismic signal. Often this noise has greater amplitudes in a limited frequency range. By cutting or filtering these frequencies it is sometimes possible to reduce noise to acceptable levels. A typical engineering seismograph (Fig. 2-32) can display ground motion on an oscillograph screen prior to energy input. Various filter values can be selected and their effect viewed on the screen to determine if filtering will be effective in reducing the noise level.

In shallow reflection work, as noted previously, we are interested in high resolution and therefore in the higher frequency portion of the seismic signal. We have already demonstrated the advantages of geophones with natural frequencies of 50 or 100 Hz. An additional strategy is to employ high-pass or low-cut filters. The capabilities of the typical equipment used for shallow seismic exploration, while impressive (especially compared to what was available just a few short years ago), do have limitations. Because the amplitudes of the lower-frequency signals are so large relative to the higher frequencies, the limits of the amplifiers are reached (saturated) by these lower frequencies, and the higher frequen-

Figure 2-33   Frequency spectrum sampled by combination of 100-Hz geophone and 200-Hz filter.

Figure 2-34   Effect of filtering on relative energies for frequency spectrum produced by 8-gauge shotgun source. Based on data in Pullan and MacAulay (1987).

cies are not recorded. However, by filtering the lower-frequency component, the higher-frequency component can be preserved. Figure 2-33 is patterned after Figure 2-31 except that here we show diagrammatically the effect of a 200-Hz high-pass filter. Figure 2-34 also illustrates the effect of filtering. This diagram plots relative energy against frequency for a filtered (300-Hz high pass) and nonfiltered (all-pass) field trial using an 8-gauge shotgun source.

Signal Recording

Virtually all seismic equipment being used in serious exploration endeavors today has some form of digital recording capability. Older equipment had only analog capability. The differences between analog and digital are straightforward. In analog recording the amplified and filtered geophone signal is essentially routed directly to a recording device that is activated by continually varying voltages. While not normally utilized in field seismographs, a strip-chart recorder is an excellent example of this process. The constantly moving ink pen draws a curve on a paper roll due to a constant rotation of the paper roll and continual deflection of the pen due to incoming voltage levels. In the early days of seismic exploration most analog records were recorded on photographic paper, which was then replaced by magnetic tape. In all cases, however, in analog recording, the detail of the signal is limited by the capabilities of the recording device.

These limitations are not present in digital recording. The main difference in digital recording occurs after signal conditioning. The conditioned signal is routed through an analog to digital converter (A/D converter), where the analog signal is converted to digital form. This is accomplished by sampling the signal at discrete intervals (perhaps 2 ms) and

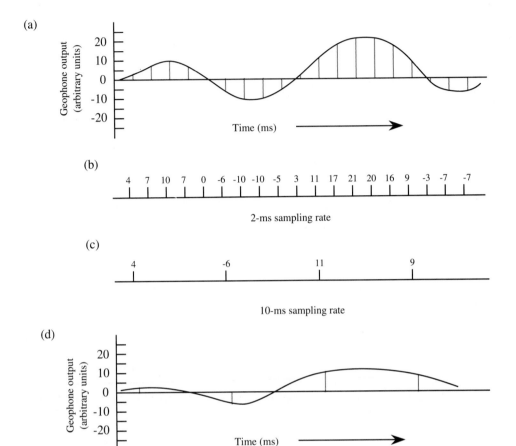

Figure 2-35 (a) Analog output of geophone (arbitrary units). (b) Digital representation of geophone response in (a) taken at 2-ms intervals. (c) Digital representation of same waveform at 10-ms sampling interval. (d) Reconstruction of waveform from data in (c) illustrates effect of inadequate sampling rate.

representing the amplitude of the signal at that moment by a signed number (Fig. 2-35). Such a process theoretically places no limit on the ratio of the greatest to least signal amplitude that can be recorded (the dynamic range), since the signal at any moment in time is represented by a number, no matter how large. The sampling rate is important, however. If the sampling intervals are too far apart, the true form of the signal is not preserved and the higher frequencies are lost (Fig. 2-35). Thus, in practice, the constraint placed on a digital system is primarily available memory in the digital recorder, as more memory is required in order to store larger numbers and to sample at higher rates (more frequently).

Several benefits accrue from digital recording in addition to greater signal fidelity. The data are stored in digital form on magnetic media (tape or disk) and can be read directly by computer for further display and processing. The data can be displayed at any time by digital to analog conversion. Also, many engineering seismographs possess signal-enhancement capabilities. Because the data is in digital form, the waveforms from several hammer blows or shots can be added together. This process tends to enhance or increase the wave amplitude of the true reflection and refraction arrivals, as their arrival pattern is always the same for the same shot point. Thus, amplitudes that are too small to be identified because of small energy sources may eventually be distinguished because their waveforms add constructively. Background noise, however, may have a substantial randomness component, such that noise interference is reduced somewhat by destructive interference.

Most modern engineering seismographs have, in addition to digital recording capability, an oscilloscope display, so that the seismic signal from each geophone can be viewed immediately after each shot (Fig. 2-36). This is invaluable in monitoring the quality of the data being recorded and also forms a critical component of a number of field procedures to be developed in subsequent chapters. Most instruments also have the capability of producing a permanent record on electrostatic or light-sensitive paper. These records (seismograms) include a trace from each geophone and timing lines spaced at equal intervals (usually 5 or 10 ms). Usually an option is present to plot the traces simply as a wiggly-line trace (Fig. 3-21) which resembles the standard analog record or a wiggly-line trace with variable-area shading superimposed (Fig. 4-9). The standard wiggly-line record tends to be used more for refraction surveys, whereas the wiggly-line/variable-area trace is used more frequently in reflection surveys.

Figure 2-36 Photograph of display on seismic unit. Horizontal lines are timing lines. Each vertical trace represents one geophone's response after amplification and filtering.

SUMMARY

You now should have some idea of how seismic waves are generated, how they are propagated, and what controls their velocities. In this context it is important to remember the velocities for compressional waves of common Earth materials and the ratio of shear wave and surface wave velocities to P-wave velocities. The relationship of frequency to wavelength also is important for understanding why certain techniques are used during data collection.

The relationships between incident and refracted and reflected rays form the foundation for the seismic refraction and reflection methods, and working with these relationships should be second nature. Having a good command of what types of seismic energy are expected along a line of geophones is the first step in interpreting field seismograms. You also should have a feel for expected amplitudes and frequencies and the connection between the latter, the energy source characteristics, and the effect of the materials through which the energy passes.

Finally, you should be comfortable in discussing the role of geophone selection and filter selection in sampling the arriving seismic waveform. If you can view a field seismogram and begin to think in terms of wave arrivals, velocities, frequencies, attenuation, energy source characteristics, and equipment sampling capabilities, you are prepared to undertake a detailed discussion of the two basic seismic exploration techniques: refraction and reflection seismology.

Before proceeding to the next chapter, however, work through the few problems that follow. This will help to reinforce your knowledge and will increase your expertise when it comes time to collect field data and interpret them.

PROBLEMS

2-1. A common compressional wave velocity for saturated, unconsolidated materials is 1400 m/s. If a seismic wave with a frequency of 10 Hz is traveling through this material, what is the wavelength of the wave? What is the wavelength of a wave with a frequency of 100 Hz?

2-2. What is the wavelength of a 10-Hz wave propagating through bedrock with a velocity of 4500 m/s? What is the wavelength of a wave with a frequency of 100 Hz?

2-3. Calculate the velocity of a compressional wave in a homogeneous rock layer with a density of 2.60 g/cm^3, a Young's modulus of 0.39×10^{11} N/m^2, and a Poisson's ratio of 0.11.

2-4. Calculate the velocity of a shear wave in a homogeneous rock layer with a density of 2.04 g/cm^3, a Young's modulus of 0.07×10^{11} N/m^2, and a Poisson's ratio of 0.14.

2-5. Estimate the velocity of Rayleigh waves in a homogeneous rock layer with a density of 2.52 g/cm^3, a Young's modulus of 0.14×10^{11} N/m^2, and a Poisson's ratio of 0.16.

2-6. In most common rocks density varies from 2.0 to 3.0 g/cm^3, Young's modulus varies from 0.12×10^{11} to 1.1×10^{11} N/m^2, and Poisson's ratio varies from 0.04 to 0.3. Using the Table 2-1 spreadsheet template, determine whether density variations or variations in elastic coefficients have the greatest effect on P-wave velocities.

2-7. A sandstone has a density of 2.28 g/cm^3, a Young's modulus of 0.14×10^{11} N/m^2, and a Poisson's ratio of 0.06. A geophone is located 50 m from a hammer blow. How long will it take the air wave, the direct compressional wave, and the surface (Rayleigh) wave to travel from the hammer blow to the geophone?

2-8. An arkose has a density of 2.62 g/cm³, a Young's modulus of 0.16×10^{11} N/m², and a Poisson's ratio of 0.29. Twelve geophones are arranged along a line at 10-m intervals. The shot point is located 5 m from the first geophone in the line (and 15 m from the second geophone, etc.). Construct a graph that illustrates time of arrival (vertical axis) against geophone position (horizontal axis) for the air wave, direct P-wave, direct S-wave, and surface wave.

2-9. Use Huygens' principle to demonstrate that plane waves encountering an opening in a barrier with a dimension similar to that of the wavelength of the waves produce the diffraction pattern illustrated in Figure 2-18. You need demonstrate only the *general* form of the diffraction pattern.

2-10. Demonstrate that the relationship specified in Eq. 2-26 between an incident P-wave and a refracted S-wave is valid (use Huygens' principle).

2-11. Demonstrate that the relationship specified in Eq. 2-26 between an incident S-wave and a refracted P-wave is valid (use Fermat's principle).

2-12. A ray from a sonic boom strikes the earth's surface at an incident angle of 20°. The material beneath the surface has a P-wave velocity of 500 m/s and an S-wave velocity of 200 m/s. Determine the angles of refraction of the refracted P- and S-waves.

2-13. An incident P-wave strikes an interface between two different rock types. The upper layer has a compressional wave velocity of 1200 m/s. The lower layer has a compressional wave velocity of 3800 m/s and a shear wave velocity of 1900 m/s. The incident angle is 18°. Calculate the angles of refraction for the refracted P- and S-waves.

2-14. A P-wave with an angle of incidence of 18° encounters a contact between two different rock types. The upper layer has a compressional wave velocity of 2500 m/s. The lower layer has a compressional wave velocity of 3800 m/s and a shear wave velocity of 1400 m/s. Calculate the angles of refraction for the refracted P- and S-waves.

2-15. A high-velocity rock unit (2500 m/s) overlies a low-velocity rock unit (1400 m/s). Calculate the angle of refraction for a P-wave incident at 30°.

2-16. Examine Snell's law (Eq. 2-25) and the relationships you calculated in problems 2-14 and 2-15. Formulate a general statement concerning the angles of incidence and refraction for compressional waves when (a) the material above an interface has a lower velocity than the material below the interface and (b) the material above an interface has a higher velocity than the material below the interface.

2-17. Given a P-wave traveling through a rock unit (velocity = 1200 m/s). This rock unit is in contact with another unit that has greater seismic wave velocities (P-wave = 3800 m/s; S-wave = 1900 m/s). At what angle must a ray strike the contact between these two units to produce a compressional head wave? a shear head wave?

2-18. Derive a general relationship to compute the minimum distance from shot to detector at which a head wave can be received. Refer to the diagram below as a guide and use the same symbols in your equation.

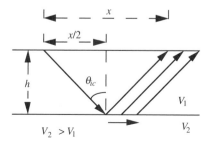

REFERENCES CITED

Birch, Francis, 1966, Compressibility; elastic constants, *in* Clark, S. P., Jr., ed., Handbook of physical constants, rev. ed.: Geological Society of America Memoir 97, p. 97–173.

Dobrin, Milton B., and Savit, Carl H., 1988, Introduction to geophysical prospecting, 4th ed.: New York, McGraw-Hill, 867 p.

Means, W. D., 1976, Stress and strain: New York, Springer-Verlag, 339 p.

Miller, R. D., Pullan, S. E., Waldner, J. S., and Haeni, F. P., 1986, Field comparison of shallow seismic sources: Geophysics, v. 51, p. 2067–2092.

Press, Frank, 1966, Seismic velocities, *in* Clark, S. P, Jr., ed., Handbook of physical constants, rev. ed.: Geological Society of America Memoir 97, p. 97–173.

Pullan, S. E., and MacAulay, H. A., 1987, An in-hole shotgun source for engineering seismic surveys: Geophysics, v. 52, p. 985-996.

Richter, Charles F., 1958, Elementary seismology: San Francisco, California, W. H. Freeman and Company, 768 p.

Slotnick, Morris Miller, 1959, Lessons in seismic computing, edited by Geyer, Richard A.: The Society of Exploration Geophysicists, p. 70–76.

Telford, W. M., Geldart, L. P., and Sheriff, R. E., 1990, Applied geophysics, 2nd ed.: Cambridge, Cambridge University Press, 770 p.

Tooley, R. D., Spencer, T. W., and Sagoci, H. F., 1965, Reflection and transmission of plane compressional waves: Geophysics, v. 30, p. 552–570.

Zoeppritz, K., 1919, Über reflexion und durchgang seismischer wellen durch Ünstetigkerlsfläschen: Über Erdbebenwellen VII B, Nachrichten der Königlichen Gesellschaft der Wissenschaften zu Göttingen, math.-phys. K1, Berlin, p. 57–84.

3

Seismic Exploration: The Refraction Method

Refraction seismology has a long and distinguished history. Although seismographs were developed in the latter part of the nineteenth century, it was not until the early twentieth century that major discoveries resulted from refraction techniques.

In 1909 A. Mohorovičić determined that a velocity discontinuity exists at the base of the crust based on his interpretation of wave arrivals from a nearby earthquake. Then, in 1913 B. Gutenberg determined the depth to the Earth's core. Thus, refraction seismology was responsible for our early knowledge of the basic structure of the Earth.

By the 1920s refraction techniques were being vigorously applied to oil exploration and were very successful. As seismic techniques and equipment improved, reflection eventually replaced refraction as the dominant method applied in oil exploration. However, refraction continues to be the most frequently applied seismic technique for shallow subsurface investigations. It is mainly for this reason that we consider it first in this text.

A HOMOGENEOUS SUBSURFACE

Before beginning to examine the refraction method, let's examine once again waves spreading throughout a homogeneous subsurface. As the hemispherical wavefront passes by a string of equally spaced geophones, each records the ground displacement due to this wave. The time of transit of this wavefront from the energy source (shot point) to each geophone can be determined from a field seismogram. From the geophone spacing (geophone inter-

Figure 3-1 Generalized diagram illustrating ray paths in a uniform material with no discontinuities. Time-distance relationships for the portion of the wave traveling directly from energy source to receiver are shown in the travel-time curve.

val) and the distance from the shot point to the first geophone (shot offset), we can construct a graph (travel-time curve) that plots time against distance (Fig. 3-1).

Because the wave is traveling at a constant velocity and the geophones are equally spaced, this time-distance plot will be a straight line. Although somewhat trivial in this case, it is instructive to begin an analysis that we will follow throughout most of the rest of our seismic studies. Let's ask ourselves if we can write an equation for this line. The form of the equation will be

$$\text{time} \;=\; \frac{\text{path}}{\text{velocity}} \qquad\qquad (3\text{-}1)$$

or

$$t \;=\; \frac{x}{V_1} \qquad\qquad (3\text{-}2)$$

For an equation this simple in form, it is fairly obvious that we can determine the velocity of the homogeneous material if we know the time and distance values. It's useful to review (or present) a bit of calculus at this point. Taking the first derivative of this equation with respect to x gives us the slope of the line. Thus,

$$\frac{dt}{dx} \;=\; \frac{1}{V_1} \qquad\qquad (3\text{-}3)$$

and

$$\text{slope} \;=\; \frac{1}{V_1} \qquad\qquad (3\text{-}4)$$

or

$$V_1 = \frac{1}{\text{slope}} \tag{3-5}$$

This means that if we have a travel-time curve similar to that in Figure 3-1, we can determine the slope (rise/run), take the inverse, and multiply by 1000 if the units for our graph are in meters and milliseconds. Following this procedure, the velocity of the direct wave determined from the travel-time curve in Figure 3-2 is 286 m/s. Although this insight into the subsurface is minimal, it does give us some information about the material near the surface because of its low velocity (see Table 2-2). However, as we soon shall see, the amount of useful information we can extract from typical travel-time curves is much greater.

A SINGLE SUBSURFACE INTERFACE

You realize, of course, that the subsurface usually is not homogeneous. We normally expect to have several interfaces present. Based on our discussion in chapter 2, it is clear that such interfaces will produce reflections, refractions, and wave conversions. At the moment we will limit our analysis to refractions and, in particular, to refractions that result in energy returning to the surface where it may be sensed and recorded. You should recall the relationship that leads to critical refraction and produces a head wave. This might be a propitious time to review Snell's law, the critical angle, and critical refraction.

Figure 3-3 illustrates the path from energy source E to geophone G that one such critically refracted ray follows. A compressional wave generated at E and traveling at V_1 strikes the interface between materials with different velocities, V_1 and V_2. The ray that strikes this interface at the critical angle, θ_{ic}, is refracted parallel to the interface and travels at V_2. A head wave is generated as energy is returned to the surface along rays such as that traveling from N to G. Several possible questions arise from an analysis of Figure 3-3. Can we detect

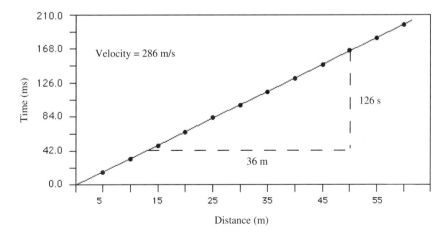

Figure 3-2 Travel-time curve (time-distance graph) illustrating only direct wave arrivals. The velocity is 286 m/s as determined from the inverse of the slope.

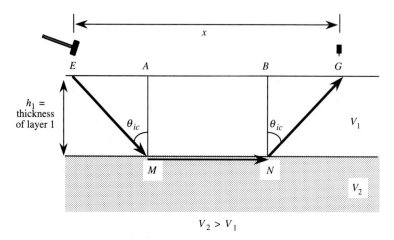

Figure 3-3   Diagram illustrating symbols used in derivation of time of travel for critically refracted ray.

the arrivals of this refracted energy? How long does it take for the energy generated at E to arrive at G? How would the head wave arrivals plot on a travel-time curve? Can we extract any information about the subsurface from the head wave assuming we can identify it on seismograms?

Derivation of Travel-Time Equation

A logical first step is to follow a procedure such as we undertook for the direct wave analysis in the preceding section. After we derive a travel-time equation for the refracted wave, we can analyze the equation and see what is possible. In deriving an equation of form similar to Eq. 3-1, we will use the relationships and quantities in Figure 3-3.

The first step is a simple one. The total time of travel must be

$$\text{time} \quad = \quad \frac{EM}{V_1} + \frac{MN}{V_2} + \frac{NG}{V_1} \tag{3-6}$$

Since

$$\cos \theta_{ic} \quad = \quad \frac{h_1}{EM} \quad \text{and} \quad EM = NG$$

then

$$EM = NG = \frac{h_1}{\cos \theta_{ic}} \tag{3-7}$$

Also,

$$EA = BG = h_1 \tan \theta_{ic} \quad \text{and} \quad MN = x - 2h_1 \tan \theta_{ic}$$

so

$$\text{time} = \frac{h_1}{V_1 \cos\theta_{ic}} + \frac{x - 2h_1\tan\theta_{ic}}{V_2} + \frac{h_1}{V_1\cos\theta_{ic}} \qquad (3\text{-}8)$$

Equation 3-8 is the basic formula for determining time of travel for a critically refracted ray at a single interface. If we know h_1, V_1, and V_2, we then can compute the travel time for any distance x from energy source to receiver. We could deduce several interesting relationships at this point, but Eq. 3-8 is more useful if we recast it in a slightly different form. Let's take care of this task before proceeding to an analysis. First, we rearrange terms to produce

$$\text{time} = \frac{2h_1}{V_1\cos\theta_{ic}} - \frac{2h_1\tan\theta_{ic}}{V_2} + \frac{x}{V_2} \qquad (3\text{-}9)$$

Next, we utilize the relationships

$$\tan\theta_{ic} = \frac{\sin\theta_{ic}}{\cos\theta_{ic}} \quad \text{and} \quad \sin\theta_{ic} = \frac{V_1}{V_2}$$

to change Eq. 3-9 to

$$\text{time} = \frac{2h_1}{V_1\cos\theta_{ic}} - \frac{2h_1\sin^2\theta_{ic}}{V_1\cos\theta_{ic}} + \frac{x}{V_2} \qquad (3\text{-}10)$$

Consolidating, we obtain

$$\text{time} = \frac{2h_1(1 - \sin^2\theta_{ic})}{V_1\cos\theta_{ic}} + \frac{x}{V_2} \qquad (3\text{-}11)$$

Using the familiar relationship

$$\sin^2\theta_{ic} + \cos^2\theta_{ic} = 1$$

and the identities noted previously, we can substitute to obtain

$$\text{time} = \frac{2h_1\cos\theta_{ic}}{V_1} + \frac{x}{V_2} \qquad (3\text{-}12)$$

$$\text{time} = \frac{2h_1(1 - (V_1^2/V_2^2))^{1/2}}{V_1} + \frac{x}{V_2} \qquad (3\text{-}13)$$

$$\text{time} = \frac{2h_1(V_2^2 - V_1^2)^{1/2}}{V_2 V_1} + \frac{x}{V_2} \qquad (3\text{-}14)$$

With the final form of our travel-time equation (3-14) directly in front of us, let's follow the same procedure as we did for the direct wave and take the first derivative of the equation with respect to x. This provides a perhaps surprisingly simple result:

$$\frac{dt}{dx} = \frac{1}{V_2} \qquad (3\text{-}15)$$

Therefore, we can say that the travel-time equation for the critically refracted ray is an equation of a straight line with a slope of $1/V_2$. Since $V_2 > V_1$, a slope of $1/V_2$ must be less than a slope of $1/V_1$.

Analysis of Arrival Times

This is sufficient information to decide if the critically refracted ray can be utilized in resolving subsurface geology. In our analysis we will use the geometry in Figure 3-4 and the values in Table 3-1. We can place the derived travel-time equations for the direct and critically refracted rays in a spreadsheet and calculate travel times for various geophone placements. Table 3-1 shows the results for one combination of geophone intervals and geology.

 If we scan the arrival times in this table, we see that the direct wave arrives first at distances up to 27 m from the shot but that the refracted wave arrives before the direct wave at all geophones at distances greater than 27 m. Thus, if we had a geophone at each distance position in Table 3-1, the first event on a field seismogram for geophones 1–10 would be the direct wave arrival, and the first event for geophones 11–24 would be the refracted wave arrival. What is a simple physical explanation of this relationship? If we examine the subsurface model in Figure 3-4, we see that the direct ray travels from left to right at 1400 m/s. The critically refracted ray travels to the interface at 1400 m/s, is refracted and travels parallel to the interface at 4500 m/s, and then returns to the surface at 1400 m/s. At significant

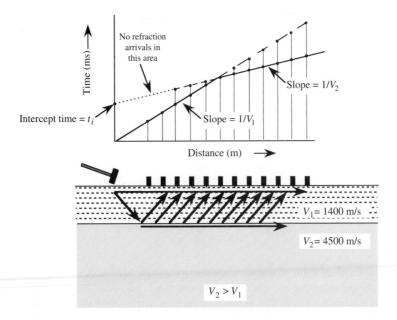

Figure 3-4 Generalized diagram illustrating ray paths in a material with one horizontal discontinuity. Time-distance relationships for both the direct and refracted rays are shown in the travel-time curve.

TABLE 3-1   Arrival Times for Direct and Refracted Waves

Distance from Shot (m)	Arrival Times (ms)	
	Direct Wave	Refracted Wave
0	0.00	13.58
3	2.14	14.24
6	4.29	14.91
9	6.43	15.58
12	8.57	16.24
15	10.71	16.91
18	12.86	17.58
21	15.00	18.24
24	17.14	18.91
27	19.29	19.58
30	21.43	20.24
33	23.57	20.91
36	25.71	21.58
39	27.86	22.24
42	30.00	22.91
45	32.14	23.58
48	34.29	24.24
51	36.43	24.91
54	38.57	25.58
57	40.71	26.24
60	42.86	26.91
63	45.00	27.58
66	47.14	28.24
69	49.29	28.91

Velocity 1-P (m/s)	**1400**
Velocity 2-P (m/s)	**4500**
Depth (m)	**10**

Shot offset (m)	**0**
Geophone interval (m)	**3**

No refracted wave intervals before (ms)	6.55

distances from the energy source, the greatest portion of the refracted wave's path is at a velocity more than three times as great as the direct wave. At some surface point, therefore, the refracted wave must arrive before the direct wave, which is traveling at a much slower velocity.

Is this information sufficient for us to reach a decision on the utility of the critically refracted wave? If we plot the arrival times in Table 3-1 against distance from the energy source (geophone positions), we produce a graph very similar to the travel-time curve in Figure 3-4. The direct wave times plot along a straight line with slope = $1/V_1$ that must pass through the origin (assuming the energy source to be at the surface), and the refracted wave times plot along a straight line with slope $1/V_2$. If we had only these times and distances (as would be the case when analyzing field data), we could deduce that we were dealing with a two-layer case (single interface) because we have two travel-time segments, and we could compute V_1 and V_2. The velocity values would provide us with some insight into the nature of the subsurface materials, but we still would not know the thickness of the upper layer. However, the travel-time equation we derived contains a layer-thickness term h_1, so there must be some way to solve for h_1 if we know V_1 and V_2.

Note that Table 3-1 also is a dynamic table and that values in bold may be changed. Experiment with the velocity and thickness values to convince yourself that the refracted wave ultimately will arrive before the direct wave. If you select velocities that are similar (remember the condition that $V_2 > V_1$), or if you select a very large depth to the interface, the times at the 24 distance positions in the table may not illustrate this relationship. The reason is that the most distant position is not far enough from the shot point for the refracted wave to arrive first. If this happens, increase the geophone interval until you are satisfied that the relationship holds. (Actually, if you have this difficulty, you have discovered one of the limitations of the refraction method, which will be discussed shortly.)

Determining Thickness

Examine Figure 3-4 once again. The straight line passing through the arrival times for the critically refracted ray can be extended until it intersects the time axis. This time is termed the *intercept time t_i*. Recall that it has no real physical significance, because no refractions arrive at the energy source ($x = 0$). However, at $x = 0$ our travel-time Eq. 3-14 reduces to

$$\text{time} = t_i = 2 h_1 \frac{(V_2^2 - V_1^2)^{1/2}}{V_2 V_1} \tag{3-16}$$

and, therefore,

$$h_1 = \frac{t_i}{2} \frac{V_2 V_1}{(V_2^2 - V_1^2)^{1/2}} \tag{3-17}$$

Thus, for a single horizontal interface, if we can determine times for direct and refracted arrivals from a field seismogram, we can calculate the thickness of the material above the interface and the velocities of the materials above and below the interface.

Crossover Distance

Before proceeding further, it is perhaps a good time to examine two other matters of interest. First, although we now have all the information necessary to solve problems related to a single horizontal interface, there is an additional approach that occasionally is used, and we now develop it for sake of completeness. If you examine the travel-time curve in Figure 3-4, you see that the straight lines for the direct wave times and refracted wave times intersect at a point. The horizontal coordinate of this point, x_{co}, is referred to as the *crossover distance*. Crossover distance can be used instead of t_i to develop a solution for h_1, the depth to the interface. At x_{co} the times of travel for the direct wave and the refracted wave are equal, and, therefore,

$$\text{time}_{\text{direct wave}} = \frac{x_{co}}{V_1} \qquad \text{(see 3-2)}$$

$$\text{time}_{\text{refracted wave}} = \frac{2h_1(V_2^2 - V_1^2)^{1/2}}{V_2 V_1} + \frac{x_{co}}{V_2} \qquad \text{(see 3-14)}$$

and

$$\frac{x_{co}}{V_1} = \frac{2h_1(V_2^2 - V_1^2)^{1/2}}{V_2 V_1} + \frac{x_{co}}{V_2} \qquad (3\text{-}18)$$

If we rearrange terms, we arrive at

$$\frac{h_1(V_2^2 - V_1^2)^{1/2}}{V_2 V_1} = \frac{x_{co}}{2}\left(\frac{V_2 - V_1}{V_2 V_1}\right) \qquad (3\text{-}19)$$

and

$$h_1 = \frac{x_{co}}{2}\left(\frac{V_2 - V_1}{V_2 V_1}\right)\left(\frac{V_2 V_1}{(V_2^2 - V_1^2)^{1/2}}\right) \qquad (3\text{-}20)$$

The $V_1 V_2$ terms cancel, and remembering that $(V_2^2 - V_1^2) = (V_2 - V_1)(V_2 + V_1)$, we can simplify Eq. 3-20 to

$$h_1 = \frac{x_{co}}{2}\left(\frac{V_2 - V_1}{V_2 + V_1}\right)^{1/2} \qquad (3\text{-}21)$$

It is mainly a matter of convenience whether we use Eq. 3-21 or Eq. 3-17 to determine layer thickness. However, because determination of x_{co} involves fitting *two* lines to field data, whereas t_i requires only one line, t_i usually can be determined with more accuracy. In the computer program RefractSolve that accompanies this book we utilize t_i (Eq. 3-17).

Critical Distance

The second item of interest is one you have seen several times before (see Problem 2-18). You should recall that, based on the geometry of the critical refraction, there is a finite distance from the energy source to the first point at which this refracted energy can be received. This distance is referred to as the *critical distance* and is equal to x_{crit} in Figure 3-5. Please note that in many refraction modeling programs and in the tables and dynamic tables used in this text, refraction times are given for distances less than the critical distance. These times obviously cannot be valid but are given to avoid the complexity of programming a spreadsheet to exclude them. You must remember, however, the geometry of critical refraction, especially when planning and interpreting field surveys.

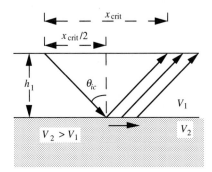

Figure 3-5 This diagram illustrates the relationships for calculating critical distance. Critical distance is the minimum distance from the energy source at which the first critical refraction can be received.

Derivation of an equation to calculate critical distance is straightforward (which you already know if you solved Problem 2-18).

$$\tan \theta_{ic} = \frac{x_{crit}/2}{h_1} \tag{3-22}$$

and, since

$$\sin \theta_{ic} = \frac{V_1}{V_2},$$

then

$$\tan \left[\sin^{-1} \left(\frac{V_1}{V_2} \right) \right] = \frac{x_{crit}/2}{h_1} \tag{3-23}$$

and

$$x_{crit} = 2 h_1 \tan \left[\sin^{-1} \left(\frac{V_1}{V_2} \right) \right] \tag{3-24}$$

Equation 3-24 is quite sufficient for our purposes in this day of inexpensive hand calculators which handle trigonometric functions with ease. However, it is possible to use the trigonometric identities we employed when deriving the travel-time equation for head waves to arrive at a different form for Eq. 3-24:

$$x_{crit} = \frac{2 h_1}{\left[\left(V_2/V_1 \right)^2 - 1 \right]^{1/2}} \tag{3-25}$$

Table 3-2 presents some critical-distance values for a range of layer thicknesses. The velocities used to calculate these values are similar to those for unsaturated sands overlying saturated sands (see Table 2-2). Note that the percentage of critical distance relative to thickness is substantial. In a case such as this the positioning of geophones is important if one is to gather the maximum amount of information in the most economical fashion.

TABLE 3-2 Critical-Distance Values for Various Thicknesses

Thickness (m)	Critical Distance (m)
2	1.53
4	3.06
6	4.59
8	6.12
10	7.65
12	9.18
14	10.71
16	12.24
18	13.76
20	15.29

Velocity 1-P (m/s)	**500**
Velocity 2-P (m/s)	**1400**

Thickness increment (m)	2

x_{crit} as % of h_1	76%

> Try a variety of velocity values to investigate the role velocity contrast has in affecting the critical distance. Examine Table 2-2 in order to use realistic velocities.

You may have noticed that the path of the first critically refracted ray to arrive at the surface is the same as a reflection which strikes the interface at the critical angle. Thus, the reflected and refracted waves arrive at the same time at x_{crit}. However, as we will later demonstrate, all other reflections arrive later than the critical refraction and so will not interfere with our analysis, which depends on the first wave arriving at each receiver.

Constructing a Travel-Time Curve from a Field Seismogram

Now that you have a good grasp of the relationships that enable us to determine velocities and thickness for a single horizontal interface separating two differing materials, let's examine a field seismogram from a site with this subsurface geology. Figure 3-6 illustrates a straightforward example. Pertinent information is contained in the caption for this figure. Recall that we are interested in the time at which the first energy arrives at each geophone. In Figure 3-6 this is indicated by the first sharp downturn in each geophone's output and is marked by arrows to help you recognize this first break.

Construct a table similar to Table 3-3 in which you record the time of each first break at each geophone. Then graph this data so as to create a time-distance plot similar to that in Figure 3-7(a). Use a sufficiently large sheet of cross-ruled paper (at least 8½ X 11 in.) so that slopes and the intercept time can be determined accurately. When measuring first-break times, it's a good idea to construct a template on a transparent sheet on which you have

Figure 3-6 Field seismogram from the Connecticut Valley, Massachusetts. Geophone traces are labeled 1–12. The first geophone is located 5 m from the energy source. The geophone interval is 3 m. First breaks for each trace are indicated by a downward directed arrow. Timing lines are at 5-ms intervals. The record encompasses 100 ms. This seismogram exhibits a classic two-layer pattern.

drawn lines at 0.5- or 1-ms intervals. Place this sheet on the seismogram in order to read values accurately and consistently.

Your next task is to interpret the data points. The first eight data points define a straight line reasonably well (Fig. 3-7(b)). The line for the direct wave must pass through the origin if the energy source was located on the surface, and this line satisfies that criterion. The last four data points also lie along a straight line. This plot looks like a familiar case in which we receive the direct wave at the closest geophones and then the head wave at the most distant geophones. Next, measure the slopes of these lines and from the slopes compute the velocities. The intercept time can be read from the vertical axis (46 ms), and the subsurface structure computed from Eq. 3-17. Our computations suggest a 9.2-m-thick layer with a velocity of 385 m/s overlying material with a velocity of 1300 m/s.

Using RefractModel and RefractSolve

Now that you have learned some of the basics of the refraction method, it seems like an appropriate time to introduce two of the computer programs included on the disk that accompanies this textbook. Instructions for using these programs are in Appendix A.

TABLE 3-3   Time-Distance Values for Seismogram in Figure 3-6

Geophone	1	2	3	4	5	6	7	8	9	10	11	12
Distance (m)	5	8	11	14	17	20	23	26	29	32	35	38
Time (ms)	13.5	21.5	28.5	37.0	43.5	51.5	59.0	62.5	68.5	71.0	73.0	75.0

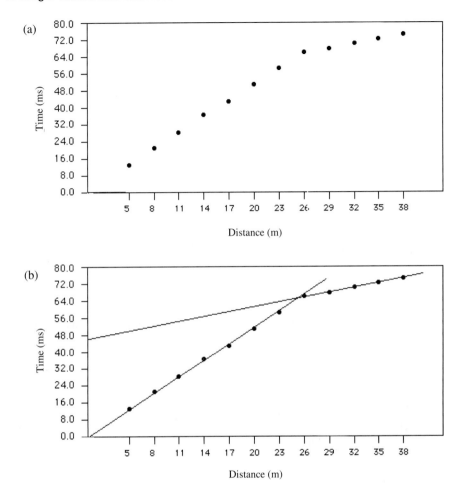

Figure 3-7   (a) Time-distance data from the seismogram in Figure 3-6. (b) Lines drawn through data points in (a).

RefractModel produces time-distance data for subsurface models that you designate. As such, it is an extremely useful tool for investigating various relationships as we discuss them in the pages that follow. Because the goal of this chapter is to enable you to interpret travel-time curves based on the arrival of refracted waves, the more experience you have with curves representing different subsurface conditions, the better your interpretations will be. So use RefractModel freely and often. A final note: RefractModel plots time-distance data as if energy sources were placed at both ends of a string of geophones. Such practice is common, but as we have not yet discussed the reasons for this procedure, study only the data points represented by the same symbols as in Figure 3-7.

RefractSolve plots time-distance data that you enter or from data files created by RefractModel. You then fit straight lines to the data by drawing with a mouse. The power of this program is in its ease of use. Because you can draw lines quickly and because subsurface structures are computed quickly, you easily can attempt several interpretations when straight lines do not perfectly pass through all field-data points. Such multiple interpretations are encouraged in order to gain some notion of the range of models that fit your data.

From time to time in the following pages, suggestions are offered for exercises utilizing these computer programs. Try to spend some time investigating these suggestions, since by doing so your interpretative abilities will improve substantially.

It is possible to save RefractModel data as a data file, transfer directly to RefractSolve, and then analyze the RefractModel data. This capability is a useful learning aid, and you are urged to try this as well.

Use RefractModel to investigate a number of subsurface variations for the single interface case. First, hold layer thickness constant and try a variety of velocity contrasts to gain an appreciation of velocities and related slopes. Remember always that we assume $V_2 > V_1$. Next hold velocities constant and vary the thickness from very small to very large. Note the manner in which the travel-time curves change. Can you formulate a number of situations in the field for which you might not recognize that two layers (one interface) are present?

The Mohorovičić Discontinuity

Virtually everyone who has taken a physical geology course is familiar with the velocity discontinuity that separates the earth's crust from the mantle. Most remember this interface as the "Moho" or M-discontinuity, a few recall that this terminology incorporates (to some extent) the name of its discoverer, Andrija Mohorovičić (1857–1936), and fewer still are aware of the evidence advanced by Mohorovičić to demonstrate the existence of the *Mohorovičić discontinuity*.

Following a strong earthquake in October 1909 that affected his place of work at the Zagreb Meteorological Observatory, Mohorovičić plotted the main deflections on seismograms recorded by 29 seismological observatories located at distances from very near the epicenter out to 2400 km from the epicenter (Bonini and Bonini, 1979). Considering only the P-wave arrivals, Mohorovičić noted that on seismograms recorded at stations close to the epicenter only one pulse appeared, but on seismograms from more distant stations there were two clearly distinguishable pulses. This same pattern held for S-wave arrivals (see Figure 3-8). Mohorovičić reasoned that the early pulses recorded at the more distant stations must be due to waves that traveled at a greater velocity over a considerable portion of their path distance. He computed velocities from the travel-time curves and calculated that

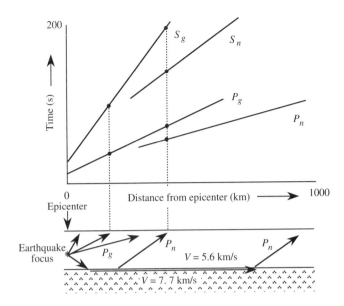

Figure 3-8 Travel-time curves simplified from Mohorovičić's original plots. P_g is the direct compressional wave, and P_n is the refracted compressional wave. Terminology for shear waves S is similar. The section below the travel-time curve illustrates Mohorovičić's interpretation of the travel-time curves.

the velocity discontinuity was located at a depth of 50 km. This was quite a remarkable achievement at such an early date in the history of seismology.

If you examine Figure 3-8, you should recognize our classic two-layer case (one horizontal interface). The direct wave curves for P- and S-waves (labeled P_g and S_g) do not pass through the origin, because the earthquake occurred not on the surface but at depth. Travel from the focus to the epicenter took a finite amount of time. Also, curves for both P- and S-waves are shown (S-ray paths are not shown). Because the distances traversed by P- and S-waves are measured in kilometers in this case (rather than in meters as in most exploration work), sufficient time is present for the disturbance due to the arrival of the P-wave to die down before the arrival of the S-wave. Therefore, it usually is possible to recognize S-wave arrivals. Although this presentation is simplified somewhat, you should be able to understand the essential elements of the evidence for the Mohorovičić discontinuity.

Depending on the objectives of your field study and the length of your geophone spread, you often will encounter a two-layer case. As you are well aware, however, most subsurface geology is more complex than a single horizontal interface. Let's take a look at the next step up in complexity.

TWO HORIZONTAL INTERFACES

It is not difficult to envisage a subsurface configuration in which there are two horizontal interfaces (often referred to as a three-layer case). For instance, consider an alluvial sequence of sands and gravels overlying bedrock. The sands and gravels near the surface normally will be above the water table, dry, and possess a relatively low velocity. The water table forms the first interface, as the saturated sands and gravels will have a substantially higher velocity than their dry counterparts. Finally, the top of the bedrock surface forms the second interface, as bedrock velocities are higher than saturated sediments. In attempting to derive a travel-time equation for this situation, we proceed in precisely the same manner as for a single interface.

Derivation of a Travel-Time Equation

Referring to Figure 3-9 we see that for this geology a direct wave will travel from E to G and that wave energy also will traverse the path $EPQG$. This is identical to the situation of the single horizontal interface. Wave energy also will strike the V_2-V_3 interface. Assuming $V_3 > V_2$, critical refraction will occur, so we are interested in a wave path that follows $EPRSQG$. We already know that $\sin \theta_{ic} = V_2/V_3$. The major question we must answer before proceeding is how to determine the θ_i required for ray PR to be critically refracted. Figure 3-9 is drawn to call your attention to the answer to this question. Study the drawing carefully before continuing.

We know from Eq. 2-25 (Snell's law) that

$$\frac{\sin \theta_i}{\sin \theta_r} = \frac{V_1}{V_2}$$

and, since $\theta_r = \theta_{ic}$, then

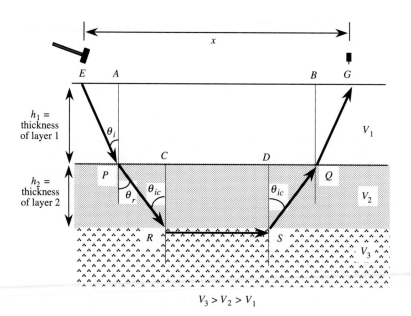

$$V_3 > V_2 > V_1$$

Figure 3-9 Diagram illustrating symbols used in derivation of time of travel for ray critically refracted along the second interface in a three-layer case.

$$\sin \theta_r \;=\; \sin \theta_{ic} \;=\; \frac{V_2}{V_3} \quad \text{and, therefore,} \quad \sin \theta_i \;=\; \frac{V_1}{V_3} \qquad (3\text{-}26)$$

Everything we now derive follows the same procedure as for the single interface. The complete travel-time equation is

$$\text{time} \;=\; \frac{EP}{V_1} + \frac{PR}{V_2} + \frac{RS}{V_3} + \frac{SQ}{V_2} + \frac{QG}{V_1} \qquad (3\text{-}27)$$

As before,

$$EP = QG = \frac{h_1}{\cos\theta_i} \quad \text{and} \quad PR = SQ = \frac{h_2}{\cos\theta_{ic}}$$

Also,

$$EA = BG = h_1 \tan\theta_i \quad \text{and} \quad PC = DQ = h_2 \tan\theta_{ic}$$

so that

$$RS \;=\; x - 2\,h_1 \tan\theta_i - 2\,h_2 \tan\theta_{ic} \qquad (3\text{-}28)$$

These relationships can be substituted into Eq. 3-27, so that our travel-time equation becomes

$$\text{time} \;=\; \frac{2\,h_1}{V_1 \cos\theta_i} + \frac{2\,h_2}{V_2 \cos\theta_{ic}} + \frac{x - 2\,h_1 \tan\theta_i - 2\,h_2 \tan\theta_{ic}}{V_3} \qquad (3\text{-}29)$$

It is worthwhile to present the major steps in simplifying this equation, but as the steps are similar to those used to arrive at Eq. 3-14, they are presented without comment. The same identities that we used previously are adequate.

$$\text{time}_{RS} \;=\; \frac{x}{V_3} - \frac{2\,h_1 \tan\theta_i}{V_3} - \frac{2\,h_2 \tan\theta_{ic}}{V_3} \qquad (3\text{-}30)$$

$$\text{time} \;=\; \frac{x}{V_3} + \frac{2\,h_1}{V_1 \cos\theta_i} + \frac{2\,h_2}{V_2 \cos\theta_{ic}} - \frac{2\,h_1 \sin^2\theta_i}{V_1 \cos\theta_i} - \frac{2\,h_2 \sin^2\theta_{ic}}{V_2 \cos\theta_{ic}} \qquad (3\text{-}31)$$

$$\text{time} \;=\; \frac{x}{V_3} + \frac{2\,h_1 - 2\,h_1 \sin^2\theta}{V_1 \cos\theta_i} + \frac{2\,h_2 - 2\,h_2 \sin^2\theta_{ic}}{V_2 \cos\theta_{ic}}$$

$$\text{time} \;=\; \frac{x}{V_3} + \frac{2\,h_1 \cos\theta_i}{V_1} + \frac{2\,h_2 \cos\theta_{ic}}{V_2} \qquad (3\text{-}32)$$

and, finally

$$\text{time} \;=\; \frac{x}{V_3} + \frac{2\,h_1 (V_3^2 - V_1^2)^{1/2}}{V_3 V_1} + \frac{2\,h_2 (V_3^2 - V_2^2)^{1/2}}{V_3 V_2} \qquad (3\text{-}33)$$

Once again we finish with an equation for a straight line. As you probably have observed by now, if we once again take a derivative—voila!

$$\frac{dt}{dx} = \frac{1}{V_3} \qquad (3\text{-}34)$$

Although it is beginning to sound a little repetitious, we can continue with the same type of analysis as we did for the single interface. Examination of Figure 3-10 should convince you that at some point the ray that has a part of its path along the second interface eventually will be the first to arrive at one of the distant geophones. We see that the velocity of the third layer can be determined by taking the inverse of the slope of the straight line passing through the arrival times from the ray that was critically refracted along the second interface.

Determining Thickness

The line corresponding to refractions from the second interface can be extended back to the vertical axis (Fig. 3-10) and is referred to as the *second intercept time*, or t_{i_2}. If we use Eq. 3-33 and solve for h_2 at $x = 0$, we arrive at

$$h_2 = \left[t_{i_2} - \frac{2h_1(V_3^2 - V_1^2)^{1/2}}{V_3V_1} \right] \frac{V_3V_2}{2(V_3^2 - V_2^2)^{1/2}} \qquad (3\text{-}35)$$

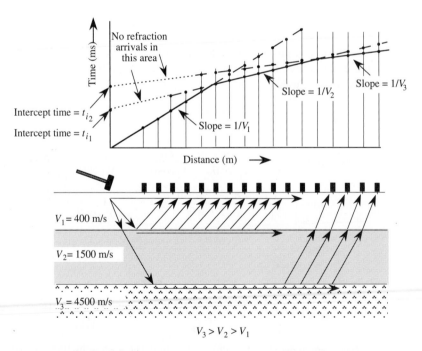

Figure 3-10 Generalized diagram illustrating ray paths in a material with two horizontal interfaces. Time-distance relationships for the direct and two critically refracted rays are shown in the travel-time curve.

V_1, V_2, and V_3 are computed from the inverse of the slopes of the three line segments on the travel-time curve in Figure 3-10. The thickness of the first layer h_1 is computed using t_{i_1} and Eq. 3-17. The second layer thickness is then calculated using t_{i_2} and Eq. 3-35.

Based on our discussions to this point, we interpret a travel-time curve such as that in Figure 3-10 as representing a three-layer case, because it has three straight-line segments, each with a lower slope than the one closer to the energy source. Velocities for each layer are available from the curve as are the thicknesses of the first two layers (and, of course, the depths to the two interfaces). From this information alone and some general knowledge of the local geology, we can arrive at a reasonably good model for the subsurface geology at the exploration site.

In the preceding analysis we used only intercept times to develop an equation for thickness of the second layer. We could develop an equation utilizing crossover distances as in the case of the single interface, but for reasons cited previously and as most computerized solutions use intercept times, it does not seem necessary to produce yet more equations. This derivation is left for those interested in such an exercise. Because of implications for field work, however, it does seem wise to once again turn to the topic of critical distance.

Critical Distance

The critical distance x_{crit} for the head wave produced by critical refraction along the second interface is EG in Figure 3-11. The ray path for this geometry is $EPRQG$. Thus, x_{crit} is equal to $EA + PC + CQ + BG$. Because $EA = BG = h_1 \tan \theta_i$ and $PC = CQ = h_2 \tan \theta_{ic}$, and because $\sin \theta_i = V_1/V_3$ and $\sin \theta_{ic} = V_2/V_3$, then

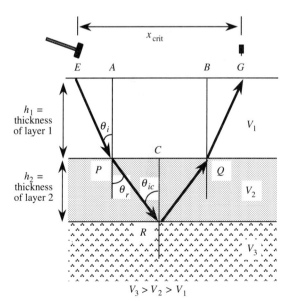

Figure 3-11 Diagram illustrating symbols used in derivation of critical distance for ray critically refracted along the second interface in a three-layer case.

$$x_{\text{crit}} = 2\left[h_1 \frac{\sin \theta_i}{\left(1 - \sin \theta_i^2\right)^{1/2}} + h_2 \frac{\sin \theta_{ic}}{\left(1 - \sin \theta_{ic}^2\right)^{1/2}} \right] \tag{3-36}$$

$$x_{\text{crit}} = 2\left\{ h_1 \frac{V_1/V_3}{\left[1 - \left(V_1/V_3\right)^2\right]^{1/2}} + h_2 \frac{V_2/V_3}{\left[1 - \left(V_2/V_3\right)^2\right]^{1/2}} \right\} \tag{3-37}$$

and

$$x_{\text{crit}} = 2\left[h_1 \frac{V_1}{\left(V_3^2 - V_1^2\right)^{1/2}} + h_2 \frac{V_2}{\left(V_3^2 - V_2^2\right)^{1/2}} \right] \tag{3-38}$$

In order to investigate the implications of this formula for critical distance, let's create a table similar to Table 3-2.

The velocities in Table 3-4 are fairly representative of what might be expected in an area with unsaturated and saturated sediments lying above bedrock. Although the magnitude of the second critical distance (the distance from the energy source to the point where refractions from the second interface can be received) is more than half the depth to the second interface (sum of h_1 and h_2), this should not be unexpected to anyone who examined Figure 3-11 closely.

TABLE 3-4 Critical Distances for Two Horizontal Interfaces

h_1 (m)	h_2 (m)	First X_{crit} (m)	Second X_{crit} (m)
1	4	0.76	2.84
2	8	1.53	5.68
3	12	2.29	8.53
4	16	3.06	11.37
5	20	3.82	14.21
6	24	4.59	17.05
7	28	5.35	19.90
8	32	6.12	22.74
9	36	6.88	25.58
10	40	7.65	28.42

Velocity 1-P (m/s)		**500**
Velocity 2-P (m/s)		**1400**
Velocity 3-P (m/s)		**4500**

h_1 increment (m)		**1**
h_2 increment (m)		**4**
Second X_{crit} as % of $h_1 + h_2$		57%

The magnitude of the second critical distance relative to the depth to the second interface varies according to the relative magnitudes of $V_1, V_2,$ and V_3. Use Table 3-4 as a dynamic table and insert realistic velocity values for a case in which the values are reasonably similar in magnitude and for a case in which the velocities are very different. Let the values in Table 3-4 represent the standard or general case for means of comparison. Do velocity ratios affect the critical distances to any significant extent?

Although critical distance is important, remember that the refraction method utilizes the first arrival at each geophone. Refer to Figure 3-11. It should be obvious that a wave traveling path *EPRQG* will take longer than a wave traveling from *E* to *G* that is refracted along the first interface. A wave refracted parallel to the second interface must travel a considerable distance at V_3 before overcoming its longer path distance and arriving prior to the head wave generated by the first interface. We can verify this impression by utilizing RefractModel. Figure 3-12 illustrates a plot using the same velocities as in Table 3-4 and layer thicknesses $h_1 = 5$ m and $h_2 = 20$ m. Table 3-4 tells us that the second critical distance for such values is 14.21 m, whereas Figure 3-12 demonstrates that it is approximately 60 m before the head wave from the second interface arrives prior to the head wave from the first interface.

Critical distance is of primary practical importance when trying to recognize later arrivals on a field seismogram. If we have a reasonable model for the subsurface, we can

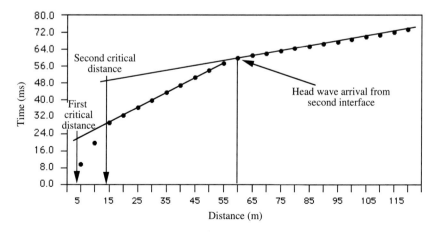

Figure 3-12   Travel-time curve based on a RefractModel plot using velocity values in Table 3-4 with $h_1 = 5$ m and $h_2 = 20$ m. Note that the first refraction from the second interface to be a first arrival is located at a distance considerably greater than the second critical distance.

calculate critical distance and compare waveforms from geophones at distances less than and greater than the critical distance. If conditions are very good, we might be able to recognize events in addition to the first arrivals. These additional data points then would enable us to refine our interpretation.

Analyzing a Second Field Seismogram

Now that we have increased our knowledge of the refraction method somewhat, let's interpret another field seismogram. The seismogram in Figure 3-13 was recorded approximately 0.5 km north of the field location for the seismogram in Figure 3-6. Note that the energy source–geophone offset and the geophone interval are different. Figure 3-13 is reasonably straightforward to interpret except that the first breaks are not quite as distinct as in Figure 3-6. Also, amplifier gains on channels 11 and 12 were quite high, so that background noise is evident and makes it a little more difficult to locate first breaks. Proceed in the same manner as before to gather your time-distance data.

When you are ready, either use RefractSolve or work by hand to interpret your data. Figure 3-14 illustrates our preferred interpretation. Energy in this case was by a hammer blow on the surface, so the direct wave should pass through the origin of the graph. As the data point from geophone 1 does not fit well on the line determined by points 2, 3, and 4 and lies below the line, we interpret it to be a direct wave arrival. Points 2, 3, and 4 are head wave arrivals from a shallow interface. Points 5–12 define a line with a clear change in slope and represent arrivals from a second interface. As is usually the case, the points do not exactly fit any straight line, so we visually fit a line that passes through or close to the maxi-

Figure 3-13 Field seismogram from the Connecticut Valley, Massachusetts. Geophone traces are labeled 1–12. The first geophone is located 3 m from the energy source. The geophone interval is 10 m. First breaks for each trace are indicated by a downward directed arrow. Timing lines are at 5-ms intervals. The record encompasses 100 ms.

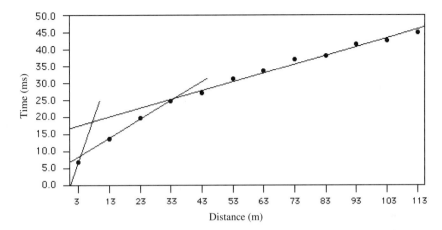

Figure 3-14   Preferred interpretation of time-distance data taken from Figure 3-13.

mum number of points. In fitting this line we also could disregard or weight less heavily points for which first breaks are not especially clear (e.g., trace 11).

Our analysis results in a thickness of 1.4 m and a velocity of 387 m/s for the first layer, a thickness of 10.0 m and a velocity of 1820 m/s for the second layer, and a velocity of 3871 m/s for the third layer. The field site is in an area where glacial sands and gravels overlie Triassic arkoses. We therefore interpret the first interface to represent the water table and the second interface to represent the contact between sands and gravels and bedrock.

Another reminder about the utility of computer programs such as RefractSolve is in order here. Fitting lines to the field data from Figure 3-13 is not completely unambiguous. Therefore, you should attempt several different interpretations in order to determine how much variation is present in the subsurface configuration computed for each interpretation. This takes only a few more minutes with RefractSolve (or a similar program) but provides you with a good sense of how much uncertainty is present in the final geologic solution you adopt.

MULTIPLE INTERFACES

Although the three-layer case represents a configuration with multiple interfaces (two), we might ask what a travel-time curve looks like for a situation in which there are more than two interfaces and velocity increases for each layer. Must we proceed through analyses like those just discussed in detail for a four-layer case, a five-layer case, and an *n*-layer case? If we reproduce Eq. 3-14 and 3-33 with the terms slightly rearranged, you can't help but notice a progressive similarity.

$$\text{time} = \frac{x}{V_2} + \frac{2 h_1 (V_2^{\,2} - V_1^{\,2})^{1/2}}{V_2 V_1} \tag{3-14}$$

$$\text{time} = \frac{x}{V_3} + \frac{2h_1(V_3^2 - V_1^2)^{1/2}}{V_3 V_1} + \frac{2h_2(V_3^2 - V_2^2)^{1/2}}{V_3 V_2} \qquad (3\text{-}33)$$

We can rewrite these equations in a slightly different form using a summation symbol and letting the letter n represent the number of layers (thus, the number of interfaces is $n - 1$). The equation represents the time of travel for a wave that is critically refracted at the $n - 1$ interface.

$$\text{time}_n = \frac{x}{V_n} + \frac{2}{V_n} \sum_{i=1}^{n-1} h_i \frac{(V_n^2 - V_i^2)^{1/2}}{V_i} \qquad (3\text{-}39)$$

Equation 3-39 has the same form as Eq. 3-14 and 3-33 and therefore represents the equation of a straight line with an intercept $t_{i_{n-1}}$ and with slope = $1/V_n$. Based on this information, we can say that a travel-time curve for a subsurface geology with n layers, each with a greater velocity than the one above, has n straight line segments with each progressive segment having a lesser slope than the preceding one. Equation 3-39 can be used to develop an equation that incorporates intercept time $t_{i_{n-1}}$ and solves for thickness h_{n-1}. Velocities are determined from the inverses of the n slopes, and thicknesses are calculated using intercept times and beginning with layer 1 and proceeding through layer $n - 1$.

Equations for thickness are not developed for $n > 3$, as the process is identical to what already has been presented. Also, for most investigations in the shallow subsurface it is not especially common for more than three line segments to be present. There are a number of reasons for this, some of which we will touch on later. One common reason is the similar velocities of saturated sediments and the velocity contrast between them and bedrock. This results in two obvious interfaces that we saw in the preceding exercise: the water table and the sediment-bedrock contact.

Fortunately, the iterative nature of the process lends itself very nicely to computer solution. The process just described is the one utilized by RefractSolve.

DIPPING INTERFACES

Although it would be simpler if the subsurface consisted only of horizontal interfaces between rocks and sediments, it would not be very interesting or challenging. In our previous discussion of the refraction method, we assumed horizontal interfaces. We all can visualize numerous natural situations in which interfaces are inclined or are irregular. First, let's examine the case of a single dipping interface and see how we might deal with this first departure from our idealized world.

Analyzing the Problem

To begin, study Figure 3-15. The geology in (a) and (b) is identical except that the interface in (b) dips at 4° down to the left. A refraction survey conducted at (a) produces a travel-time curve with two straight-line segments, which is exactly the same type of plot that a survey conducted at (b) would produce. There is absolutely no way to distinguish an inclined inter-

Figure 3-15   (a) Horizontal interface at a depth of 20 m with velocities above and below the interface of 500 m/s and 1500 m/s, respectively. (b) A dipping interface with identical depth to that in (a) at the site of the hammer impact and with identical velocities to (a). (c) Travel-time curves for (a) and (b).

face from a horizontal interface based on this type of field survey and the resulting travel-time curves. How then should we proceed? Although Figure 3-15 does not solve our problem, part (c) does contain a clue. The slope of the line labeled "horizontal interface" is equal to $1/V_2$, as you already know. However, the slope of the lined labeled "dipping interface" is less than the prior value and therefore gives a velocity that is greater than the true velocity. We refer to this as an *apparent velocity*.

Our clue lies in the reason for the different arrival times for the head wave in the two subsurface configurations. Notice in Figure 3-15(c) that the arrival of the refracted wave at each geophone is early relative to similar geophone positions in the case of the horizontal interface. The reason for this is evident when comparing (a) to (b). All rays in (a) returning to the surface from the interface have identical path distances to traverse, and all are longer than those in (b). Because the surface in (b) is inclined upward to the right, each ray returning to the surface through V_1 traverses a shorter distance than the ray immediately to its left. Thus, the gap between critically refracted wave *travel-times* from the horizontal and inclined interfaces increases up-dip. This accounts for the different slopes in (c). Of course, this observation does not solve our problem, but you might ask, What would a travel-time curve look like if we reversed the location of the energy source from the left edge of the diagram to the right edge? You should be able to predict that it will appear different from the horizontal interface case. How will it appear when compared to the curve for the dipping interface that we already examined?

Before attacking this question, we first need to compare the results of an energy source position switch for the case of the horizontal interface. Figure 3-16 illustrates such a situation. Geophones are distributed along the surface, and an energy source is placed as shown at the left edge of the diagram. After travel times are recorded from this position, the

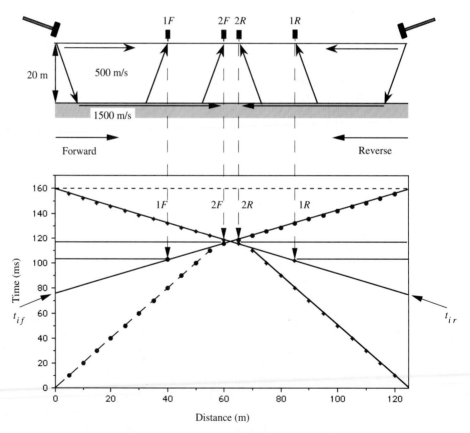

Figure 3-16 Correlation of a travel-time curve with wave paths to geophones at equal distances from an energy source for a forward and a reverse traverse.

energy source is moved to the other end of the geophone string as shown at the right edge of the diagram, and times are again recorded. The first position is called the *forward* traverse because it is the initial position. The second position is termed a *reverse* traverse because the energy source position is reversed relative to the first traverse. The main purpose of Figure 3-16 is to demonstrate that path distances and, hence, travel times are identical from either energy source position to a given geophone interval. For instance, the surface distance from the energy source for the forward traverse to geophone 1F (40 m) is the same as the surface distance from the reverse energy source to geophone 1R. The path distances for both refracted waves are identical, as are their travel times. Such geometry results in travel-time curves for each traverse that are identical.

It is customary to plot data from forward and reverse traverses in the manner of Figure 3-16. When the curves are plotted in this manner, they are mirror images of each other. Note, however, that if the data from both traverses were plotted with the origin at the lower left of the graph, the curves would fit precisely on top of each other. Also observe that because the curves are identical, the slopes of the respective direct and refracted wave segments are the same, as are the intercept times.

There is one additional important feature of this travel-time curve that illustrates a critical relationship to remember when we investigate dipping interfaces. Refer to the time it takes for wave energy that is refracted along the interface to travel from the position of the energy source for the forward traverse to the position of the energy source for the reverse traverse as T_F. T_R is the travel time from the reverse energy source to the position of the forward energy source. T_F and T_R must be equal because the paths of travel are precisely the same. This equivalence is known as *reciprocity*. That reciprocity holds in Figure 3-16 is evident because the uppermost horizontal line on the travel-time curve indicates that both the reverse and forward head wave curve segments intersect the vertical axes (coincident with the energy source locations) at 160 ms.

Now we can return to one of our original questions: What would a travel-time curve look like if we reversed the location of the energy source from the left edge of the diagram to the right edge? Figure 3-17 illustrates the same situation as Figure 3-15 except that a reverse traverse is added. As you likely guessed by now, the curves for a forward and reverse traverse are not the same when an interface is dipping. Therefore, we *always* collect data for both a forward and reverse traverse, as this is the only way to determine if an inclined interface is present. If the curves are not symmetrical, as in Figure 3-17 (as opposed to Figure 3-16), then the interface is inclined.

The reason that the curves are not symmetrical should be clear from Figure 3-17. Geophones 1F and 1R are the same distances from their respective energy sources (40 m) as are geophones 2F and 2R (60 m). Examine the path traveled by the incident and critically refracted wave. If we measure the distance traveled in the V_1 layer to geophone 1F, it is greater than a similar measurement for geophone 1R. This produces a greater travel time, as shown on the travel-time curve. This greater travel time for the forward traverse relative to the reverse traverse holds for all source-geophone positions except one. The differential between the forward and reverse times decreases as the source to geophone distance increases. What is the one case for which the times are equal? Recall the requirement of reciprocity. Examine Figure 3-17 and determine to your satisfaction that T_F and T_R, when

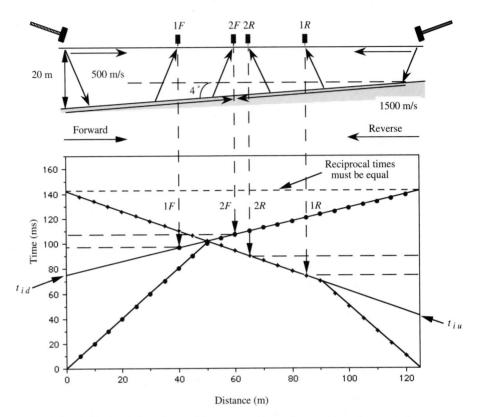

Figure 3-17   Correlation of a travel-time curve with geophone positions above a single dipping interface. The purpose of this diagram is to demonstrate the different path distances and arrival times for geophones located at identical offsets for a forward and reverse traverse.

defined as in our initial discussion of reciprocity, must be equal. Once again these particular path differences are the same, and head wave curve segments intersect the vertical axes at 142 ms for both forward and reverse plots.

In our discussion of Figure 3-15 we pointed out that the apparent velocity for layer 2 was greater than the true velocity. If we calculate a velocity from the reverse traverse segment, we discover that the apparent velocity is *less* than the true velocity. Actually, if dips are low (<10°) the average of the two apparent velocities is very nearly equal to the true velocity.

Before we begin a derivation of a travel-time equation for the single dipping interface, perhaps a brief review is in order. The main points to keep in mind are summarized in Figure 3-18. Forward and reverse travel-time curves are not symmetrical when a dipping interface is present. Velocity V_1 for the layer above the interface is determined in the standard way. V_1 values for the forward and reverse traverses must agree. Reciprocal times must be equal. Inverses of slopes for the head wave arrivals do not give V_2. The average of V_{2F} and V_{2R} approximates V_2 when the dip is low. Slope values are equated to m_u and m_d fol-

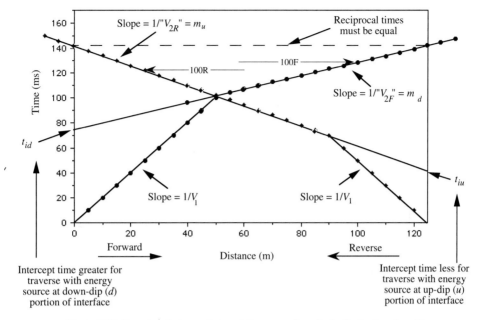

Figure 3-18 Important features of a travel-time curve for a single dipping interface. Note that reciprocal times must be equal. However, the times from forward and reverse points to a geophone at a given distance are not equal. This is illustrated by the arrows labeled 100F and 100R, which designate travel times to geophones located 100 m from the forward (F) and reverse (R) shots.

lowing the convention for the equation of a straight line: $y = mx + b$, where m designates slope. We can determine the direction of dip of the interface because the intercept time is less for the traverse with its energy source at the up-dip portion of the interface (where the vertical distance to the interface is least). Although this seems like a lot to keep straight, everything is important when interpreting curves related to dipping interfaces.

In order to become more adept at working with curves of this type, enter several configurations of velocity, depth, and dip into RefractModel. At first work only with two layers, but when you feel at ease with these curves, attempt some three-layer situations with both interfaces dipping. Following the rules you already know, you should be able to work out similar relationships for this more complicated situation. The convention used in this text for dip angles is: a positive angle represents a surface inclined upward toward the right; a negative angle represents a surface inclined downward toward the right.

Derivation of a Travel-Time Equation

The derivation of a travel-time equation for the single dipping interface essentially is the same as for the single horizontal interface. The initial equation has a few more terms, and reducing the equation to a more compact form requires a bit more effort. Otherwise everything is familiar.

The symbols employed in the following derivation are identified in Figure 3-19. The quantities h_d and j_d are located at the *down-dip* portion of the interface. In the travel-time equation time$_d$ represents the travel time when the energy source is located in the down-dip position. The intercept time determined from this equation is t_{id}, and the slope determined from the derivative of the equation is labeled m_d. Similarly, when the energy source is located at the up-dip portion of the interface, all these symbols take on the "u" subscript, as in h_u. You should be aware that this convention is the opposite of that used in most exploration geophysics textbooks. However, we believe that this terminology is easier to keep straight if you relate the subscript to the position of the energy source.

Note also that h is the vertical distance to the interface. This uses h in the same manner as in the case of horizontal interfaces. RefractSolve and RefractModel (as well as many other computer programs) refer to this quantity as thickness. This is not thickness in the stratigraphic sense but is usually thickness in the sense used by field explorationists. Thickness as measured perpendicular to layer boundaries always can be computed when the dip is known.

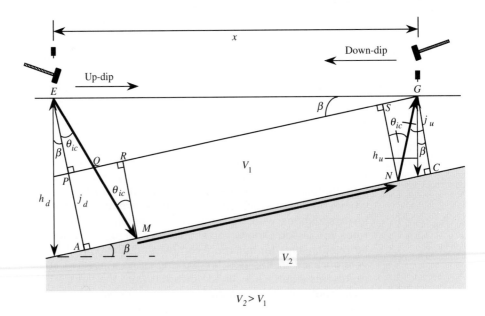

Figure 3-19 Diagram illustrating symbols used in derivation of travel-time equation for a single dipping interface.

In examining Figure 3-19, recall that θ_{ic} is measured from the perpendicular to the interface. First, we tackle the case when the energy source is located in the down-dip position (we are "shooting" up-dip).

$$time_d = \frac{EM}{V_1} + \frac{MN}{V_2} + \frac{NG}{V_1} \qquad (3\text{-}40)$$

$$EM = \frac{j_d}{\cos\theta_{ic}}, \quad NG = \frac{j_u}{\cos\theta_{ic}}$$

and since

$$j_u = j_d - EP = j_d - x\sin\beta,$$

$$NG = \frac{j_d - x\sin\beta}{\cos\theta_{ic}}$$

Since $MN = PG - AM - NC$, we determine the relevant quantities

$$PG = x\cos\beta, \qquad AM = j_d\tan\theta_{ic}, \quad \text{and} \quad NC = j_u\tan\theta_{ic} = (j_d - x\sin\beta)\tan\theta_{ic}$$

This gives us the basic travel-time equation

$$time_d = \frac{j_d}{V_1\cos\theta_{ic}} + \frac{x\cos\beta - j_d\tan\theta_{ic} - (j_d - x\sin\beta)\tan\theta_{ic}}{V_2} + \frac{j_d - x\sin\beta}{V_1\cos\theta_{ic}} \qquad (3\text{-}41)$$

We can reduce this equation to a simpler form by using the trigonometric identities we used in our previous derivations and by remembering

$$\sin(\theta_{ic} - \beta) = \sin\theta_{ic}\cos\beta - \cos\theta_{ic}\sin\beta$$

$$\sin(\theta_{ic} + \beta) = \sin\theta_{ic}\cos\beta + \cos\theta_{ic}\sin\beta$$

Finally, we arrive at

$$time_d = \frac{2j_d\cos\theta_{ic}}{V_1} + \frac{x}{V_1}\sin(\theta_{ic} - \beta) \qquad (3\text{-}42)$$

Our next step is to move the energy source to the up-dip part of the interface to shoot down-dip. The equation we arrive at looks remarkably similar to Eq. 3-41. After simplifying once again, we have

$$time_u = \frac{2j_u\cos\theta_{ic}}{V_1} + \frac{x}{V_1}\sin(\theta_{ic} + \beta) \qquad (3\text{-}43)$$

These two equations are not only very similar to each other but are not as different from the horizontal single-interface case as you might have predicted. Our next step, by now almost a ritual, is to determine the slopes of the straight lines that these equations define. Taking the derivatives of Eq. 3-42 and 3-43 results in

$$\frac{d(time_d)}{dx} = \frac{\sin(\theta_{ic} - \beta)}{V_1} \quad \text{and} \quad \frac{d(time_u)}{dx} = \frac{\sin(\theta_{ic} + \beta)}{V_1} \qquad (3\text{-}44)$$

Following the terminology established in Figure 3-18 for slopes and intercept times, we begin by rewriting Eq. 3-44 in the form

$$m_d = \frac{\sin(\theta_{ic} - \beta)}{V_1} \quad \text{and} \quad m_u = \frac{\sin(\theta_{ic} + \beta)}{V_1} \tag{3-45}$$

Remembering that $\sin \theta_{ic} = V_1/V_2$, we see that we can determine V_2 if we know θ_{ic}. Of course, we also want to know the dip β of the interface. Noting the + and - signs in Eq. 3-45, we see that we can solve these slope equations for both β and θ_{ic}. Therefore, we put these equations in the form

$$\theta_{ic} - \beta = \sin^{-1}(V_1 m_d) \quad \text{and} \quad \theta_{ic} + \beta = \sin^{-1}(V_1 m_u) \tag{3-46}$$

Because m_u, m_d, and V_1 are determined directly from the travel-time curve, we solve for θ_{ic}:

$$\theta_{ic} = \sin^{-1}(V_1 m_d) + \beta \quad \text{and} \quad \theta_{ic} = \sin^{-1}(V_1 m_u) - \beta$$

$$2\theta_{ic} = \sin^{-1}(V_1 m_d) + \sin^{-1}(V_1 m_u)$$

and, finally,

$$\theta_{ic} = \frac{\sin^{-1}(V_1 m_d) + \sin^{-1}(V_1 m_u)}{2} \tag{3-47}$$

Similarly, we determine the relationship for β using

$$\beta = \theta_{ic} - \sin^{-1}(V_1 m_d) \quad \text{and} \quad \beta = -\theta_{ic} + \sin^{-1}(V_1 m_u)$$

to arrive at

$$2\beta = \sin^{-1}(V_1 m_u) - \sin^{-1}(V_1 m_d)$$

and

$$\beta = \frac{\sin^{-1}(V_1 m_u) - \sin^{-1}(V_1 m_d)}{2} \tag{3-48}$$

Determining Thickness

However, we are not yet quite finished. We still need to derive an equation that provides a solution for thickness, at both the up-dip and down-dip portions of the interface. At this point we know the critical angle and the velocity of the first layer, so we can express Eqs. 3-42 and 3-43 in terms of intercept times:

$$t_{id} = \frac{2 j_d \cos\theta_{ic}}{V_1} \tag{3-49}$$

and

$$t_{iu} = \frac{2 j_u \cos\theta_{ic}}{V_1} \tag{3-50}$$

Noting from Figure 3-19 that $\cos \beta = j_d/h_d$, and $\cos \beta = j_u/h_u$, we complete our task for the single dipping interface

$$j_d = \frac{t_{id}V_1}{2\cos\theta_{ic}} \tag{3-51}$$

$$j_u = \frac{t_{iu}V_1}{2\cos\theta_{ic}} \tag{3-52}$$

$$h_d = \frac{j_d}{\cos\beta} \tag{3-53}$$

and

$$h_u = \frac{j_u}{\cos\beta} \tag{3-54}$$

MULTIPLE DIPPING INTERFACES

We now turn to the development of travel-time equations for a situation where more than one interface is present and at least one interface is inclined to the horizontal. The geometric technique follows the same pattern we utilized several times already but is much more complex. You would gain relatively little new knowledge by pursuing this in detail, so we simply present the results and encourage you to consult original sources if you are interested in details.

Travel-Time Equation

A clear exposition of the development of travel-time equations for several dipping interfaces is presented by Adachi (1954). Figure 3-20 is based on his work, but the symbols are changed to be consistent with those used in this text. Note that angles θ_i and ϕ_i are measured from the perpendicular to the various interfaces and therefore represent angles of incidence and refraction. Angles u_i and d_i are measured from the vertical. These symbols as well as other subscripts follow the same usage as in the case of the single dipping interface.

The equations developed by Adachi are

$$t_d = \frac{x\sin u_1}{V_1} + \sum_{i=1}^{n-1} \frac{h_{d_i}}{V_i}\left(\cos d_i + \cos u_i\right) \tag{3-55}$$

$$t_u = \frac{x\sin d_1}{V_1} + \sum_{i=1}^{n-1} \frac{h_{u_i}}{V_i}\left(\cos d_i + \cos u_i\right) \tag{3-56}$$

where n = the number of interfaces. Solutions to these equations are obtained by a combination of the procedures for the single dipping interface and multiple horizontal interfaces. Intercept times are determined for each straight line on the travel-time curve. The same

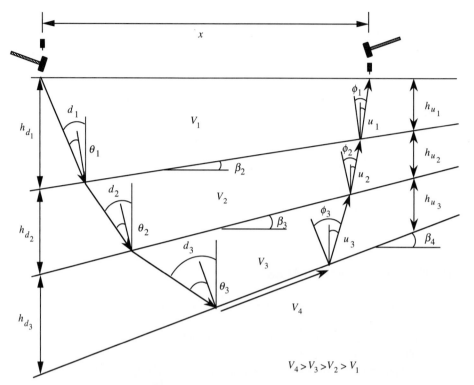

Figure 3-20   Diagram illustrating symbols used in travel-time equation for multiple dipping interfaces (after Adachi, 1954).

techniques are applied to determine slopes, angle of dip, and critical angle. These quantities then are used to solve for V_2, h_{d_1}, and h_{u_1}. Once all quantities related to the shallowest interface are known, Eqs. 3-55 and 3-56 are expanded with $i = 2$, and the process begins again. As before, this technique is an especially appropriate computer task, and RefractSolve is based on these equations.

> In order to demonstrate the applicability of Eqs. 3-55 and 3-56 for all cases, rewrite them for the case of $\beta = 0$ and $n = 1$, and compare the result to Eq. 3-2.

In practice Adachi's equations usually need to be expanded for only one or two dipping interfaces. The many complexities of the subsurface and the limitations of the refraction method combine to make it generally impracticable to apply this approach to more complicated situations. On the positive side are the typical goals of shallow refraction surveying. In the majority of cases, the objectives of the survey are such that these equations are sufficient for our purposes.

Analyzing Field Seismograms

You now have the knowledge to interpret the majority of routine refraction surveys. In order to review procedures and to practice with more complicated seismograms, let's examine a representative survey.

Figure 3-21 illustrates four field seismograms of the wiggle-trace variety. Two are from a forward traverse, and two are from an accompanying reverse traverse. Field parameters are noted in the figure caption. Because the instrument used in this survey has only 12-channel capability, long traverses requiring 24 or 36 geophones must be recorded in 12-geophone segments. Hence, the need for two records for the forward and reverse traverses.

These records are instructive for a variety of reasons. First, conditions at the field site are fairly typical. First breaks on the geophones near the Buffalo-gun ("shotgun") energy source are reasonably clear, but first breaks on the more distant geophones are sometimes very difficult to identify. This is not at all unusual at localities that do not have particularly efficient energy-coupling characteristics (dry sand). Computer programs such as RefractSolve require first-break times for every geophone, but questionable times can be ignored when drawing travel-time curves. You therefore should keep careful records so that you know which first-break times you have more confidence in. These should guide you during your interpretation. Once again, remember that obtaining a solution from the travel-time curves is only a matter of a few minutes, so be sure to try several possible fits to your data to determine to what extent your computed subsurface model varies depending upon line fits.

Consult the travel-time curve based on our first-break picks (Fig. 3-22). Segments $1F$ and $1R$ should have the same slopes if the first layer is homogeneous. Their slopes are slightly different but lead to only very slightly different velocity values. Unfortunately, each curve is defined only by one point, so it is difficult to know which line is more reliable. Here is a case where experimenting with interpretations could provide useful information. If layer thickness are little affected by the possible extremes in the V_1 values, additional information probably is not warranted. However, if you cannot tolerate the error in thickness interpretations due to this uncertainty, the obvious solution is to place a string of geophones

You may wish to analyze the seismograms from Figure 3-21 independently. Our first-break picks are in the RefractSolve file: K82688021-24

Do not hesitate to substitute your own interpretations of arrival times or to draw different fits to the data points. In fact, you are encouraged to do so.

Figure 3-21 (a) Field seismogram from the Connecticut Valley, Massachusetts. Forward traverse. Geophone traces are labeled 1–24. The first geophone is located 3 m from the energy source. The geophone interval is 10 m. First breaks for each trace are indicated by a downward directed arrow. Timing lines are at 5-ms intervals. The record encompasses 185 ms.

with a smaller interval at the beginning of each traverse in order to record more direct wave arrivals and hence to determine velocities to closer tolerances.

Although some scattering of points exists, segments $2F$ and $2R$ are reasonably well defined. Because t_{if_1} and t_{ir_1} are almost equal, the first interface is essentially horizontal. Note that these two segments must satisfy the principle of reciprocity, even though the

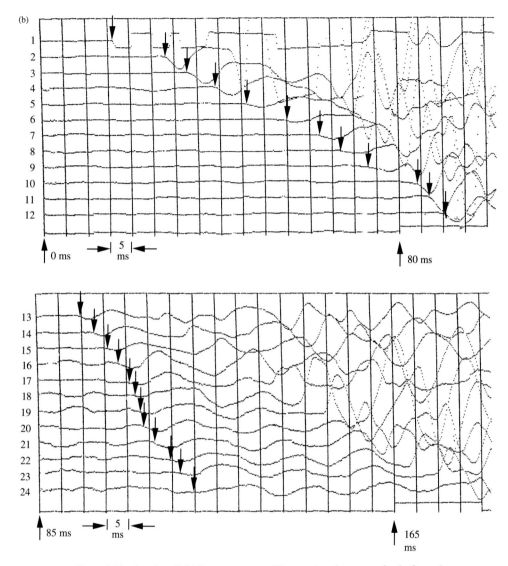

Figure 3-21 *(continued)* (b) Reverse traverse. All parameters the same as for the forward traverse.

intersections of their extensions with the vertical axes are not shown. The data points defining segments 3F and 3R have considerable scatter, but the majority fit on the lines as drawn in Figure 3-22. These two lines satisfy reciprocity, as indicated by R_r and R_f and the horizontal dotted line joining these points. However, 3F and 3R have different slopes, and t_{if_2} is less than t_{ir_2}. This indicates that the second interface is dipping and that the up-dip or shallowest part is located at the energy source position used for the forward traverse. Note, however, that if a third interface is present and is dipping, we cannot use this simple check

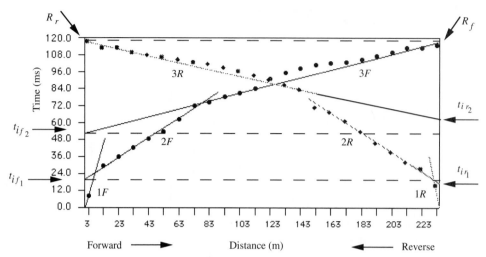

Figure 3-22   Travel-time curve based on data from Figure 3-21.

to determine its direction of dip because the dip of the second interface affects the intercept times for the third interface.

If you examine the structure computed from this positioning of curve segments (Fig. 3-23), you will see that our analysis is correct. We hope that by now you are able to study curves of this complexity and to arrive at the same general conclusions. This is a valuable ability, for it enables you to plot data and make a very generalized interpretation while still in the field. This is the most appropriate time to try and resolve uncertainties and/or to gather data for a more complete analysis.

Figure 3-23   Structure computed by RefractSolve from curve segments illustrated in Figure 3-22.

THE NONIDEAL SUBSURFACE

After the previous lengthy introduction to the seismic refraction method, you may well presume that we have covered most contingencies. Unfortunately, that is nowhere near the true situation. Fortunately, what you have learned is sufficient to cover a great many cases, especially if you are not concerned with detailed subsurface structure but need only one item of information such as depth to the water table or bedrock surface. Can you think of some fairly common and simple subsurface configurations that we might add to our repertoire?

One that comes to mind immediately is a lateral change of material. All our models to date have assumed lateral homogeneity. Another that should occur to you is the assumption of velocity increasing with depth. Although this is true in a general way, especially in shallow subsurface investigations, it often is possible to have a layer with a velocity lower than the one immediately above it. All our derivations also assume planar interfaces, whether horizontal or inclined. What happens if an interface is faulted? Can we deal with cases in which an interface has undulations? Finally, certain combinations of layer thicknesses and velocities can result in critical refractions from a deeper layer arriving before critical refractions from a shallow layer. If this happens, no data points will be available for this shallow layer, and quantities computed from such a travel-time curve will contain errors. We're confident you can think of other situations we should discuss, but those mentioned, either alone or in combination, cover much common subsurface geology. For the sake of time and space, these are the ones we analyze in the following pages.

Hidden Zones: The Low-Velocity Layer

In all our derivations to this point we assumed that each layer had a higher velocity than the layer immediately above it. Although we mentioned that this often is the case, especially in many routine shallow investigations, there is no guarantee that this assumption holds. Most likely you can create reasonable models for which a layer has a lower velocity than the layer above. In such a case with two layers, the consequences are obvious. Critical refraction does not occur, no refracted energy is returned to the surface, first breaks are due to the direct wave, the travel-time curve has one segment only, and we see no evidence that an interface is present.

Three-layer situations provide more combinations for analysis. In the somewhat rare case that a surface layer has a higher-velocity than those below, the direct wave traveling through the higher velocity medium always will arrive before refractions coming from the deeper layers. The basic refraction method will not "see" these layers, and their presence will go undetected unless additional information is available (e.g., from well logs). Similarly, the deepest layer will not be revealed in a configuration in which $V_1 < V_2 > V_3$. This situation produces a normal travel-time curve for V_1 and V_2, but no critical refraction occurs at the second interface. Although the third layer is undetected, the depth to the first interface will be correct.

Let's now consider the effect of a low velocity layer sandwiched between layers with higher velocities. Based on the preceding discussion, we can understand why the low-velocity layer in the middle of the stack will go undetected. But what is the appearance of a travel-time curve for this arrangement, and what types of errors are introduced into our analysis? The ray paths in Figure 3-24 illustrate that no critical refraction occurs at the first interface, because the angle of refraction must be smaller than the angle of incidence (Snell's law). Critical refraction does occur at the second interface, so a plot of time-distance data for this model produces a travel-time curve with two segments (Fig. 3-24). The first segment arises from direct wave arrivals. As wave energy is refracted at the first interface, some energy will arrive at the second interface at the appropriate angle for critical refraction. Equation 3-33 still is valid, and therefore the slope of the second line segment must be equal to $1/V_3$. If the travel-time curve in Figure 3-24 was the only information available to an interpreter, the person would have no evidence for the presence of the low-velocity layer and would compute a two-layer model with a depth to the top of the V_3 layer

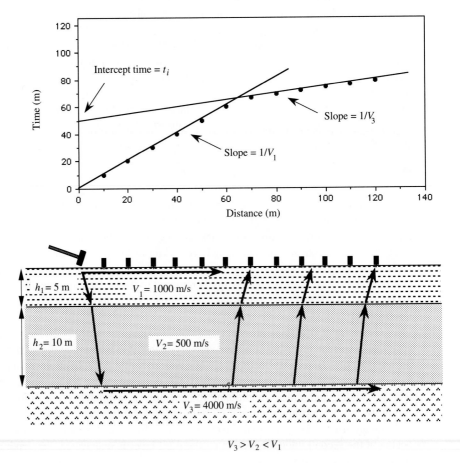

Figure 3-24   A plot of time-distance data from Table 3-5 correlated with a diagram of the model used to calculate the data. Note that there is no head wave generated at the V_1 -V_2 interface, since $V_2 < V_1$. Therefore, there is no indication of the V_2 - layer on the travel-time curve.

of 25.5 m. This depth obviously is in error, as the presence of the V_2-layer did not enter into the computation. But what is the magnitude of the error, and is the depth too large or too small?

> Note the convention for referring to travel-time curve seg-
> ments in this and the following section. We refer to these
> segments relative to the existing geologic subsurface. Thus, if
> a curve has two segments, but one segment is due to head
> wave arrivals from the top of the V_3-layer, we refer to this as
> the V_3-segment. Normally, however, if a curve has only two
> segments, we refer to the one with the lower slope (higher
> velocity) as the V_2-segment.

RefractModel assumes increasing velocity with depth, so we cannot use this program to create a travel-time curve to analyze the problem. However, Table 3-5 provides the nec-essary information. Table 3-5 contains the model and the time-distance data illustrated in Figure 3-24. The depth to the top of the V_3-layer is the sum of h_1 and h_2, which is 15 m. Thus, the depth based on the two-segment curve is greater than the true depth to the top of the V_3-layer. You could continue to investigate this problem by utilizing Table 3-5 as a dynamic table. Enter a model in Table 3-5, type the resulting time-distance data into RefractSolve, use RefractSolve to compute a model, and compare the result with the origi-

TABLE 3-5 Refraction Times for a Three-Layer Case with a Low-Velocity Intermediate Layer

Distance (m)	Direct Wave (ms)	Interface 2 Refraction (ms)
10	10.0	51.9
20	20.0	54.4
30	30.0	56.9
40	40.0	59.4
50	50.0	61.9
60	60.0	64.4
70	70.0	66.9
80	80.0	69.4
90	90.0	71.9
100	100.0	74.4
110	110.0	76.9
120	120.0	79.4

V_1 (m/s)	**1000**
V_2 (m/s)	**500**
V_3 (m/s)	**4000**
h_1 (m)	**5**
h_2 (m)	**10**
Interval (m)	**10**

Intercept time (ms)	49.4

nal model you entered into Table 3-5. In each case you investigate, you will find that the incorrect depth always is greater than the actual depth. The reason for the sense of the error is straightforward. A two-segment travel-time curve warrants use of Eq. 3-17. The vertical thickness of the first layer, h_1, is directly proportional to t_i. However, the travel time of the critically refracted wave is slowed by passage through V_2, and these arrival times are later than they would be if the entire distance was through V_1 material. Hence, the intercept time is greater, and it follows that the calculated depth to the interface is greater than the actual depth.

Unless some additional evidence is available, the low-velocity layer will remain undetected. If, however, stream cuts, road cuts, well logs, or other geophysical methods demonstrate the presence of a low-velocity layer in the geologic section, it is possible to approximate the depth to the second interface using a travel-time curve such as the one in Figure 3-24. This travel-time curve provides values for V_1 and V_3. We demonstrated that Eq. 3-33 remains valid in the case of the "sandwiched" low-velocity layer. This substantiates that the intercept time t_i in Figure 3-24 is correct and is equivalent to t_{i_2} in Eq. 3-35. If our independent evidence provides a good estimate for h_1 and V_2, we need only solve Eq. 3-35 to determine h_2. Table 3-6 provides a means for testing this assertion. It is a dynamic table and uses the values just discussed to calculate h_2. Values used in the table are the same as those used in Table 3-5 and Figure 3-24.

Explore the sensitivity of h_2 to the magnitudes of V_2 and h_1. Note that V_1, V_3, and the intercept time are interrelated and should not be changed independently. If you want to investigate additional values, use Table 3-5 as a dynamic table to determine values for a new model. Then insert these into Table 3-6.

TABLE 3-6   Computation of Thickness for Low-Velocity Second Layer

V_1 (m/s)	1000
V_2 (m/s) (assumed)	500
V_3 (m/s)	4000
h_1 (m) (assumed)	5
Intercept time (ms)	49.4
h_2 (m)	10.0

Because the presence of a low-velocity layer is a common possibility, every effort should be made to obtain independent information about the subsurface in the area in which you are working. The inability of the refraction method to detect low-velocity layers is one of its major shortcomings.

Hidden Zones: The Thin Layer

Unfortunately, a low-velocity layer is not the only circumstance that results in a layer escaping detection by seismic refraction methods. Thin layers or layers that have a low-velocity contrast with a layer below also will escape detection. Let's try to determine why this occurs and what type error results. A dynamic table (Table 3-7), RefractSolve, and RefractModel are very useful for our purposes. First enter the following model into RefractModel or into Table 3-7: $V_1 = 500$ m/s, $V_2 = 1400$ m/s, $V_3 = 4500$ m/s, $h_1 = 10$ m, and $h_2 = 15$ m. RefractModel produces a classic three-segment travel-time curve, and a solution to this data using RefractSolve yields values that are identical to the starting model. Now change the value of h_2 to 5 m. This time the result is a two-segment curve even though a three-layer model was entered.

First study Table 3-7 and then inspect Figure 3-25. All head waves from the first interface arrive later than the head waves from the second interface. Usually we are able to identify only the first energy to arrive at a geophone, so the later-arriving waves will not be identified. Hence, we observe a two-layer case when we have a three-layer geologic section. If we use the time-distance data from our second model in RefractSolve, we find that the computed vertical thickness to the top of the V_3-layer is 11.7 m. This is less than the true value of 15 m.

The hidden-layer problem due to a thin layer yields a depth value less than the true depth, whereas the low-velocity layer resulted in a depth value greater than the true value.

TABLE 3-7 Refraction Times for a Thin Intermediate Layer

Distance (m)	Direct Wave (ms)	Interface 1 Refraction (ms)	Interface 2 Refraction (ms)		
5	10	44.50	47.65		
10	20	51.65	48.76		
15	30	58.79	49.87		
20	40	65.93	50.99		
25	50	73.08	52.10	V_1 (m/s)	500
30	60	80.22	53.21	V_2 (m/s)	1400
35	70	87.36	54.32	V_3 (m/s)	4500
40	80	94.50	55.43	h_1 (m)	10
45	90	101.65	56.54	h_2 (m)	5
50	100	108.79	57.65		
55	110	115.93	58.76	Interval (m)	5
60	120	123.08	59.87		

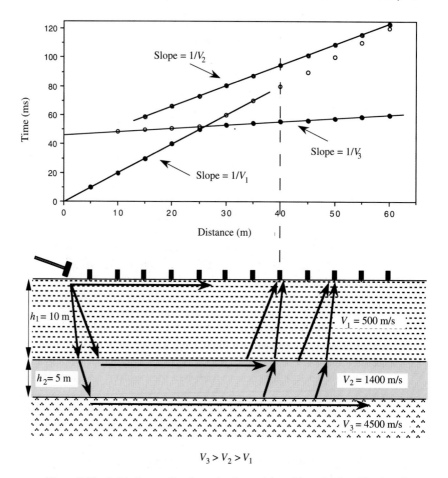

Figure 3-25   A thin intermediate layer that is not detected by refraction. The time-distance data from Table 3-7 are correlated with a diagram of the model used to calculate the data. No head wave arrivals for the first interface arrive before the direct wave or the head wave from the second interface due to the V_2 -V_3 velocity contrast and the thinness of the V_2 -layer.

Why? Once again, a two-segment travel-time curve warrants use of equation 3-17. The vertical thickness of the first layer, h_1, is directly proportional to t_i. However, the travel time of the critically refracted wave (second interface) is speeded by passage through V_2, and these arrival times are earlier than they would be if the entire distance was through V_1-material. Hence, the intercept time is less, and it follows that the calculated depth to the interface is less than the actual depth.

Do we have any recourse as we did in the case of the low-velocity layer? If we have no independent information that the V_2 -layer may be present, then the layer will be missed, and the computed depth will be accepted as accurate when it actually is too shallow. If independent information indicates that we should be dealing with a three-layer situation, we can

Figure 3-26 How to utilize the travel-time curve in Figure 3-25 to calculate a maximum depth to the second interface. Procedure explained in the text.

calculate a maximum depth. As we already have a minimum depth, this process at least provides some limits. We have values for V_1, V_3, and the intercept time for the line through head wave arrivals from the second interface. If independent information indicates that another layer, V_2, is present, we likely have a reasonably good constraint on the value of V_2 and therefore can draw a line with the correct slope on our curve. This line is adjusted as illustrated in Figure 3-26 until it just touches the intersection of the V_1 and V_3 lines (we are assuming that the V_2 arrivals were just a bit too slow to arrive before the V_3 arrivals that were recorded). This gives us the lowest possible intercept time for the head wave from the first interface.

You can use this approach in RefractSolve. Be sure to draw the V_2 line before drawing the V_3 line, because lines are assigned to layers in the order drawn. You then can adjust the slope of the V_2 line until the velocity recorded in a solution is correct.

Laterally Varying Velocity

Lateral variations in materials at the surface, at least in a minor way, may well be the rule rather than the exception. If these variations are not extreme, surface velocities will not vary significantly, and calculated depths to interfaces will have only small errors. You can gain a sense of the effect of small variations by changing the slope of the direct wave arrivals in RefractSolve.

In this section we analyze the effect of an abrupt vertical change in surface materials (Fig. 3-27). The travel-time curve in Figure 3-27 illustrates that a field survey shooting in the direction of the higher-velocity unit sees a curve analogous to that produced by a single horizontal interface. However, a reverse traverse produces a very unusual curve, because the first segment has a lower slope than the second segment. It should be clear why the slopes for the first segments of the forward and reverse traverses have the values that they do. However, it may not be as evident in the case of the second segments.

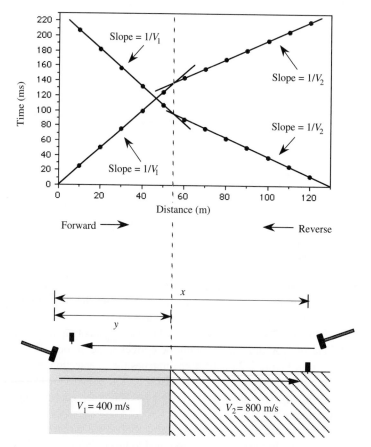

Figure 3-27   The effect of a vertical velocity discontinuity on forward and reverse travel-time curves. Note the distinct pattern of slope reversals depending on traverse direction. The position of the vertical discontinuity is marked by the crossover point.

Once again, we can understand the result more easily if we develop a travel-time equation and determine the slope (refer to Fig. 3-27):

$$\text{time} = \frac{y}{V_1} + \frac{x - y}{V_2} \qquad (\text{for } x \geq y) \qquad (3\text{-}57)$$

$$\frac{dt}{dx} = \frac{1}{V_2} \qquad\qquad\qquad\qquad\qquad\qquad (3\text{-}58)$$

Therefore, in such a case you can recognize the presence of a vertical surface discontinuity by the special pattern of the travel-time curves, determine the velocities of the materials, and locate the position of the discontinuity by the crossover distances of the forward and reverse curve segments.

Will such a unique pattern be produced if the discontinuity is buried beneath a surface layer? If you examine Figure 3-28, you will conclude that the answer is yes. As in the

Use Table 3-7 and/or RefractModel to investigate a hidden layer due to a low velocity-contrast between the first and second layers in a three-layer sequence. Once you obtain a model that produces a two-segment curve, use RefractSolve to determine the error in the depth value to the second interface.

case of the vertical discontinuity at the surface, a reversal in the slope patterns is produced by the sense of the change in velocities between a forward traverse and a reverse traverse. In this case a three-segment curve arises due to the layer above the vertical discontinuity. Once again, the slopes of the travel-time curve segments yield all velocities illustrated in the model. The general position of the discontinuity is indicated by the change in slope of the second and third segments of each traverse direction. Note, however, that the discontinuity position is only bracketed by the slope changes and also that it is not midway between these points.

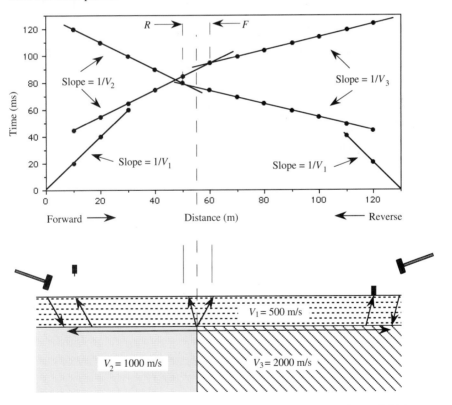

Figure 3-28 Travel-time curves for a vertical discontinuity beneath a surficial layer. Slope reversals are present for head wave arrivals from the horizontal interface.

If you are interested in substantiating the values of the slopes of the travel-time curve segments or in calculating the positions of the crossover points, examine Eq. 3-8 and rewrite it for the case of the vertical discontinuity. Don't worry about simplifying the equation beyond the form of Eq. 3-8.

Interface Discontinuities

In our discussion of interface discontinuities, we confine ourselves to two cases: (1) a sharp vertical offset in a horizontal discontinuity, and (2) an irregular surface rather than a flat surface. First, let's examine the case of the vertical offset, which might represent an erosional step or fault scarp.

Figure 3-29 illustrates ray paths for forward and reverse traverses over a vertical step and resulting travel-time curves. The area of interest is the region surrounding the vertical step, for we essentially are dealing with a two-layer situation at significant distances from the step in either direction. First concentrate on the forward traverse. Early arrivals consist of the direct and head waves in the classic pattern producing a two-segment travel-time curve. The last ray path to follow the normal state of affairs is QA. The position of A on the surface is controlled by h_1 and the critical angle, but it is extremely unlikely that a geophone will be located at this exact spot. In Figure 3-29 the last geophone to record head wave energy from the more shallow portion of the interface is immediately to the left of A.

Some wave energy follows the path PR, encounters the bottom of the step, produces diffractions, and travels along RS. That energy traveling along the deeper portion of the interface generates waves following such paths as RB and SG, which also will return at the critical angle. Once again the position of B on the surface is controlled by the critical angle and $h_1 + z$. At some point to the right of B this energy arrives first. These arrival times plot along a straight line with slope $1/V_2$. Note that this line must be displaced later in time because ray paths such as SG are longer than ones such as QA. This later line has an intercept time t_{i_2}, whereas the earlier line has an intercept time t_{i_1}.

If we compare QA and RB, we see that the difference in intercept times, $t_{i_2} - t_{i_1}$, is due to the extra distance, RD, traveled by RB at V_1. RD is related to the critical angle and the height of the step z:

$$RD = z\cos\theta_{ic} \tag{3-59}$$

so that the intercept time difference can be computed and z determined by the following:

$$t_{i_2} - t_{i_1} = \frac{z\cos\theta_{ic}}{V_1} \tag{3-60}$$

and

$$z = \frac{\left(t_{i_2} - t_{i_1}\right)V_1}{\cos\theta_{ic}} \tag{3-61}$$

or

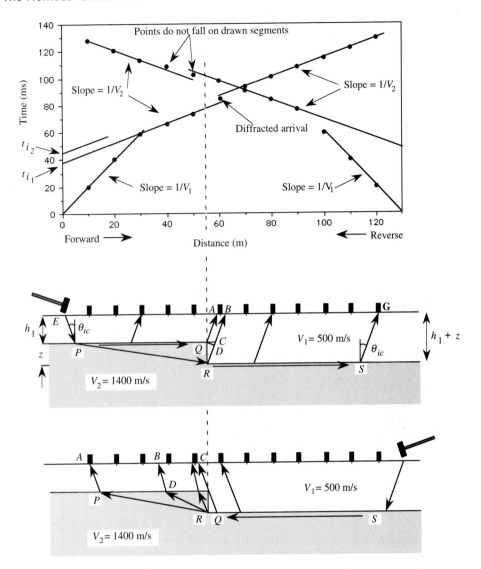

Figure 3-29 Ray paths and resulting travel-time curves for forward and reverse traverses over a vertical step. A detailed explanation is given in the text.

$$z = \frac{\left(t_{i_2} - t_{i_1}\right)V_2 V_1}{\left(V_2^2 - V_1^2\right)^{1/2}} \qquad (3\text{-}62)$$

If the reason for this computation is not yet obvious, note that the time *delay* at *B* due to the step can be envisaged as follows. If the step was not present, the time delay difference between positions A and B would be due to travel along QC at V_2. Because of the step the time *difference* is due to RC at V_1. However, the time to travel along QC at V_2 is equal to travel along DC at V_1, so that the *delay* at B is the time to travel RD at V_1.

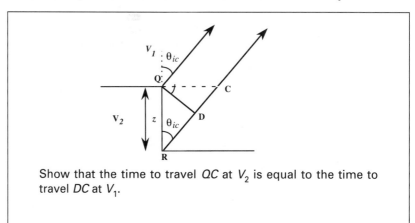

Show that the time to travel QC at V_2 is equal to the time to travel DC at V_1.

Before proceeding to an analysis of the reverse traverse, we need to point out that the actual arrival pattern is a bit more complex than presented. Diffraction occurs at point Q. These diffracted waves arrive everywhere at a later time than that of ray QA, but they will arrive earlier at B than ray RB, as both are traveling at V_1. At some point to the right of B, waves refracted from the deeper portion of the interface will arrive before the diffracted waves. This distance is determined by the velocity ratio and z. Although in Figure 3-29 only one diffracted arrival midway between the early and late refraction segments is illustrated, in practice several such points may be present and may well make it more difficult to discern the deceptively simple pattern of Figure 3-29. Here we have a clear-cut case where diffracted waves play a key role in the form of the travel-time curve, albeit not a clarifying one.

Naturally, information from a reverse traverse is collected. In the case illustrated in Figure 3-29, it appears that the step geometry yields two curve segments with slope equal to $1/V_2$ in addition to the direct wave segment. The first of the $1/V_2$ segments is due to head wave arrivals from the deeper portion of the interface. The last such path received at the surface is QC. Wave energy will travel up toward the surface from point R along paths such as RDB and RPA. This energy is refracted upward from the shallower part of the interface, but the angles of refraction do not equal the critical angle. Rays close to the step, such as RDB, will arrive earlier than QC, as they follow a somewhat similar path length but with a portion of travel at V_2. Rays such as RPA that travel to a considerable distance to the left of R have a portion of their path (RP) at a shallow angle to the interface. The angle of refraction of such rays begins to approximate the critical angle, so that a plot of their arrival times versus distance approximates $1/V_2$.

The offset of the $1/V_2$ segments for the reverse traverse is the opposite of the offset for the forward traverse. Thus, the critical criteria for recognizing the presence of the step or fault scarp are the offset $1/V_2$ segments, with the offset sense switched for forward and reverse traverses. Remembering these relationships is important, not only for detecting the presence of the step, but for selecting the correct traverse for computing the height of the step. The preceding suggests that under good field conditions the traverse with the shot point located above the shallower portion of the interface is likely to yield superior data.

In all the preceding discussions of the refraction method applied to various subsurface geometries, planar interfaces were assumed. These interfaces produce time-distance data to which straight lines are fit, and from which parameters are determined and geology is deduced. Even in the case of a lateral velocity change or an offset interface, the only adjustment to this approach is a single vertical contact. By this time you may be beginning to question how one copes with situations when an interface is not planar but possesses an irregular surface with highs and lows—a more realistic geologic scenario. Such situations are considered in the following section.

Before proceeding to the next section, we call your attention to a valuable resource. Due primarily to space considerations, many variations on the subsurface geometries discussed here are not included. With some careful consideration you should be able to engage in analyses of more complicated geometries. What type of travel-time curve would result from a dipping interface with a vertical step? How are time-distance data affected when an interface changes inclination from horizontal to dipping? A good variety of subsurface structures and their associated travel-time curves as well as analyses of a number of curves is presented in Mooney (1977, p. 15-1 to 15-18 and 16-1 to 16-9).

THE DELAY-TIME METHOD

In this section we take a slightly different approach that ultimately will allow us to map irregular refractor surfaces. Study Figure 3-30(a). Of course, the path *EBCG* is a familiar one. However, if we concentrate our attention on the interface between V_1 and V_2, which is the refractor surface we are interested in mapping, we might think in terms of the difference in time it takes a wave to travel *EBCG* as opposed to the time to travel the same distance between source and receiver y in the V_2-layer (path *AD*). Such a consideration leads to the concept of *delay time*.

Delay time can be defined as the time spent by a wave to travel up or down through the V_1-layer (the slant path) compared to the time the wave would spend if traveling along the projection of the slant path on the refractor (*AB* or *CD* in Figure 3-30(a)) (Redpath, 1973, p. 10). For the illustration in Figure 3-30(a) the *total* delay time is

$$T_{EG} = t_R - \frac{y}{V_2} \tag{3-63}$$

where t_R is the total travel time along *EBCG* or *GCBE* (recall the requirement of reciprocity). In more advanced analyses this total travel time is referred to as *reciprocal time*. Hence, we use the symbol t_R to refer to this quantity. In the development that follows uppercase T is used to represent delay times, whereas lowercase t is used to refer to travel times that can be measured from time-distance plots.

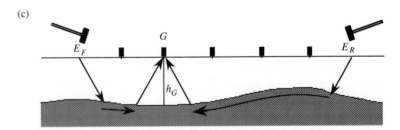

Figure 3-30 (a) Symbols used in derivation of delay-time relationships. (b) Schematic of field setup for determining delay times and depths to refractors. (c) Mapping irregular refractors with the delay-time approach.

Since, by definition,

$$T_{EG} = T_E + T_G \qquad (3\text{-}64)$$

we can derive an expression for the delay time at a geophone:

$$T_E + T_G = t_R - \frac{y}{V_2}$$

and

$$T_G = t_R - \frac{y}{V_2} - T_E \tag{3-65}$$

Also, by definition,

$$T_G = \frac{CG}{V_1} - \frac{CD}{V_2} \tag{3-66}$$

Using simple trigonometry and the same relationships we used when deriving Eq. 3-14, we can demonstrate that

$$T_G = \frac{h_G}{V_1 \cos\theta_{ic}} - \frac{h_G \tan\theta_{ic}}{V_2}$$

$$T_G = h_G \frac{\left(V_2^2 - V_1^2\right)^{1/2}}{V_1 V_2} \tag{3-67}$$

and, finally,

$$h_G = T_G \frac{V_1 V_2}{\left(V_2^2 - V_1^2\right)^{1/2}} \tag{3-68}$$

Thus, we can calculate the depth of the V_1-V_2 interface beneath the geophone if we can determine the delay time at the geophone (and, of course, know V_1 and V_2). How can we determine T_G if we can't measure it directly?

Once again, the reverse traverse comes to our rescue. Consider Figure 3-30(b). A geophone is located at G, a forward source is at E_F, and a reverse source is at E_R. We rewrite Eq. 3-65 as an expression for reciprocal time:

$$t_R = T_G + T_E + \frac{y}{V_2} \tag{3-69}$$

Following our earlier convention, $t_{E_F G}$ represents the travel time from E_F to G which is measured on a field seismogram and plotted on a time-distance plot. Using the basic form of Eq. 3-69, we see that

$$t_{E_F G} = T_{E_F} + T_{G_F} + \frac{x}{V_2} \tag{3-70}$$

and

$$t_{E_R G} = T_{E_R} + T_{G_R} + \frac{y - x}{V_2} \tag{3-71}$$

If we assume that the refractor surface is planar beneath G, then

$$T_{G_F} = T_{G_R}$$

and

$$t_{E_F G} + t_{E_R G} = T_{E_F} + T_{E_R} + 2T_G + \frac{x}{V_2} + \frac{y - x}{V_2} \tag{3-72}$$

Because of the requirement of reciprocity,

$$t_R = T_{E_F} + T_{E_R} + \frac{y}{V_2}$$

and, therefore,

$$t_{E_F G} + t_{E_R G} = t_R - \frac{y}{V_2} + 2T_G + \frac{y}{V_2}$$

$$t_{E_F G} + t_{E_R G} = t_R + 2T_G$$

and

$$T_G = \frac{t_{E_F G} + t_{E_R G} - t_R}{2} \tag{3-73}$$

Since we now have an expression for T_G that is composed entirely of quantities that are obtainable from field observations, we can use Eq. 3-68 to calculate h_G. This same procedure can be repeated for each geophone along a typical spread, so that, in theory at least, we can map the form of a refractor's surface by using as many geophones as required data points (Fig. 3-30(c)).

Several assumptions are included in the development of the delay-time method that limit its applicability, namely (1) that there was a planar refractor surface immediately beneath the geophone and the energy source when deriving expressions for delay times and (2) that distances along the refractor were equivalent to surface distances x and y. The first assumption effectively limits the application of this method to irregular refractor surfaces only if the relief on the refractor is small in amplitude relative to the average refractor depth (Kearey and Brooks, 1984, p. 124; Redpath, 1973, p. 26). The second assumption limits the method's application to refractor dips of less than 10° (Kearey and Brooks, 1984, p. 125).

Perhaps the best way to visualize the application of the delay-time method is to work with a specific example. Table 3-8 (a dynamic table) lists forward and reverse times for an irregular refractor similar to that illustrated in Figure 3-30(c). Each set of forward and reverse times for a specific geophone distance is equivalent to the two energy source–to-geophone times in Eq. 3-73. Reciprocal time is the time from the forward source position to the reverse source position. These times and V_1 and V_2 are entered into Table 3-9 (also a dynamic table) which uses Eq. 3-73 to compute delay times and Eq. 3-68 to calculate depths. Because Table 3-9 is a dynamic table, you also can use it to perform depth calculations for actual field examples. Although Tables 3-8 and 3-9 aid in visualizing the steps in depth computations, you should apply the delay-time approach to an actual travel-time curve.

Figure 3-31(a) is a time-distance plot of the times calculated in Table 3-8, but only the times which would appear as first arrivals are plotted. The only positions where depths can be determined are enclosed in a dashed outline, as these are the only geophone positions where travel times are available from both the forward and reverse sources. These few points limit the amount of information about the refractor. Usually, we can obtain addi-

TABLE 3-8 Head Wave Arrival Times for an Irregular Refractor

Position	Distance (m)	Depth (m)	Direct Wave (ms)	Forward Time (ms)	Reverse Time (ms)
Forward source	0	15	0.00		55.20
Geophone 1	10	14	7.14	21.91	52.14
Geophone 2	20	13	14.29	23.62	49.09
Geophone 3	30	12	21.43	25.33	46.03
Geophone 4	40	13	28.57	28.38	44.33
Geophone 5	50	13	35.71	30.76	41.94
Geophone 6	60	14	42.86	33.82	40.24
Geophone 7	70	15	50.00	36.87	38.53
Geophone 8	80	16	57.14	39.92	36.82
Geophone 9	90	17	64.29	42.98	35.11
Geophone 10	100	19	71.43	46.71	34.08
Geophone 11	110	20	78.57	49.76	32.37
Geophone 12	120	21	85.71	52.82	30.67
Reverse source	130	21	92.86	55.20	

V_1 (m/s)	1400
V_2 (m/s)	4200
Geophone interval (m)	10

Reciprocal time (ms)	55.20

TABLE 3.9 Refractor Depths Computed Using Delay Times

Position	Forward Time (ms)	Reverse Time (ms)	Delay Time (ms)	Depth (m)
Geophone 1	21.9	52.1	9.4	14
Geophone 2	23.6	49.1	8.8	13
Geophone 3	25.3	46.0	8.1	12
Geophone 4	28.4	44.3	8.8	13
Geophone 5	30.8	42.0	8.8	13
Geophone 6	33.8	40.2	9.4	14
Geophone 7	36.9	38.5	10.1	15
Geophone 8	39.9	36.8	10.8	16
Geophone 9	43.0	35.1	11.4	17
Geophone 10	46.7	34.1	12.8	19
Geophone 11	49.8	32.4	13.5	20
Geophone 12	52.8	30.7	14.1	21

V_1 (m/s)	1400
V_2 (m/s)	4200
Reciprocal time (ms)	55.20

(a)

(b)

Figure 3-31   (a) A plot of the direct wave and forward and reverse head wave times listed in Table 3-8. (b) A graph of the difference between forward and reverse travel times for each geophone plotted against distance. The slope of the line fit to these points is used to calculate V_2.

tional points by increasing the number of geophones, but we must be careful not to use arrivals from a deeper refractor. Strategies for dealing with these problems are covered clearly and in detail in Redpath (1973).

Recall that in the delay-time method the quantity h_G is defined as perpendicular to the refractor surface. Therefore, once depths beneath each geophone are determined, the refractor can be mapped by drawing arcs with a radius equal to the calculated depths and constructing a smooth curve that is tangent to these arcs (Fig. 3-32).

If you have entertained a nagging suspicion that something is slightly amiss, congratulations! Head wave times from an irregular refractor will not lie along a straight line in a time-distance plot and, therefore, V_2 cannot be determined accurately. Lacking a good

Figure 3-32   A refractor surface is constructed from delay-time depths by drawing an arc from each geophone location. The arc's radius is equal to the calculated depth at that position.

value for V_2, we cannot calculate h_G. How can we circumvent this problem? If we rewrite Eqs. 3-70 and 3-71, we can arrive at the form

$$t_{E_F G} - t_{E_R G} = T_{E_F} + T_{G_F} + \frac{x}{V_2} - T_{E_R} - T_{G_R} - \frac{y - x}{V_2} \qquad (3\text{-}74)$$

and

$$t_{E_F G} - t_{E_R G} = T_{E_F} - T_{E_R} + \frac{2x}{V_2} - \frac{y}{V_2} \qquad (3\text{-}75)$$

Equation 3-75 demonstrates that travel-time differences plotted against distance yield a line with a slope of $2/V_2$. Figure 3-31(b) is a graph of this type using the times and distances in Table 3-8. Using the calculated slope of 0.476, we arrive at a value for V_2 of 4202 m/s.

The delay-time method represents a somewhat different approach to using typical refraction data and under good conditions provides a much more detailed picture of a refracting interface than the basic approach discussed first in this chapter. However, the delay-time method is fraught with the same difficulties as most commonly employed refraction methods, the most obvious of which is the hidden-layer problem.

In the next section we briefly mention other more advanced refraction methods. Although these may circumvent some of the limitations of the approaches already discussed, they are more complex in design and implementation. They are presented primarily to provide an idea of additional possibilities.

OTHER METHODS

Wavefront Method

The wavefront method was developed by Thornburgh (1930) and utilizes wavefront construction to locate points on refracting surfaces. The method is simple to apply in the case of horizontal or dipping planar interfaces but is most useful for delineating irregular refractor surfaces. However, its application in the case of more complex surfaces is tedious. Because of the time involved in constructing these wavefronts, the method has been replaced by other more powerful methods such as the generalized reciprocal method mentioned later.

Nevertheless, it is instructive to illustrate how the wavefront method is applied in a simple case, as this illustrates yet an additional way to view arrival times and to work with seismic data. Consider the planar horizontal interface illustrated in Figure 3-33(a). Refracted energy arriving at the geophone G travels a path $E_F ABG$. The other end of the planar wavefront arriving at G is located on the refractor at point P. By using the definition of the critical angle and simple geometry, it is straightforward to show that the time for a wave to travel from B to G at V_1 is equal to the time for a wave to travel from B to P at V_2:

$$\sin \theta_{ic} = \frac{V_1}{V_2} = \frac{BG}{BP} \tag{3-76}$$

and

$$\text{time}_{BG} = \frac{BG}{V_1} = \frac{BP}{V_2} = \text{time}_{BP} \tag{3-77}$$

A similar wavefront associated with point P resulting from a reverse shot at E_R defines point H on the surface. As in the previous derivation,

$$\text{time}_{CH} = \text{time}_{CP} \tag{3-78}$$

The time to travel $E_F ABP$ plus the time to travel $E_R DCP$ must be equal to the time to travel $E_F ADE_R$ which is reciprocal time t_R. From this relationship and the identities in Eqs. 3-77 and 3-78, it becomes apparent that

$$t_{E_F G} + t_{E_R H} = t_R \tag{3-79}$$

Knowing V_1 and V_2, we calculate θ_{ic}. If we have a geophone position with a head wave arrival (G), we can construct a planar wavefront such as the one at 40 m in Figure 3-33(b). Given a reciprocal time, we need only subtract to find the unknown time, and hence position H, relative to the energy source at E_R. We then construct a second wavefront which locates point P (Fig. 3-33(b)). This procedure can be duplicated for every geophone that records an arrival from the refractor of interest. Although this would be a waste of effort for a horizontal surface, two or more points would delimit a planar, dipping refractor, and many points would map in some detail an irregular refractor. In these latter cases the wavefront construction becomes more complicated and may not be warranted in light of other methods which produce similar results with less effort or superior results with equivalent effort.

Ray Tracing and the Generalized Reciprocal Method

Although computer capabilities were applied for much of the material in this chapter, the processing capabilities of computers barely were challenged. The approaches of ray tracing (Cerveny et al., 1974) and the generalized reciprocal method (Palmer, 1980) utilize computer capabilities more fully in dealing with refraction data from more complicated subsurface situations than have been treated in this text.

In applying the ray-tracing method an investigator proposes a subsurface model based on whatever information is available including local surface geology, well data, and even refraction data. Ray paths for this model then are calculated by computer and travel

(a)

(b)

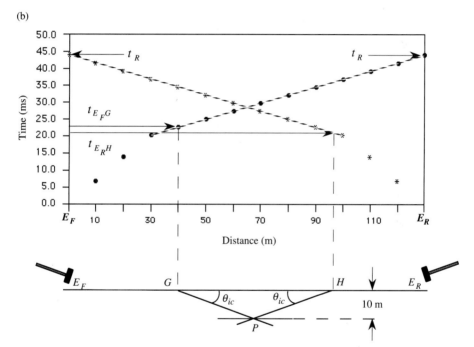

Figure 3-33 Diagram illustrating the basic procedure for the wavefront method. (a) Relationships used to derive method. (b) Locating point on refractor from travel-time curve.

times determined. These times are compared with observed times, the model is adjusted with the goal of improving the correspondence between observed and computed times, and new times are computed. This process is continued until a realistic model is in hand that conforms to known geologic information and that produces travel times which have a close correspondence to measured times.

The generalized reciprocal method (GRM) is a powerful, yet relatively straightforward, approach. It utilizes reciprocal time and the arrival times at two geophones as does the delay-time method. However, the extensions in analysis that characterize the GRM enable it to deal especially well with irregular refractors, as the method is able to deal with dip angles up to 20°. In addition, the GRM can indicate the presence of hidden layers and their probable nature (velocity inversion or thin layer), which provides an indication of the nature of depth errors. It also is possible to calculate depths to interfaces below a hidden layer in situations where good control of critical parameters can be achieved. Especially good reviews of the generalized reciprocal method as compared to "classic" refraction techniques are presented in Lankston (1989, 1990).

FIELD PROCEDURES

Site Selection and Planning Considerations

Once you identify a locality that you will explore using seismic refraction techniques, there are a number of considerations that you should keep in mind. First, you must exhaust all likely sources in order to determine as much as possible about the local geology, both surface and subsurface, *before* undertaking a refraction survey. Although such advice may seem self-evident, it is amazing how many individuals give this preparation short shrift in order to get started with seismic data collection. Some may believe that geology can be consulted after geophysical data is in hand, but more effective surveys result from thorough premeasurement preparation.

Assume for a minute that you are exploring for buried valleys cut into bedrock and covered by glacial sands and gravels. If you have 24 geophones available, should you use a cable with 3-m, 5-m, or 10-m take-outs? Your preliminary geologic studies reveal recorded bedrock depths vary from surface exposures to 70 m. Because you are interested in areas with the greater depths, you might plan to use spread lengths capable of detecting bedrock at depths of from 50 to 70 m. A quick input into RefractModel using a model of 5 m of unsaturated sands and gravels (500 m/s), 65 m of saturated sands and gravels (1400 m/s), and a bedrock velocity of 4500 m/s reveals that the first recorded bedrock refraction is 190 m from the energy source. Because you want to record a minimum of three or four refractions, your line length must be at least 210 m, and, therefore, you decide to use a cable with 10-m take-outs. A good rule of thumb for situations with these velocities is to use spread lengths at least three times the target's depth. Although somewhat simplified, this example is intended to illustrate the importance of presurvey planning. In most seismic work, time and funds are limited. You must work as efficiently as possible and be as confident as possible that you are collecting useful data.

Other considerations enter into the placement of geophones. Geophones must be well coupled with the ground. Firm placement of the spike is essential. Often, the time and trouble of augering 1 m deep holes, filling them with water to saturate the sediment, and then placing the geophones in the holes yields such improved results that the effort is well worthwhile. If at all possible, lay cable along a line of equal elevation. If geophones are at various elevations, corrections must be made (see following section).

Wind and continual traffic must be avoided. Both cause ground motion, which often is of similar or greater amplitude than the waves you are trying to record. Be flexible and return on a nonwindy day or during the night, when traffic is at a minimum. Because of some of the inherent uncertainties in refraction surveys, reliable data are essential. Obtaining reliable data under the preceding conditions is next to impossible.

Equipment Considerations

In the majority of what often are termed *shallow* refraction surveys, target depths typically are less than 100 m. Many surveys must operate in populated areas. Results usually influence engineering and/or planning decisions. Contracts tend to be for specific, short-term objectives and cover a limited area of investigation, thus favoring individual and small consulting firms. All these factors directly influence the type of survey that can be conducted, including energy sources, equipment, computing facilities, and time.

At this point it would be a good idea to read once again the material in Chapter 2 dealing with energy sources and seismic equipment. When using a signal-enhancement seismograph, a hammer source is adequate for many refraction surveys if target depths do not exceed 50 m and saturated sediments are close to the surface. Under more difficult conditions and for greater depths, a minimum of a shotgun or weight-drop source is mandatory. Explosive charges likely would be preferred, but legal, safety, and expense considerations often rule them out.

As demonstrated previously in this section, line lengths for refraction surveys must be at least three times the target's depth. This means that ray paths are long, that much attenuation occurs, and that first-break amplitudes are likely to be small. These long path lengths also result in the disappearance of higher frequencies. Even though low-frequency geophones are employed in refraction surveys, the waveform tends to be smoother and more rounded, making the first-break onset even more difficult to recognize. These factors only increase the need for careful planning, site selection, and patience in gathering quality data.

If test records contain objectionable noise, and the problem cannot be solved by returning to the site at a more propitious time, filters may save the day. Filters are not used as frequently in refraction surveys as they are in reflection surveys, but at sites where noise is a problem, filters may reduce the effect of the noise to some extent. Most modern field seismographs have an oscilloscope display with an option permitting real-time observation of ground motion. Viewing the display while switching filters may help in narrowing the choice. Another option is to measure dominant noise wavelengths and determine noise frequencies from a test record. Of course, utilizing filters also removes frequency components from signals of interest, which may be more deleterious than noise interference.

Geophone-Spread Geometries and Placements

The standard geophone spread is referred to as the *in-line spread* and is illustrated in Figure 3-34(a). The energy source, or "shot" point, is placed along the same line as the geophone cable and usually is placed at a distance beyond the first geophone that is similar to the take-out interval. The reverse shot is placed at an equivalent position beyond the other end

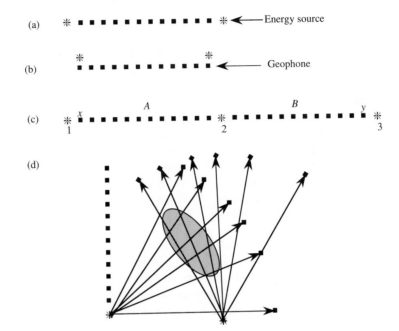

Figure 3-34　Geophone-spread arrangements. (a) In-line spread. (b) Offset spread. (c) In-line with center shot. (d) Fan shooting.

of the spread. This is the arrangement used in many diagrams in this chapter and in RefractModel and RefractSolve.

In cases where a single cable length is not sufficient to record arrivals from the target of interest, the geometry illustrated in Figure 3-34(c) is used. The cable is placed at A and shot points taken at position 1, then position 2, and then position 3. The cable is moved to B and shots taken at positions 1, 2, and 3 once again. Such a plan provides sufficient geophone distribution to record deep arrivals, contains forward and reverse traverses, and has a center shot which is useful in checking variations in velocity and thickness of the first layer along the line of investigation. Of course, this basic plan can be extended to greater lengths if necessary, provided sufficient energy is available to record first arrivals at the most distant geophones.

Note that (a) and (c) in Figure 3-34 do not give exact total travel times. Reciprocal times can be estimated from these arrangements, but in cases such as the delay-time method, where t_R is included in calculations, Redpath (1973, p. 33) recommends the offset spread (Fig. 3-34(b)). Another viable option, especially where 24 or more geophones are used and the energy source produces minimum damage to the ground (e.g., a hammer blow), is to disconnect geophone x in Figure 3-34(c) and place shot 1 at that position and then to reconnect geophone x, disconnect geophone y, and place shot 3 at that position.

Several other spread geometries are used and are detailed in advanced texts dealing with refraction methods (see Suggested Reading). As an example of other possibilities, consider fan shooting (Fig. 3-34(d)). Such an arrangement is used when searching for a subsur-

face form with anomalous velocity (the stippled oval in Fig. 3-34(d)). Examples include salt
domes in earlier, deeper refraction studies (Nettleton, 1940) or buried bedrock hills in shallow exploration. A normal refraction line is shot to determine local velocities and depths.
This is followed by one or more shots with geophones arranged in a fan shape at equal distances from the shot point. Anomalous increases or decreases in velocity along some shot-
geophone lines can aid in defining the general area of the target.

In addition to the geophone spread chosen, placement of the line is important, especially in complex regions. Also, one always should undertake preliminary analysis of data
as it is being collected. In order to illustrate the importance of both of these points, consider
the subsurface structure diagrammed in Figure 3-29. Plotting time-distance data in the field
would reveal the likely presence of a fault or step. A more complete analysis of the structure might be possible by removing the spread to the left of its present position (in the diagram), taking a shot, and then repeating the process by moving to the right of the diagram.

In the last section of this chapter the "weathering" correction is mentioned. In order to
make this correction reliably, good control of thickness and velocity variations in the surface layer is essential. This can be achieved only by several traverses with relatively small
geophone intervals. The main advice here is to plan carefully, continually monitor data,
have objectives and data requirements carefully worked out and clearly in mind, and be
flexible.

In some cases energy constraints, site access, or other factors may preclude the long
spread lengths required to receive refractions from the target. This may not spell disaster
depending on the survey requirements. If you have a good grasp of the methods detailed in
this chapter, some information always can be salvaged. For example, consider the travel-
time curve in Figure 3-35.

Imagine you obtained this curve in an investigation where your goal was to deter-

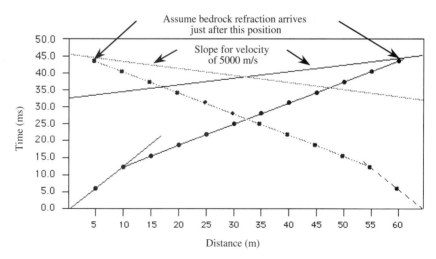

Figure 3-35 Travel-time curve illustrating method to determine minimum depth to
bedrock when no bedrock refractions are received at geophones. This approach assumes
that a bedrock refraction is present just beyond the last position sampled in the survey.

mine bedrock depth. Analysis of the curve suggests a two-layer case with velocities of 800 m/s and 1500 m/s. No bedrock refractions are present, and you have no time in which to run a longer spread. Can you deduce anything about bedrock depth? The answer is yes. You can obtain a minimum figure for bedrock depth if you have a reasonably good value for bedrock velocity.

Assume a bedrock refraction immediately after the last geophone in the spread in Figure 3-35. Draw a line with the correct slope for bedrock velocity such that the line barely touches the last point on the curve. Repeat for the reverse traverse. Your solution now will be for a three-layer case, and you will have a depth which is the minimum depth at which bedrock could be present. In the case of Figure 3-35 the minimum depth is 24 m. This information could well be all you require if your assignment was to determine if any bedrock in the area is shallower than 15 m.

> Use RefractModel to create data for a two-layer case. Transfer to RefractSolve, saving this data set. Pretend this data set represents field data from a consulting assignment. Assuming a bedrock velocity of 4500 m/s, determine the minimum depth at which bedrock could be present.

Because computer programs such as RefractSolve provide easy data input, line fitting, and speedy computations, use them creatively to glean as much information as the data supports.

Corrections to Data

If a geophone spread cannot be arranged so that all the detectors are at the same elevation, a portion of the observed variations in arrival times will be due to these elevation differences and not to subsurface structure. In order to understand how corrections for this situation are derived, consult Figure 3-36. Because of the variation in topography along the spread, the energy source and geophones are projected to lie along a common datum. For example, we assume that E is located at A, a vertical distance of Z_E below the surface, and that G is at D. The actual wave path, $EQRG$, is transformed to $APSD$. Travel times in the surface layer are reduced by EB/V_1 and CG/V_1 (denoted by the minus sign in Figure 3-36) and increased by PQ/V_2 and RS/V_2 (denoted by the plus sign). As $AB=PQ$ and $CD=RS$, we can confine our derivation to triangles EAB and GCD.

The correction time at the energy source is

$$t_{c_E} = \frac{Z_E \tan\theta_{ic}}{V_2} - \frac{Z_E}{V_1 \cos\theta_{i_c}} \tag{3-80}$$

By the critical-angle equation, Eq. 3-80 simplifies to

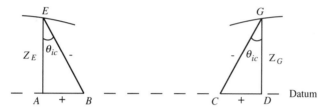

Figure 3-36   Symbols and geometry used in the text to derive the elevation correction.

$$t_{c_E} = Z_E \frac{\sin^2 \theta_{ic} - 1}{V_1 \cos \theta_{ic}} \tag{3-81}$$

which can be reduced further to

$$t_{c_E} = Z_E \frac{\left(V_2^2 - V_1^2\right)^{1/2}}{V_2 V_1} \tag{3-82}$$

The same relationship holds at the geophone position except that Z_G substitutes for Z_E. Thus, the total elevation correction is

$$t_{elevation} = \left(Z_E + Z_G\right) \frac{\left(V_2^2 - V_1^2\right)^{1/2}}{V_2 V_1} \tag{3-83}$$

If a buried shot is used, its depth simply is subtracted from Z_E. If the energy source or geophone is below the datum, the relationship remains the same except that Z_E or Z_G takes on a negative value.

   If the surface layer possesses significant variations in thickness, this can introduce unacceptable time variations due to the fact that this layer's velocity is typically very low. In such cases a correction must be made (often, though misleadingly, referred to as the *weathering correction*). Before such a correction can be introduced, however, good control on surface layer thickness and velocity must be in hand. As this requires a number of closely spaced geophone spreads, such an expenditure of effort is made only when one is interested in detailed mapping of a refractor surface. Dobrin and Savit (1988, p. 494) provide an explanation for this correction.

APPLICATIONS USING SEISMIC REFRACTION

Whately, Massachusetts

The town of Whately is a rural community located in west-central Massachusetts. Although a number of its population of more than 1300 individuals depend on vegetable truck farming for their livelihood, many commute to jobs in larger communities to the south. Over a period of several months in 1983 and 1984, pesticide contamination was discovered in the wells of many homeowners. Because most of the water supply for the community is derived from shallow wells which penetrate an unconfined aquifer, the contami-

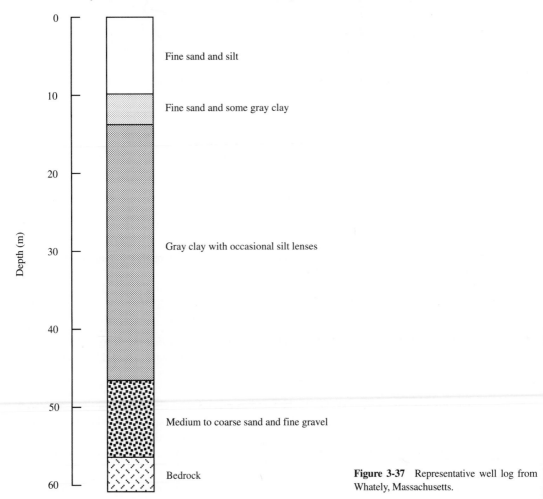

Figure 3-37   Representative well log from Whately, Massachusetts.

nation was understandably viewed with great alarm. The level of concern ultimately resulted in state funding for a study of the contamination problem and possible remedies.

The study was undertaken by a Boston-based engineering firm, which in turn sub-
contracted some of the subsurface investigations to a geophysical consulting company. Due
to the extent of the contamination, it soon became obvious that an alternate water source
was mandatory. Fortunately, the local geologic data base was good and some well logs were
available. These logs (see Figure 3-37) demonstrated the presence of a sand and gravel
layer sandwiched between a thick sequence of glacial-lake clays above and compacted
glacial till and/or arkosic bedrock beneath. The thick clay layer forms an impermeable barrier between the deeper sands and gravels and the surface sands that form the contaminated
aquifer. The consulting firms reasoned that the deeper sands and gravels might constitute a
confined aquifer if areas of sufficient thickness could be discovered. Some prior, but spotty,
geophysical investigations suggested the presence of anomalously deep bedrock areas
which might be a prime location for the aquifer thicknesses required.

Seismic refraction and electrical resistivity investigations were initiated in October
1984 and were supplemented by carefully selected drilling sites. Neither geophysical
method was able to differentiate between the clay layer and underlying sands and gravels,
but both were able to locate the bedrock surface at all occupied sites. Ultimately, the geophysical survey delineated a buried river valley filled with the lower sequence of sands and
gravels to an extent sufficient to constitute a viable water source for the community. This
source now is developed and, with sufficient caution, is likely to remain uncontaminated
due to the impermeable clay layer.

A typical travel-time curve from the Whately area is presented in Figure 3-38.
Although this time-distance plot is from a site where bedrock is much shallower than the

Figure 3-38 Field seismogram from Whately, Massachusetts. Geophone traces are
labeled 1–12. The first geophone is located 3 m from the energy source. The geophone
interval is 10 m. First breaks for each trace are indicated by a downward directed arrow.
Timing lines are at 10-ms intervals. The record encompasses 200 ms.

60–70-m depths associated with the buried valley, the curve exhibits the classic three-layer case typical throughout the area. The first layer is unsaturated sands and alluvial deposits. The second layer consists of saturated sand, gravels, silts, and the lake clays. The third layer typically is bedrock.

Southeastern New Hampshire

A study conducted in New Hampshire in 1974 (Birch, 1976) illustrates how seismic refraction studies can be used to map water-table elevations as well as bedrock elevations. This study is noteworthy because it was conducted in a short period of time at low cost and produced useful results. Field work outlined a bedrock trough which could be correlated with a meta-sedimentary unit with relatively low erosional resistance. Analysis of water-table elevations led to the conclusion that these elevations correlate with bedrock elevations and not surface elevations in the study area. This finding contains implications for water flow-paths, which in turn contain implications for planning to avoid contamination of local aquifers.

Waste-Disposal Site

Typical information necessary for waste-disposal site evaluations are depth to bedrock, geometry of the bedrock surface, and thickness and velocity characteristics of the overburden. This information often is best gathered by seismic refraction, and where quite detailed information is necessary, the GRM should be the method of choice.

Lankston (1990) describes in detail such a study for a municipality considering a waste-disposal site at an isolated location near a major river. Figure 3-39 illustrates the information obtained using the GRM. In this study three seismic lines were oriented in a north-south direction, and three were oriented in an east-west direction. Interpretations of depths at line intersections vary by less than 5 percent. Especially striking is the degree of detail obtained concerning the basalt surface as well as the position of lateral velocity variations in the basalt (interpreted by Lankston as layering in the basalt).

Maricopa Area, Arizona

The area near the Maricopa Mountains in Arizona was one of several locations in the United States proposed as a site for the Superconducting Super Collider and, as such, required that detailed geologic and geophysical information be collected. Gravity, seismic, and electrical methods in conjunction with surface geology and well logs were used to determine the subsurface geology at specific sites within the Maricopa area.

Figure 3-40 illustrates a seismic refraction study from one location at the Maricopa site (Sternberg et al., 1990). The refraction travel-time curve (Fig. 3-40(a)) is relatively straightforward. The subsurface geology derived from this seismic information is illustrated in Figure 3-40(b)). The location of a logged well also is shown. The results of logging cuttings as the well was drilled and the drill rate are diagrammed in Figure 3-40(c)). The seismic-based location of the 1000 m/s–2130 m/s interface is 35 m at the well location. The

Figure 3-39   An example of a geologic section derived from a detailed refraction study using the Generalized Reciprocal Method (GRM). Note that this method provides much detail on the surface of the basalt refractor and is able to determine lateral velocity changes which are interpreted as layering in the basalt. (Modified from Lankston, Robert W., 1990, High-resolution refraction seismic data acquisition and interpretation *in* Ward, Stanley H., ed., Geotechnical and environmental geophysics; Volume 1: Review and tutorial: Society of Exploration Geophysicists Investigations in Geophysics, no. 5, p. 45–73).

lithology log shows this velocity change (sand versus compacted fanglomerate) to be significantly deeper (at about 42 m). However, the drill rate indicates a pronounced change from soft to hard material at 37 m. This likely is the true location of the interface, whereas the lithology log is misleading due to inaccuracies inherent in working with cuttings from loose, soft material traveling to the surface.

The well bottomed in very hard material encountered at 62 m. Seismic bedrock is much deeper at this location (approximately 82 m), and it is likely drilling was terminated after striking a large boulder. In general, well information is essential for aiding and constraining geophysical interpretations. In this case the drill-rate log provides excellent confirmation of the 1000 m/s–2130 m/s interface depth. However, the lithology log is misleading and provides a cautionary note. One always should use well logs as an additional, but not infallible, source of information. Well logs, especially those from unconsolidated deposits, are only as good as the expertise of the drilling company and the person logging the well.

As we conclude this refraction seismology chapter, we hope you now have an appreciation for the strengths of the method as well as some of its weaknesses. In favorable geologic conditions this approach to exploring the subsurface returns valuable data at reasonable cost. Many of the uncertainties inherent in applying seismic refraction disappear when the method is used in conjunction with other geophysical approaches. Seismic reflection often can discern the presence of low-velocity layers, for example. Your work will best be served if you keep the inherent problems of the method in mind, learn the standard field approaches well enough to design flexible surveys, and always attempt to employ addition-

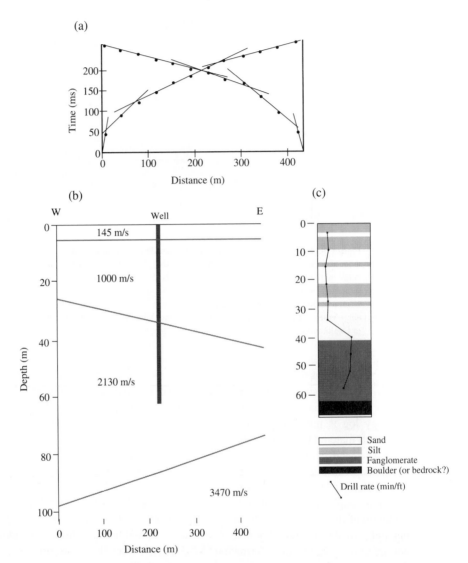

Figure 3-40   Seismic investigation of a proposed site for the Superconducting Super Collider. (a) Travel-time curves at site MR1. (b) Interpretation of seismic data. (c) Lithology log and drill-rate log at the well location identified in (b). (Modified from Sternberg, Ben K., Poulton, Mary M., and Thomas, Scott J., 1990, Geophysical investigations in support of the Arizona SSC Project *in* Ward, Stanley H., ed., Geotechnical and environmental geophysics; Volume 3: Geotechnical: Society of Exploration Geophysicists Investigations in Geophysics, no. 5, p. 211–228).

al geophysical methods. In the next chapter we continue to utilize seismic techniques via the reflection method. This approach has its own strengths and weaknesses, but many of these tend to complement the refraction approach, such that the two, used in conjunction, constitute a formidable pair.

PROBLEMS

3-1. Refer to Figure 3-38. Assuming horizontal interfaces, determine the subsurface geology at this site.

3-2. Figure 3-41 is a field seismograph from a region with horizontal interfaces. Interpret the data.

3-3. Using the information in Figure 3-42, determine as much as possible about the subsurface geology at the site where the information was recorded. Be sure to include comments concerning sources of possible error.

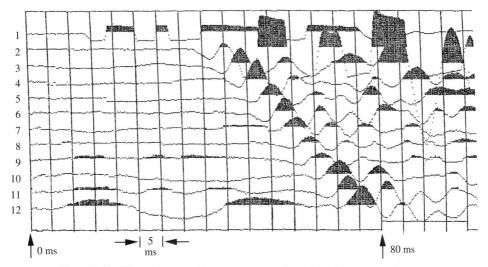

Figure 3-41 Field seismogram. Geophone traces are labeled 1–12. The first geophone is located 3 m from the energy source. The geophone interval is 10 m. Timing lines are at 5-ms intervals.

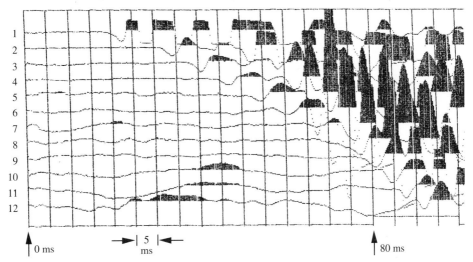

Figure 3-42 Field seismogram. Geophone traces are labeled 1–12. The first geophone is located 3 m from the energy source. The geophone interval is 10 m. Timing lines are at 5-ms intervals.

3-4. Calculate the critical distances for the first and second interfaces in the accompanying diagram.

12 m   800 m/s

15 m   1800 m/s

6000 m/s

3-5. Interpret the following refraction data:

Distance from Shot (m)	Forward Traverse (ms)	Reverse Traverse (ms)
5	7.1	7.1
10	14.3	14.3
15	21.4	18.6
20	28.6	22.0
25	35.7	25.4
30	42.9	28.8
35	48.5	32.2
40	51.0	35.6
45	53.5	39.0
50	56.0	42.3
55	58.4	45.7
60	60.9	49.1
65	63.4	52.5
70	65.9	55.9
75	68.4	59.3
80	70.8	62.7
85	73.3	66.1
90	75.8	69.5
95	77.5	72.8
100	78.3	76.2
105	79.1	77.9
110	80.0	79.0
115	80.8	80.2
120	81.6	81.3

3-6. Provide a qualitative interpretation for the following travel-time curve, but be as specific as possible.

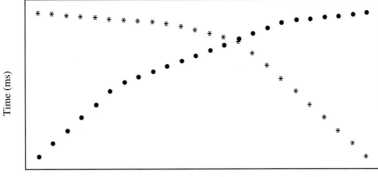

Distance (m)

3-7. In the area where the following time-distance data were collected, bedrock velocities average 5000 m/s. What is the minimum depth to bedrock?

Distance from Shot (m)	Forward Traverse (ms)	Reverse Traverse (ms)
10	16.7	16.7
20	33.3	33.3
30	50.0	50.0
40	66.7	66.7
50	77.6	77.6
60	83.9	83.9
70	90.1	90.1
80	96.4	96.4
90	102.6	102.6
100	108.9	108.9
110	115.1	115.1
120	121.0	121.0

3-8. Assume that a vertical contact separates two layers of moderate velocities. These layers are overlain by a layer with low velocity. Using the following time-distance data, determine the velocities of the materials, the thickness of the top layer, and, as closely as possible, the location of the vertical discontinuity.

Distance from Shot (m)	Forward Traverse (ms)	Reverse Traverse (ms)
10	33.3	33.3
20	66.7	66.7
30	100.0	100.0
40	111.4	119.8
50	114.3	125.1

Distance from Shot (m) *(cont.)*	Forward Traverse (ms) *(cont.)*	Reverse Traverse (ms) *(cont.)*
60	117.3	134.1
70	120.2	137.0
80	123.1	140.0
90	133.6	142.9
100	138.9	145.8
110	144.1	148.8
120	149.4	151.7

3-9. The refraction data in file RM11990001 on the computer disk included with this text illustrates a classic three-layer pattern. The second layer in this sequence thins to the east at a rate of 0.5 m for every 10 m of ground distance traversed. For simplicity, assume horizontal interfaces and that the thickness of the first layer remains constant. Approximately how far to the east will the second layer no longer be detected?

3-10. These data are taken from a traverse across a buried vertical discontinuity. Explain why the travel-time curve has the appearance it does, then determine as much as possible about velocities, the thickness of the overburden, and the location of the vertical discontinuity.

Distance from Shot (m)	Forward Traverse (ms)	Reverse Traverse (ms)
10	12.5	12.5
20	25.0	25.0
30	37.5	37.5
40	50.0	50.0
50	64.2	62.5
60	69.2	75.0
70	74.2	80.8
80	79.2	85.8
90	84.2	90.8
100	89.2	96.7
110	94.2	99.2
120	99.2	101.7

3-11. The following refraction data produce a classic two-layer pattern. However, well logs indicate that the typical sequence consists of three layers with a low-velocity layer occupying an intermediate position. The depth to the top of the low-velocity layer remains essentially constant at 3 m. The velocity of the layer is 1200 m/s. Determine the depth to the top of the third layer (bedrock) using only the information on the travel-time curve. Next determine the actual depth to bedrock based on the additional information from well logs. What is the percent error involved when using only the information on the travel-time curve?

Distance from Shot (m)	Forward Traverse (ms)
5	2.8
10	5.6
15	8.3

Distance from Shot (m) *(cont.)*	Forward Traverse (ms) *(cont.)*
20	11.1
25	13.9
30	16.7
35	19.4
40	22.2
45	25.0
50	27.8
55	28.7
60	29.5

3-12. Data file RM11990003 is from a seismic traverse in an area where supplementary information indicates that a 2200 m/s layer underlies a 1400 m/s layer. Utilizing this information, what constraints can you place on the depth to bedrock (average velocity = 3900 m/s) in the area?

3-13. The time-distance data presented here were recorded in an area where steep fault scarps in bedrock are overlain by alluvial debris. Deduce as much information as possible from the data.

Distance from Shot (m)	Forward Traverse (ms)	Reverse Traverse (ms)
5	3.6	3.6
10	7.1	7.1
15	10.7	10.7
20	14.3	14.3
25	17.9	17.9
30	21.4	21.4
35	23.0	25.0
40	24.0	28.6
45	24.9	30.4
50	25.8	25.8
55	26.7	26.7
60	27.7	27.7
65	28.6	28.6
70	29.5	29.5
75	30.4	30.4
80	31.4	31.4
85	37.8	32.3
90	38.7	33.2
95	39.7	34.1
100	40.6	35.1
105	41.5	36.0
110	42.4	36.9
115	43.4	37.9
120	44.3	38.8

3-14. The following time-distance values were recorded from forward and reverse traverses. Because your contract requires detailed information on a bedrock refractor, use the delay-time method to obtain depths and then construct the bedrock surface. Reciprocal time is 40.7 ms.

Distance from Shot (m)	Forward Traverse (ms)	Reverse Traverse (ms)
10	8.3	8.3
20	14.2	16.7
30	16.8	25.0
40	20.3	30.0
50	21.3	30.2
60	25.6	33.7
70	28.2	34.7
80	28.4	34.1
90	31.8	36.7
100	34.5	36.9
110	35.5	37.9
120	39.7	40.5

REFERENCES CITED

Adachi, R., 1954, On a proof of fundamental formula concerning refraction method of geophysical prospecting and some remarks: Kumamoto Journal of Science, series A, v. 2, p. 18–23.

Birch, Francis S., 1976, A seismic ground-water survey in New Hampshire: Ground Water, v. 14, p. 94–100.

Bonini, W. E., and Bonini, R. R., 1979, Andrija Mohorovičić: Seventy years ago an earthquake shook Zagreb: EOS, v. 60, no. 41, p. 699–701.

Cerveny, V., Langer, J., and Psencik, I., 1974, Computation of geometric spreading of seismic body waves in laterally inhomogeneous media with curved interfaces: Geophysical Journal of the Royal Astronomical Society, v. 38, p. 9–19.

Dobrin, Milton B., and Savit, Carl H., 1988, Introduction to geophysical prospecting, 4th ed.: New York, McGraw-Hill, 867 p.

Kearey, P., and Brooks, M., 1984, An introduction of geophysical exploration: London, Blackwell Scientific Publications, 296 p.

Lankston, Robert W., 1989, The seismic refraction method: A viable tool for mapping shallow targets into the 1990s: Geophysics, v. 54, p. 1535–1542.

Lankston, Robert W., 1990, High-resolution refraction seismic data acquisition and interpretation *in* Ward, Stanley H., ed., Geotechnical and environmental geophysics; Volume 1: Review and tutorial: Society of Exploration Geophysicists Investigations in Geophysics, no. 5, p. 45–73.

Mooney, Harold M., 1977, Handbook of engineering geophysics: Minneapolis, Minnesota, Bison Instruments, Inc.

Nettleton, L. L., 1940, Geophysical prospecting for oil: New York, McGraw-Hill, 442 p.

Palmer, Derecke, 1980, The generalized reciprocal method of seismic refraction interpretation: Tulsa, Oklahoma, Society of Exploration Geophysicists, 104 p.

Redpath, Bruce B., 1973, Seismic refraction exploration for engineering site investigations: Technical Report E-73-4, Explosive Excavation Research Laboratory, Livermore, California, 51 p.

Sternberg, Ben K., Poulton, Mary M., and Thomas, Scott J., 1990, Geophysical investigations in support of the Arizona SSC Project *in* Ward, Stanley H., ed., Geotechnical and Environmental Geophysics; Volume 3: Geotechnical: Society of Exploration Geophysicists Investigations in Geophysics, no. 5, p. 211–228.

Thornburgh, H. R., 1930, Wave-front diagrams in seismic interpretation: Bulletin of the American Association of Petroleum Geologists, v. 14, p. 185–200.

SUGGESTED READING

Lankston, Robert W., 1990, High-resolution refraction seismic data acquisition and interpretation *in* Ward, Stanley H., ed., Geotechnical and Environmental Geophysics; Volume 1: Review and tutorial: Society of Exploration Geophysicists Investigations in Geophysics, no. 5, p. 45–73.

Musgrave, A. W., ed., 1967, Seismic refraction prospecting: Tulsa, Oklahoma, Society of Exploration Geophysicists, 604 p.

Palmer, Derecke, 1986, Refraction seismics: London, Geophysical Press, 269 p.

Redpath, Bruce B., 1973, Seismic refraction exploration for engineering site investigations: Technical Report E-73-4, Explosive Excavation Research Laboratory, Livermore, California, 51 p.

Robinson, Edwin S., and Coruh, Cahit, 1988, Basic exploration geophysics: New York, John Wiley, 562 p.

Sheriff, R. E. and Geldart, L. P., 1982, Exploration seismology, Volume 1: History, theory, and data acquisition: Cambridge, Cambridge University Press, 253 p.

4

Seismic Exploration: The Reflection Method

The 1920s witnessed a great deal of fundamental effort in applying seismic theory to exploration for natural resources, particularly petroleum. During this decade developments in equipment and field methods were rapid, so that a technique that was relatively unproven at the beginning of the decade was well on its way to assuming the overwhelming prominence that is so evident today. In 1921 the first reflection was received and identified in the field from a shale-limestone contact (Sheriff and Geldart, 1982, p. 6–7), but most of the early efforts concentrated mainly on the refraction method and were directed at the discovery of saltdomes. This early dominance of refraction soon declined as the reflection method recorded a number of successes. By the beginning of the 1930s reflection began to replace refraction as the dominant seismic method in petroleum exploration (Sheriff and Geldart, 1982, p. 12). If you are interested in a vivid account of this early history, especially in the details of instrument development and application, we highly recommend the Sheriff and Geldart (1982) account.

You already are aware of some of the shortcomings of the refraction method: (1) the hidden-layer problem, which can be acute when dealing with the geologic sequences encountered in petroleum exploration; (2) the requirement for geophone cable lengths of three to four times the depth of the target horizon—a requirement which obviously becomes untenable when probing to depths on the order of kilometers; and (3) the energy source needed to create recognizable head wave arrivals at distant geophones. In view of these difficulties and due to the dominance of reflection in the petroleum industry, you might expect the same history to hold for shallow seismic exploration. However, we have already noted that refraction still is the most widely used technique when seeking information about shallow targets. What is the reason for this different history in the deep and shallow realms?

If you recall the discussion of energy partitioning in Chapter 2, you already may realize that the difficulties caused by the low-velocity layer disappear in reflection work (see the following). Also, since reflected waves are received at very short shot-receiver distances (even with distance equal to zero), long line lengths will not be required. This suggests that the size of the energy source compared to refraction can be much smaller. Nothing in these relationships answers our question, but perhaps a clue lurks in Figure 2-21. Note that reflections always arrive after several other waves. This suggests that reflected energy might be more difficult to recognize than the first arrivals associated with the refraction method. And, indeed, this is the case. The vast computer resources provided by petroleum exploration companies are used primarily for data processing which is directed at enhancing reflection arrivals so that they may be recognized and interpreted.

It is only within the very recent past that the combination of a relatively low cost multichannel signal-enhancement seismograph with the personal computer and its increased processing capabilities has made shallow-target reflection surveys a viable option to refraction. The reason, as before, is that these capabilities provided the impetus to develop field and other techniques to further clarify the arrival of reflected waves from shallow targets. Our strategy in this chapter is somewhat similar to that of the preceding chapter. We begin with geometric considerations and work toward situations of increasing complexity. After developing these fundamentals, we turn to field techniques and design matters. However, throughout both discussions, we continue to call attention to relationships that ultimately will facilitate our recognition of reflections from shallow interfaces.

A SINGLE SUBSURFACE INTERFACE

In beginning our analysis of reflected waves, we follow the same general plan as when we discussed refracted waves. First, we treat the simple case of a single horizontal interface and develop a travel-time equation for that special case. We graph this equation on a time-distance plot to investigate how the travel-time curve might aid our analysis. We then analyze this equation in order to determine what it might tell us about the subsurface, specifically the depth and velocity to an interface. Finally, we examine seismograms in order to apply the methods we developed to actual field data.

Using ReflectModel

A modeling program such as ReflectModel is useful for exploring a number of the parameters involved in understanding patterns of reflections and their relationships to other wave arrivals. ReflectModel is similar to RefractModel in that you enter a subsurface model for which arrival times then are calculated. Whereas RefractModel plots results as data points on a time-distance curve (since this is the basic presentation for refraction analysis), ReflectModel produces a simulated field seismogram. This mode of presentation was chosen in order to begin the process of recognizing reflections on records. It is possible to plot reflections from several interfaces as well as to add arrivals of additional wave types (direct wave, refraction from the first interface, etc.) thereby creating patterns that bear some relationship to those typically encountered on field records or that are useful in elucidating certain strategies employed during field work.

      Detailed instructions for using all reflection programs (ReflectModel, ReflectSolve, and ReflectIVel) are in Appendix B. Although computation times are generally longer for ReflectModel than for RefractModel, we hope the ability to create these simulated seismograms and their usefulness in reinforcing concepts discussed in the text will encourage you to use this program frequently. A number of simplifying assumptions were adopted while developing ReflectModel in order to keep the size of the program manageable as well as to limit computation time. These simplifications are mentioned in Appendix B but, in general, have no bearing on your utilization of the program for most purposes suggested in this text.

Derivation of a Travel-Time Equation

The derivation of time of travel for a wave reflected from a single horizontal interface is straightforward. Referring to Figure 4-1, we see that

$$\text{time} = \frac{EA + AG}{V_1}$$

$$EA = AG = \left[\left(\frac{x}{2} \right)^2 + h_1^2 \right]^{1/2}$$

and

$$\text{time} = \frac{\left(x^2 + 4h_1^2 \right)^{1/2}}{V_1} \tag{4-1}$$

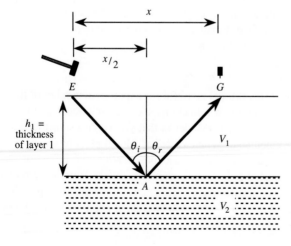

Figure 4-1 Diagram illustrating symbols used in derivation of time of travel for reflection.

TABLE 4-1 Travel Times for Reflected Waves and Normal Move-Out

Geophone	Distance (m)	Time (ms)	NMO (ms)
Source	0	21.4	00.0
1	5	21.7	00.3
2	10	22.6	01.2
3	15	24.0	02.5
4	20	25.8	04.3
5	25	27.9	06.5
6	30	30.3	08.9
7	35	32.9	11.5
8	40	35.7	14.3
9	45	38.6	17.2
10	50	41.6	20.2
11	55	44.7	23.3
12	60	47.9	26.5

Velocity (m/s)	**1400**
Thickness (m)	**15**
Increment (m)	**5**

If we take the derivative of this equation, we see that we will not be so fortunate as we were in refraction because we don't have a simple way to determine velocity. In order to understand the form of Eq. 4-1 more completely, let's create a table relating reflection travel-times to source-geophone distances for a given subsurface model. If we take the values in Table 4-1 and plot them in a standard time-distance graph, we arrive at the travel-time curve in Figure 4-2(a). Additional points are added for small values of x in the graph than are present in Table 4-1 in order to define more completely the form of the curve in this region.

The curve form in Figure 4-2(a) is reminiscent of a hyperbola. An equation for a hyperbola that is symmetric about $x = 0$ is

$$\frac{y^2}{a^2} - \frac{x^2}{b^2} = 1 \tag{4-2}$$

Equation 4-1 can be rearranged in the form of Eq. 4-2, so our initial impression is correct. Of course, the relationship would be even more apparent if we placed geophones along a line symmetrically about $x = 0$ so that the curve in Figure 4-2(a) also extended to the left of the energy source (see Figure 4-20).

> Show that Eq. 4-1 can be placed in the form of Eq. 4-2. *Hint*: Square both sides and then divide through by one of the terms that contains only constants.

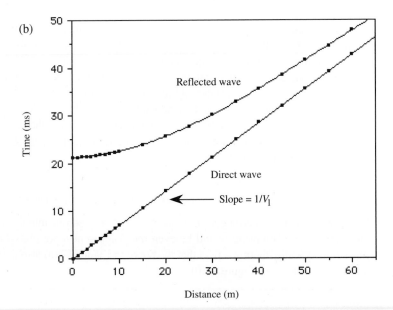

Figure 4-2 A plot of reflection time-distance data from Table 4-1. More points are added at short distances from the energy source in order to establish more clearly the form of the travel-time curve. (a) Reflection arrivals only. (b) Direct wave arrivals added to demonstrate the asymptotic relationship to the hyperbolic form of the reflection arrivals. (c) Head wave arrivals added to (b).

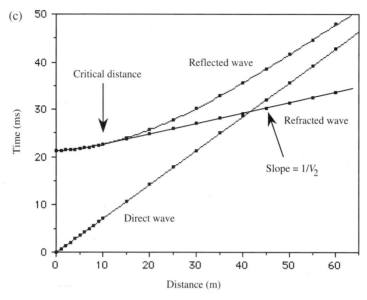

Figure 4-2   *(continued)*

We can see how the reflections in Table 4-1 would look on a field seismogram by placing the same model into ReflectModel (Fig. 4-3(a)). The time and distance coordinates for ReflectModel plots are arranged to be the same as those on field seismograms generated by most seismographs used in shallow-refraction and -reflection surveying. In order to more directly compare the pattern of arriving reflections with the arrangement of standard time-distance curves, we simply rotate the simulated seismogram in Figure 4-3(a) through 90° to arrive at Figure 4-3(b). Although the coordinates are slightly different, the similarity between Figures 4-2(a) and 4-3(b) is evident. Later in our discussion of reflection surveying, we apply corrections to the arrival of reflected pulses in order to generate a seismic "picture" of the subsurface. In geologic sections depth increases downward, so seismic sections are arranged with time increasing downward. We can obtain a sense of how these appear by flipping Figure 4-3(b) about a horizontal axis to obtain Figure 4-3(c). Note that no corrections have been applied to these data. Also, we've simply rotated a graphic image on a computer screen rather than plot the digital information recorded at a geophone position. Thus, while Figure 4-3(c) suffices for the moment as an example, the polarity of the waveforms is 180° out of phase.

In Figure 4-1 no note is made of the relationship between V_1 and V_2 for the simple reason that reflection of wave energy will occur at the interface if $V_1 < V_2$ or if $V_1 > V_2$. Actually, the factor controlling the partitioning of energy at an interface is known as *acoustical impedance*, which is the product of velocity and density. Table 2-7 lists reflection coefficients for values of density and compressional wave velocity at normal incidence. If we simply reverse the density and velocity values assigned to the two layers, the only change in the value of the reflection coefficient is one of sign. Thus, equivalent energy is reflected in either case, but waveforms exhibit a 180° phase change.

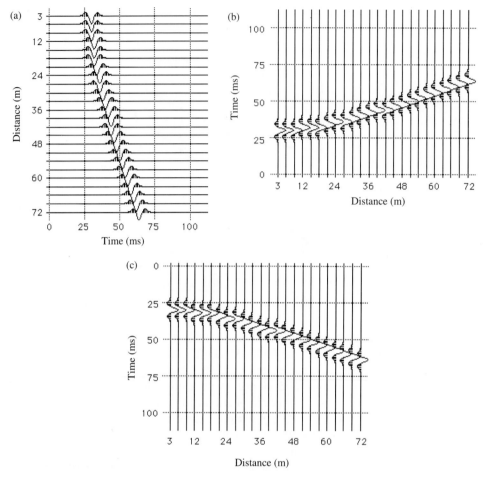

Figure 4-3 Three variations in presentation of reflection arrivals on seismogram. (a) A portion of a simulated seismogram produced by ReflectModel. (b) Seismogram rotated so that axes correspond to conventional time-distance curve. (c) Similar to (b) except time increases downward (as does depth). This is the arrangement used when seismic sections are prepared.

Analysis of Arrival Times

We intimated at the beginning of this chapter that a problem with the reflection method was identifying reflections on seismograms, so it seems important to ask about the relation of reflection arrivals to those of other waves.

    If you remember some analytic geometry, you might recall that hyperbolas have asymptotes. The equation for the asymptote of a hyperbola of the form given in Eq. 4-2 is

$$y = \frac{a}{b} x \qquad\qquad (4\text{-}3)$$

If you substitute the values for *a* and *b* you determined placing Eq. 4-1 in the form of Eq. 4-2, you find that Eq. 4-3 becomes

$$y = \frac{1}{V_1} x \qquad (4\text{-}4)$$

which is the equation of a straight line passing through the origin with a slope of $1/V_1$. In other words, our asymptote is the travel-time curve for the direct wave. This relationship between the reflected and direct waves is illustrated in Figure 4-2(b). The physical reason for the relationship is easy to visualize. Both the direct wave and the reflected wave travel at V_1. Thus, travel time will be directly related to path. At small source-detector distances the reflected wave travels a much longer path than the direct wave. Although the reflected wave always travels a longer path, at large source-detector distances the reflected path becomes closer and closer in length to the direct wave path, and, therefore, the travel times become more and more similar.

In Figure 4-2(c) head wave arrivals also are added. Notice the rapid convergence of reflected and refracted arrivals near the critical distance for refraction (at about 10 m). Clearly, this would not be a region in which to try to identify reflections if the head wave pulse has any amplitude at all. We now have at least one reason why it might be difficult to sort out and identify reflections. In order to illustrate a bit more vividly the relationship graphed in Figure 4-2(b), we add direct wave arrivals to this simulated seismogram (Fig. 4-4). Although the reflection still is discernible, its pattern is not nearly so clear due to inter-

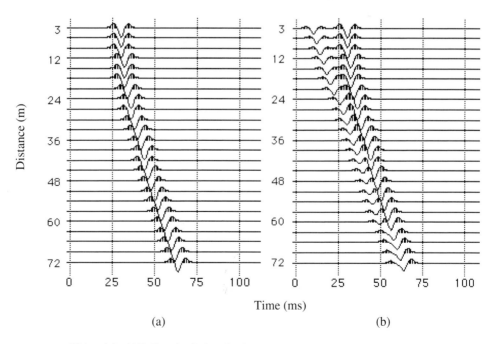

Figure 4-4   (a) Reflected arrivals as in Figure 4-3(a). (b) Direct wave pulse added to pattern in (a).

(a)

(b)

Figure 4-5 Time-distance plots demonstrating the effect of depth and velocity on curvature of reflection travel-time data. (a) Effect of depth with velocity constant. (b) Effect of velocity with depth constant.

ference with the waveform of the direct wave arrival. Add a bit of noise and a few other complicating factors, and the reflection might no longer be identified.

Let us examine Eq. 4-1 once again. It should be clear that velocity and depth affect the form of the reflection travel-time curve. Because we can expect reflections from various depths traveling at a range of velocities to appear on our field seismograms, it seems wise to determine how these factors affect curve form. Figure 4-5(a) illustrates the effect of depth differences. All curves were calculated using a velocity of 1400 m/s. The degree of curvature is greatest for the shallow reflections, with curvature decreasing with increasing depth.

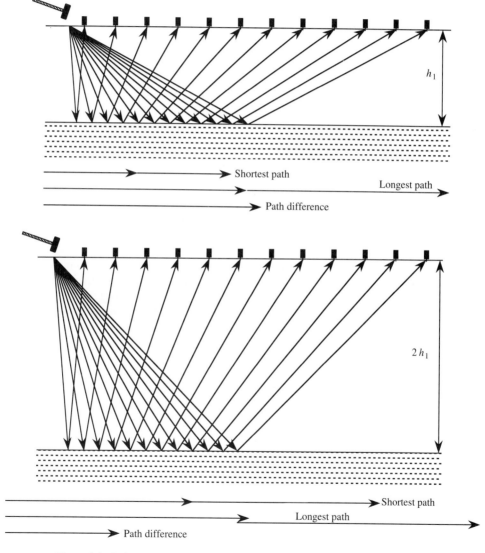

Figure 4-6   Reflection paths for two layer thicknesses: h_1 and $2h_1$. Geophone spread is the same in each case. Differences in path lengths for the two cases are evident. The difference between the longest and shortest path for each case are shown diagrammatically.

The reason for this change is the different path lengths followed by the reflected energy. In Figure 4-6 path lengths to 12 geophones are illustrated for two different layer thicknesses, one of which is twice the thickness of the other. For constant velocity, the degree of curvature of a time-distance plot will depend on the differences in path lengths. Even though the reflection paths for the thicker layer in Figure 4-6 are almost twice the length of those in the thinner layer, the differences in path lengths in the thinner layer are almost twice those of the thicker layer. Thus, we see the reason for the difference in curvature for cases such as h_1 and $2h_1$ (15 m and 30 m) in Figure 4-5(a).

Increasing velocity also reduces the curvature of reflection travel-time curves for a constant h_1. This effect is diagramed in Figure 4-5(b). Take a moment to clarify these relationships in your mind as they are of major importance in reflection surveying.

Normal Move-Out

The forgoing discussion leads to an important relationship that is used extensively in reflection work, namely, *normal move-out*. Normal move-out (NMO) is defined as the difference in reflection travel-times from a horizontal reflecting surface due to variations in the source-geophone distance. NMO is determined by subtracting the travel time at the source ($x = 0$), t_0, from the travel time at the receiver distance t_x. Thus, NMO is equal to $t_x - t_0$, or, in terms of our travel-time equation

$$T_{\text{NMO}} = \frac{\left(x^2 + 4h_1^2\right)^{1/2}}{V_1} - \frac{2h_1}{V_1} \tag{4-5}$$

Dynamic Table 4-2 calculates normal move-out times for three different combinations of velocity and depth. Figure 4-7(a) presents this information in graphical form. Because we are subtracting a constant term (for a given geology), t_0, from reflection travel-times to calculate NMO, NMO values should exhibit the same behavior as reflection travel-time curves with respect to curvature. Referring to Eq. 4-5 or Figure 4-7(a), we observe that NMO decreases with increasing depth to the reflecting surface. NMO also decreases with increasing velocity. Of course, NMO also increases with increasing source-geophone distances.

In particular, Figure 4-7(a) plots NMO against source-geophone distance for three geologic models. Figure 4-7(b) repeats Figure 4-5(a) by plotting reflection travel-times against source-geophone distance. Two normal move-out values are illustrated. The models used are the same as for Figure 4-7(a).

> Establish that normal move-out decreases with increasing velocity by utilizing Table 4-2 or Eq. 4-5.

We will return to normal move-out in later sections of this chapter. One of the ultimate goals in reflection surveying is to produce a seismic section such that all reflections on

TABLE 4-2 Normal Move-Out Comparisons for Different Velocities and/or Depths

Geophone	Distance (m)	NMO-1 (ms)	NMO-2 (ms)	NMO-3 (ms)
Source	0	00.0	00.0	00.0
1	3	00.1	00.1	00.0
2	6	00.4	00.2	00.1
3	9	00.9	00.5	00.3
4	12	01.7	00.8	00.6
5	15	02.5	01.3	00.9
6	18	03.6	01.9	01.3
7	21	04.7	02.5	01.7
8	24	06.0	03.3	02.2
9	27	07.4	04.1	02.8
10	30	08.9	05.1	03.5
11	33	10.4	06.1	04.2
12	36	12.0	07.1	05.0
13	39	13.7	08.3	05.8
14	42	15.4	09.5	06.7
15	45	17.2	10.7	07.6
16	48	19.0	12.0	08.6
17	51	20.8	13.4	09.6
18	54	22.7	14.8	10.7
19	57	24.6	16.3	11.8
20	60	26.5	17.8	13.0
21	63	28.4	19.3	14.2
22	66	30.4	20.9	15.4
23	69	32.3	22.5	16.7
24	72	34.3	24.1	18.0

Increment (m)	3

Velocity-1 (m/s)	**1400**	Thickness-1 (m)	**15**
Velocity-2 (m/s)	**1400**	Thickness-2 (m)	**30**
Velocity-3 (m/s)	**1400**	Thickness-3 (m)	**45**

each geophone trace are presented as if the trace were recorded at $x = 0$ (i.e., such that the reflecting points are directly beneath the source point). In order to accomplish such a presentation, all reflecting times must be corrected for NMO. Normal move-out considerations also come into play when determining velocities and other parameters entirely by computer processing techniques. When undertaking these discussions, we also derive an alternate equation for calculating NMO. Before proceeding to these more esoteric analyses, however, it might be prudent first to investigate how to determine velocity and interface depth for the simple two-layer case.

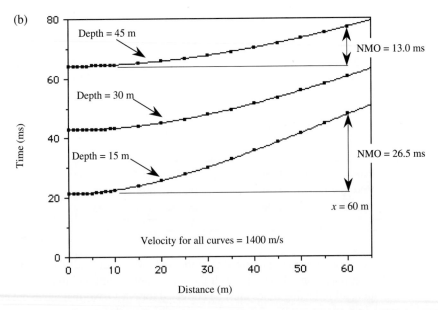

Figure 4-7 Illustration of normal move-out. (a) NMO curves for various interface depths at constant velocity. NMO for shallowest and deepest interface at $x = 60$ m shown by vertical arrows. (b) A repeat of reflection times in Figure 4-5(a) with NMO values at $x = 60$ m designated.

Determining Velocity and Thickness

The most straightforward method for determining velocity and thickness when dealing with a single interface was first proposed by Green (1938). This method, referred to as the x^2 - t^2 method, is very appropriate for working with the type of data typically available when exploring shallow targets.

If we square both sides of the basic reflection travel-time equation,

$$\text{time} = \frac{\left(x^2 + 4h_1^2\right)^{1/2}}{V_1} \tag{4-1}$$

we arrive at an equation that can be written in the form of the equation for a straight line, $y = mx + b$,

$$t^2 = \frac{x^2 + 4h_1^2}{V_1^2} = \frac{1}{V_1^2}x^2 + \frac{4h_1^2}{V_1^2} \tag{4-6}$$

Thus, on an x^2 - t^2 graph, the data points plot as a straight line with slope equal to the inverse of V_1^2 (see Fig. 4-8(a)). This line is projected back to $x = 0$, which gives us t_0^2 (see Fig. 4-8(a)). Velocity and t_0 now are known. If we examine the form of Eq. 4-6 at $x = 0$, we see that

$$t_0^2 = \frac{4h_1^2}{V_1^2}$$

$$t_0 = \frac{2h_1}{V_1}$$

and

$$h_1 = \frac{t_0 V_1}{2} \tag{4-7}$$

If our initial conditions of constant velocity and a horizontal interface are met, the x^2 - t^2 method is straightforward to apply. Because of the requirement that data be squared, it is easiest to use some type of computer aid if possible rather than simply a hand calculator. The dynamic table, Table 4-3, illustrates how a spreadsheet program can fulfill this requirement. Source-geophone distances and reflection arrival times are typed in. Ideally, the x^2- and t^2-values are graphed by the spreadsheet program, which provides the slope and t_0^2-values. Alternatively, a separate graphing program can be used. These two quantities then are placed back in the spreadsheet template for final determination of the velocity and thickness of the layer about the reflecting horizon.

The time-distance values in Table 4-3 are taken from the simulated seismogram illustrated in Figure 4-8(b). The x^2- and t^2-values in Figure 4-8(a) are from Table 4-3.

Another alternative is to use a program, such as ReflectSolve, designed expressly for the x^2-t^2 method. Time-distance data created in ReflectModel can be transferred directly to ReflectSolve, as detailed in Appendix B. You should utilize this capability to solidify your

(a)

(b)

Figure 4-8 An example of the x^2 - t^2 method. (a) An x^2 - t^2 graph. (b) The seismogram from which the data points for (a) were taken.

TABLE 4-3 x^2-t^2 Method

Distance (m)	Time (ms)	x^2	t^2		
15	**50.7**	225	2,570.49		
25	**51.9**	625	2,693.61	slope	**0.3086**
35	**53.6**	1,225	2,872.96	t-squared at x = 0	**2500**
45	**55.9**	2,025	3,124.81		
55	**58.6**	3,025	3,433.96		
65	**61.7**	4,225	3,806.89		
75	**65.1**	5,625	4,238.01		
85	**68.8**	7,225	4,733.44		
95	**72.7**	9,025	5,285.29	Velocity (m/s)	1800
105	**76.8**	11,025	5,898.24	Thickness (m)	45
115	**81.1**	13,225	6,577.21		
125	**85.6**	15,625	7,327.36		

Shot offset (m)	**15**		Interval (m)	**10**

comprehension of this approach to deriving subsurface information from reflection data. As always, it is a good idea to work through a method once or twice by hand (using Table 4-3 to provide the squared data) before relying entirely on computer automation.

Applying the x^2-t^2 Method to a Field Seismogram

Now let's apply the x^2 - t^2 method to real field data. Figure 4-9 illustrates a seismogram recorded on the floodplain of the Connecticut River in Massachusetts during a survey attempting to map the bedrock surface. The survey was designed especially to record reflections. Details of survey design will be presented in a later section of this chapter. Although there are suggestions of earlier reflections, the first distinct reflection is denoted by arrows. For purposes of discussion, we will assume this is the first reflection on the record and, therefore, will treat the data as in a two-layer case.

We also will postpone a detailed discussion of how to recognize reflections on records. However, two good criteria are normal move-out and waveform similarity. If our source-geophone interval is not large relative to the reflector's depth, all paths will be similar in length, and the energy at each geophone should be similar. The reflection on the record exhibits only very slight NMO but does possess good waveform similarity.

Usually it is difficult to identify the arrival of the onset of the reflected energy, and the seismogram in Figure 4-9 is no exception. In such a situation the first trough or peak of the reflection that can be identified from trace to trace is selected for analysis. Clearly, the rule here is consistency. In order to practice identifying reflection arrival times, record time-distance data much as you did in analyzing refraction seismograms. Using ReflectSolve, Table 4-3, or some other means, determine the velocity and thickness of material above the reflecting horizon.

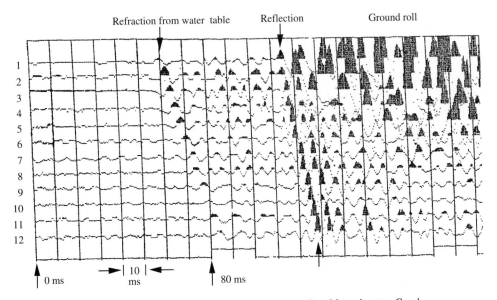

Figure 4-9 Field seismogram from the Connecticut Valley, Massachusetts. Geophone traces are labeled 1–12. The first geophone is located 30 m from the energy source. The geophone interval is 3 m. The reflection selected for analysis is marked by arrows. Timing lines are at 10-ms intervals. The record encompasses 200 ms. Other wave arrivals of interest are noted and will be discussed in detail at a later point in this chapter.

An x^2-t^2 graph constructed from the reflections in Figure 4-9 is presented in Figure 4-10. The computed thickness of 53 m agrees well with bedrock depths in the area determined from refraction studies. Refraction profiles also identify an unsaturated layer that averages 7

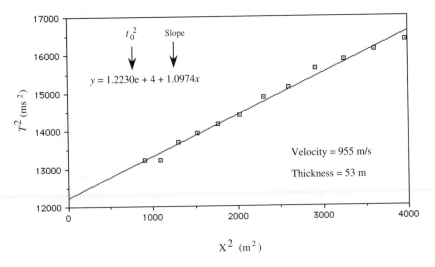

Figure 4-10 An x^2-t^2 plot for the reflection identified in Figure 4-9.

m in thickness (velocity = 350 m/s) and saturated sediment velocities of 1100 m/s. These figures suggest an average velocity to bedrock of 1000 m/s, which is close to that calculated from the x^2-t^2 graph.

Finally, in order to satisfy our curiosity concerning the amount of NMO that should be exhibited by reflection from an interface at 53 m, a velocity of 955 m/s, and source-geophone offsets used in our survey, we place these values into ReflectModel. The simulated seismogram is illustrated in Figure 4-11(a). The curvature of the reflection arrivals is essen-

Figure 4-11 Enhancing the normal move-out exhibited by reflections from an interface by adjusting the geophone interval. The geophone interval for (a) is 3 m; the interval for (b) is 6 m. (c) 24 traces with an interval of 3 m.

tially the same as in the field seismogram. Although this anticipates a bit our discussion of field survey design, we might ask if we would be able to identify this reflection in the absence of obvious NMO if the amplitudes were not so great. We can adjust the shot offset value in ReflectModel, but this only moves the pattern in Figure 4-11(a) earlier in time. Increasing the geophone interval from 3 m to 6 m does result in significantly greater NMO values, which are easily seen in Figure 4-11(b). Perhaps the best solution is to keep a 3-m interval and to use 24 geophones or two 12-geophone arrays. The main lesson from this brief discussion (which will be expanded later) is that survey design is especially critical in providing an optimum situation in which to identify reflections.

MULTIPLE HORIZONTAL INTERFACES

When we developed the refraction method, the step from the single interface to multiple interfaces was not particularly difficult. Although the trigonometric complexity increased somewhat, the equations were of similar form, and the required unknown quantities could be obtained from travel-time curves. If we worked our way through the equations, obtaining information from layer 1, then layer 2, and so forth, both velocities and thicknesses for every layer could be calculated. This capability to determine all velocities and thicknesses in the refraction method ultimately depends on the fact that we are dealing with critically refracted rays and hence know the angle of refraction (90°) at each interface.

Unfortunately, the same jump from the single interface to multiple interfaces in the reflection method involves more complexities than were encountered in the refraction approach. If we consider Figure 4-12 for a moment, it should be obvious that we cannot apply the straight-line ray path that was used to derive Eq. 4-1 to a reflection from, for example, the third interface. Refraction occurs at the first and second interfaces, so that the actual path of the reflected ray is quite different from the straight-line path. In addition, each source-receiver reflection path from a given reflecting horizon is different from all others.

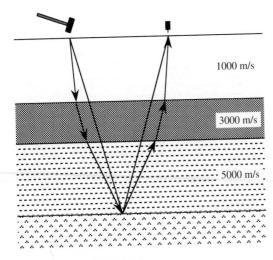

1000 m/s

3000 m/s

5000 m/s

Figure 4-12 Actual reflection ray paths in a multilayered sequence compared with straight-line ray path.

Of course, given a subsurface model and an angle of inclination for a ray at the energy source, we can calculate the precise path of that ray. However, when we do not possess knowledge of the subsurface and have only time-distance data for a given reflection, there simply are too many variables to permit the derivation of a unique equation that does not involve some type of approximation. What are our options?

One option is to neglect refraction at interface boundaries and to assume that the straight-line path is a permissible approximation, especially if source-receiver distances are kept small. When using this option (which we will refer to as the *Green method*), we simply create an x^2-t^2 plot for each reflection event on a seismogram. Since each line through the squared time-distance values results in a velocity value, an h-value for the thickness of material to a reflecting interface, and a t_0-value, we can obtain velocity and thickness values for each layer in the sequence. The values for velocity and thickness of the first layer are correct, because the ray paths involved are straight lines (this is the one-interface case). All values for deeper interfaces clearly are only approximations, because the actual ray paths are not straight lines. Because the next method (*Dix method*) we discuss provides much better values, we will not discuss the Green option further. Consult Table 4-5 to obtain an idea of how values computed by the Green method compare with those obtained via the Dix method.

The Dix Equation

A second option was derived by C. Hewitt Dix (1955). Dix demonstrated that in a case where there are n horizontal beds, travel-times can be related to actual paths traversed by employing a special velocity value known as the root-mean-square velocity V_{rms}. In the general case where there are n horizontal beds and Δt_i is the *one-way vertical travel-time* through bed i, the Dix equation states that

$$V_{rms}^2 \approx \frac{\sum_{i=1}^{n} V_i^2 \Delta t_i}{\sum_{i=1}^{n} \Delta t_i} \tag{4-8}$$

For example, if we want to determine the rms velocity to the second in a series of reflecting interfaces, we expand Eq. 4-8 to arrive at

$$V_{rms} = \left(\frac{V_2^2 \Delta t_2 + V_1^2 \Delta t_1}{\Delta t_2 + \Delta t_1} \right)^{1/2} \tag{4-9}$$

Careful inspection of Eq. 4-9 should lead to an understanding of why this velocity is termed the root-mean-square velocity.

Although we soon will see that the Dix equation provides very good results, note that it is an approximation and assumes that source-receiver distances are kept small relative to the distances to reflecting interfaces. In addition to Dix's article, good discussions relating to the derivation of this equation are in Sheriff and Geldart (1982, p. 86–87) and Robinson and Coruth (1988, p. 102–105).

Determining Velocities

Dix's analysis allows us to continue to use an x^2-t^2 plot to determine layer velocities and thicknesses. Since travel-times are related to actual paths by the rms velocity, Eq. 4-6 still is valid if velocity terms are replaced by V_{rms}.

$$t^2 = \frac{1}{V_{rms}^2} x^2 + t_0^2 \tag{4-10}$$

Recall that we still are working with an approximation. In a following section we will take a more detailed look at the nature of this approximation and try to gain some insights into how good it is. For the moment we see that little has changed with respect to the x^2-t^2 plot except that now slope = $1/V_{rms}^2$.

Let's consider an actual interpretation problem. For simplicity, we consider the seismogram in Figure 4-13(a). Three reflection events are present. If we prepare an x^2-t^2 plot from this information, we arrive at the three lines in Figure 4-13(b). Lines were fit to data points using a least-squares approach, and the equations of these fits are given below and to the right of each line. Beginning at the shallowest reflection, the slope of each line is 0.11113, 0.21007, and 0.13243. These slopes result in the rms velocities listed along the left portion of each line. Since the ray paths connected with the first reflecting horizon are straight lines, $V_{rms_1} = V_1$. Of course, what we also seek are V_2 and V_3. Can we obtain these from the rms velocities?

We begin with Eq. 4-8. Given n reflecting horizons, let V_{rms_n} represent the rms velocity to the nth reflector and let $V_{rms_{n-1}}$ represent the rms velocity to the $(n-1)$st reflector. Then

$$V_{rms_n}^2 = \frac{\sum\limits_{i=1}^{n} V_i^2 \Delta t_i}{\sum\limits_{i=1}^{n} \Delta t_i} \tag{4-11}$$

and

$$V_{rms_n}^2 \sum_{i=1}^{n} \Delta t_i = \sum_{i=1}^{n} V_i^2 \Delta t_i \tag{4-12}$$

Next we expand the summation on the right of the equal sign in Eq. 4-12 and write an equation for $V_{rms_{n-1}}$.

$$V_{rms_n}^2 \sum_{i=1}^{n} \Delta t_i = \sum_{i=1}^{n-1} V_i^2 \Delta t_i + V_n^2 \Delta t_n \tag{4-13}$$

$$V_{rms_{n-1}}^2 = \frac{\sum\limits_{i=1}^{n-1} V_i^2 \Delta t_i}{\sum\limits_{i=1}^{n-1} \Delta t_i} \tag{4-14}$$

Figure 4-13   (a) Seismogram showing reflections from multilayered model. (b) x^2-t^2 plot of time-distance values taken from (a). Rms velocities noted. Results of Dix analysis given in text.

and

$$V_{\mathrm{rms}_{n-1}}^2 \sum_{i=1}^{n-1} \Delta t_i = \sum_{i=1}^{n-1} V_i^2 \Delta t_i \qquad (4\text{-}15)$$

Finally, we substitute the equivalency established in Eq. 4-15 into Eq. 4-13, divide by Δt_n, and solve for V_n^2:

$$V_n^2 = \frac{V_{rms_n}^2 \sum_{i=1}^{n} \Delta t_i - V_{rms_{n-1}}^2 \sum_{i=1}^{n-1} \Delta t_i}{\Delta t_n}$$ (4-16)

All that is left is to recast Eq. 4-16 into a form that uses quantities directly obtainable from an x^2 - t^2 diagram. Remembering that Δt_i is the *one-way vertical travel-time* through bed i and that t_{0_n} is the two-way vertical travel time from reflector n (at $x^2 = 0$ on an x^2 - t_2 diagram), we can write

$$\sum_{i=1}^{n} \Delta t_i = \Delta t_1 + \Delta t_2 + \Delta t_3 \ldots$$

$$= \frac{t_{0_1} + \left(t_{0_2} - t_{0_1} \right) + \left(t_{0_3} - t_{0_2} \right) \ldots}{2}$$ (4-17)

$$= \frac{t_{0_n}}{2}$$

and also,

$$\Delta t_n = \frac{t_{0_n} - t_{0_{n-1}}}{2}$$ (4-18)

Therefore, we finally arrive at,

$$V_n^2 = \frac{V_{rms_n}^2 \dfrac{t_{0_n}}{2} - V_{rms_{n-1}}^2 \dfrac{t_{0_{n-1}}}{2}}{\dfrac{\left(t_{0_n} - t_{0_{n-1}} \right)}{2}}$$ (4-19)

or

$$V_n^2 = \frac{V_{rms_n}^2 t_{0_n} - V_{rms_{n-1}}^2 t_{0_{n-1}}}{t_{0_n} - t_{0_{n-1}}}$$ (4-20)

Equation 4-20 allows us to determine the velocity of any unit for which we have reflections from the interfaces bounding the unit on its upper and lower surfaces. For this reason V_n is referred to as an *interval* velocity. Equation 4-20, in addition to Eq. 4-8, often is referred to as the *Dix equation*. Now, let's return to Figure 4-13(b) and apply Eq. 4-20 to the task of determining interval velocities.

Rms velocities from the three reflecting horizons and square roots of the t^2-values at $x^2 = 0$ for the three lines also are noted in Figure 4-13(b). These are the t_0 values which represent two-way vertical travel times from each reflector. In determining interval velocities we simply select the interval of interest and substitute the appropriate quantities into Eq. 4-20. For purposes of illustration, let's assume we want to determine V_3. We write

$$V_3^2 = \frac{V_{rms_3}^2 t_{0_3} - V_{rms_2}^2 t_{0_2}}{t_{0_3} - t_{0_2}}$$

$$V_3^2 = \frac{(2748 \ \text{m/s})^2 \ (0.22561 \ \text{s}) - (2182 \ \text{m/s})^2 (0.14667 \ \text{s})}{(0.22561 \ \text{s} \ - \ 0.14667 \ \text{s})}$$

and

$$V_3 = 3569 \ \text{m/s}$$

In many presentations of determining interval velocities the subscripts L for *lower* and U for *upper* are substituted for n and $n - 1$. Here lower is synonymous with deeper, and upper is synonymous with shallower. Equation 4-20 then becomes

$$V_L^2 = \frac{V_{\text{rms}_L}^2 t_{0_L} - V_{\text{rms}_U}^2 t_{0_U}}{t_{0_L} - t_{0_U}} \tag{4-21}$$

ReflectIVel is one of the computer programs supplied with this text, and it uses the preceding terminology. Although the computations in determining interval velocities are straightforward using a hand calculator, ReflectIVel speeds the process and tends to minimize errors.

Calculate the interval velocity for the second unit from the data in Figure 4-13(b).

Determining Thicknesses

Once we have interval velocities in hand, calculating thicknesses is trivial. Because travel time equals path distance divided by velocity, vertical thickness (h_n) is equal to velocity times the one-way vertical travel time, or

$$h_n = V_n \left(\frac{t_{0_n} - t_{0_{n-1}}}{2} \right) \tag{4-22}$$

If we use this equation to find the thickness of the third unit in our three-unit sequence of Figure 4-13, we have

$$h_3 = V_3 \left(\frac{t_{0_3} - t_{0_2}}{2} \right)$$

and

$$h_3 = (3569 \ \text{m/s}) \left(\frac{0.22561 \ \text{s} - 0.14667 \ \text{s}}{2} \right) = 141 \ \text{m}$$

Further Discussion of the Dix Method

In the past the Dix method was used extensively in reflection analysis. Presently, most large exploration firms use the computational power of computers to automate the determination of velocities, and the Dix method has fallen into disuse in these circles. Although we will discuss these more sophisticated procedures in a following section, many small-scale operations still utilize the Dix approach. For this reason, it seems prudent to take a slightly more detailed look at the assumptions of the method and how its results compare with known values.

We have already called attention to the role of refraction at layer boundaries in shaping the form of the reflection ray path. This is emphasized once again in Figure 4-14. Actually, a valuable procedure in analyzing reflection travel-times from such a multilayered situation is to prepare a computer program or a spreadsheet template to calculate these times for a given subsurface model. For reasons explained in Appendix B, ReflectModel calculates times based on straight-line paths, so this program does not help us meet our present objectives. However, the dynamic table, Table 4-4, does supply accurate times. Given a geologic model, this table calculates reflection time-distance values from each of three interfaces for a range of incidence angles at the first interface. *Theta* 1 represents the angle of incidence at the first interface, *Theta* 2 represents the angle of refraction at the first interface and the angle of incidence at the second interface, and so on. *X1R* and *T1R* represent horizontal distance and time values for rays reflected from the first interface. *X2R*, *X3R*, *T2R*, and *T3R* follow the same pattern. A wide range of time-distance data from each inter-

Figure 4-14   Reflection ray paths for source-receiver offsets of 30 m, 60 m, and 120 m for the model parameters noted in the diagram. A straight-line raypath is shown for the 30-m offset.

TABLE 4-4 Reflection Time-Distance Values for Refracted Ray Paths in a Multilayered Sequence

| Increment (m) | **0.40** | Velocity 1 (m/s) | **400** | Velocity 2 (m/s) | **1800** | Velocity 3 (m/s) | **3500** |
| | | Thickness 1 (m) | **10** | Thickness 2 (m) | **40** | Thickness 3 (m) | **10** |

Theta 1 (°)	X1R (m)	T1R (ms)	Theta 2 (°)	X2R (m)	T2R (ms)	Theta 3 (°)	X3R (m)	T3R (ms)
0.50	0.17	50.00	2.25	3.32	94.48	4.38	4.85	100.21
0.90	0.31	50.01	4.05	5.98	94.56	7.90	8.76	100.33
1.30	0.45	50.01	5.86	8.66	94.69	11.45	12.71	100.52
1.70	0.59	50.02	7.67	11.37	94.87	15.04	16.74	100.78
2.10	0.73	50.03	9.49	14.11	95.09	18.70	20.87	101.13
2.50	0.87	50.05	11.32	16.88	95.37	22.43	25.14	101.56
2.90	1.01	50.06	13.16	19.71	95.71	26.27	29.59	102.08
3.30	1.15	50.08	15.01	22.60	96.10	30.24	34.26	102.71
3.70	1.29	50.10	16.88	25.57	96.55	34.37	39.25	103.47
4.10	1.43	50.13	18.76	28.61	97.07	38.72	44.65	104.39
4.50	1.57	50.15	20.67	31.76	97.66	43.34	50.63	105.51
4.90	1.71	50.18	22.60	35.02	98.32	48.35	57.51	106.92
5.30	1.85	50.21	24.56	38.41	99.08	53.91	65.85	108.78
5.70	2.00	50.25	26.54	41.96	99.93	60.33	77.06	111.47
6.10	2.14	50.28	28.56	45.68	100.89	68.38	96.14	116.39
*********	*********	*********	*********	*********	*********	*********	*********	*********

face can be obtained by varying the *Theta* 1 increment value (only a portion of the actual table is illustrated in the text).

Time-distance values of reflections that undergo refraction at overlying interface(s) no longer lie along a hyperbolic curve, as time-distance values do in the case of a single reflecting horizon. It follows that in an x^2-t^2 plot, data points no longer will lie along a straight line. These points actually define a curve that is slightly convex toward greater t^2-values. Recall that Dix's analysis assumed small x-values (source-receiver offsets). Actually, in determining rms velocity values, we should be drawing a straight line that is tangent to the curve defined by points near $x^2 = 0$. Because the x^2 - t^2 curve is nearly straight in this region, however, for all practical purposes the tangent is the same as the line that best fits the data points.

Another approach to the preceding discussion is summarized in Sheriff and Geldart (1983, p. 19). They state, "Note that the Dix formula implies that the travel paths to the $(n-1)$st and nth reflectors are essentially identical except for the additional travel between the two reflectors." Refer to Figure 4-14. Assume that the geophone interval is 2.5 m for the 30-m spread, 5 m for the 60-m spread, and 10 m for the 120-m spread. You should be able to picture that geophones at 27. 5 and 30 m meet this criterion, that those at 55 and 60 m come close, and that those at 110 and 120 m likely do not.

In order to appreciate more fully the practical implications of the preceding discussion and to review some generalizations, let's formulate a model and undertake a Dix analysis on three quite different variations in geophone spread length. The model we will use is defined in Figure 4-14. Since the total depth to the third reflecting surface is 60 m, let's investigate reflection time-distance values in the ranges 0–30 m, 0–60 m, and 0–120 m. Ray paths to the maximum of each of these ranges are shown in Figure 4-14 to illustrate how the shape of the path varies with increasing source-receiver distances.

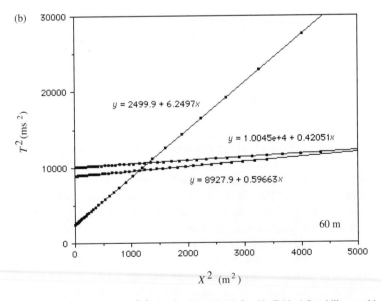

Figure 4-15 Comparison of x^2-t^2 plots for the model defined in Table 4-5 and illustrated in Figure 4-14. (a) Data points for various source-geophone distances up to a maximum of 30 m. (b) Data points for various source-geophone distances up to a maximum of 60 m. (c) Data points for various source-geophone distances up to a maximum of 120 m.

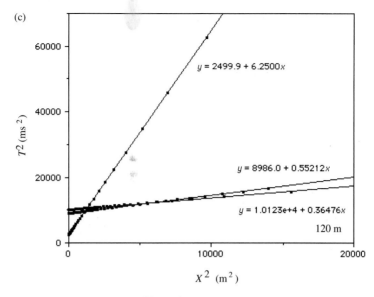

(c)

$y = 2499.9 + 6.2500x$

$y = 8986.0 + 0.55212x$

$y = 1.0123e+4 + 0.36476x$

120 m

Figure 4-15 *(continued)*

Figure 4-15 illustrates three separate x^2-t^2 diagrams. Each contains lines fit by a least-squares analysis to reflection time-distance values from each of the reflectors in Figure 4-14. Each diagram corresponds to one of the ranges selected for analysis; therefore, they are labeled 30 m, 60 m, and 120 m. Equations for each of the lines are included, and the diagram illustrates that orientations of the lines are different in each case.

Results of a Dix analysis are summarized in Table 4-5. Green method results are included also for informational purposes. They clearly are not nearly as reliable as the Dix values. As expected, results for the 30-m spread correspond very well to model parameters. The 60-m values are acceptably close. At 120 m, however, we begin to see a percentage error that is unacceptable given the perfect data we are working with. Remember that these results are only for this one model. Other subsurface configurations (thicknesses and veloci-

TABLE 4-5 Comparison of Model Parameters with Green and Dix Solutions for Three Geophone Spreads

Parameters	Model	Green			Dix		
		30 m	60 m	120 m	30 m	60 m	120 m
Velocity 1 (m/s)	400	400	400	400	400	400	400
Velocity 2 (m/s)	1800	2250	2293	2411	1812	1839	1912
Velocity 3 (m/s)	3500	5254	5585	6529	3542	3736	4234
Thickness 1 (m)	10	10	10	10	10	10	10
Thickness 2 (m)	40	50	51	54	40	41	43
Thickness 3 (m)	10	15	16	19	10	11	12

ty contrasts) would produce different results. In some cases the divergences would be greater and in some cases less than are presented here. However, we would observe the same trends from small offsets to large offsets.

It also is instructive to examine the details in the variation of these data points. Figure

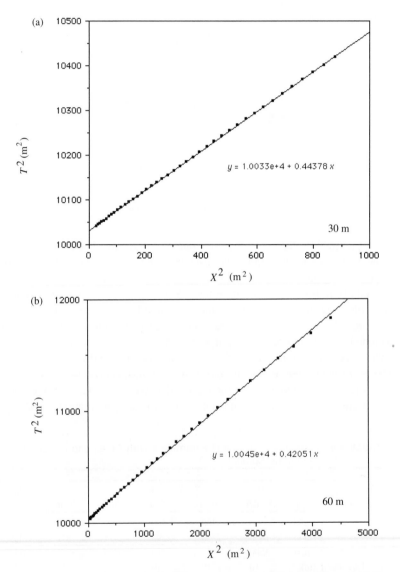

Figure 4-16   A more detailed presentation of selected material from Figure 4-15. Data points are squared time-distance values for reflections from the third and deepest interface defined in Table 4-5. (a) 30-m geophone spread. (b) 60-m geophone spread. (c) 120-m geophone spread.

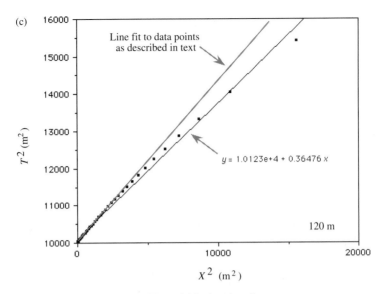

Figure 4-16 *(continued)*

4-16 provides us with 30-m, 60-m, and 120-m plots for data from the third and deepest reflecting horizon. In the 30-m case, Figure 4-16(a), the points define a nearly straight line, since source-receiver offsets are small relative to the depth of the interface. This line produces the best results (Table 4-5). In Figure 4-16(b) the points clearly do not lie on a straight line. This divergence from a straight line is most marked for the 120-m case, Figure 4-16(c), and this line yields the poorest results. Since the least-squares fit gives all points equal weight, points with large source-receiver values are included and lead to increased error. Note that if we examine the lines going from the 30-m to the 60-m to the 120-m spreads, the slopes are decreasing and t_0^2-values are increasing, which leads to the increased velocities and thickness catalogued in Table 4-5.

Let us examine Figure 4-16(c) once again. Even in this case where some of the larger values clearly are unacceptable, we could arrive at much better values by drawing a tangent to the data points at small x^2-values or by fitting a line to data points in a way that gives greater weight to points at small x^2-values. Such a line is identified in Figure 4-16(c). It has a greater slope and lower t_0^2-values, which are precisely what is required to produce a solution that corresponds more closely to our initial model.

We hope that this discussion has made you more aware of the need to keep source-receiver offsets small relative to the depths at which exploration targets are assumed to lie. Ideally, spreads should not exceed target depths. If this is not possible, lines drawn on x^2-t^2 plots should give greater weight to points at small x^2-values. Admittedly, the quality of average reflection data and the difficulty of accurately picking the beginning of the reflection pulse often will generate a considerable spread in data points. Nevertheless, if you employ the Dix method for resolving reflection data, you should follow these guidelines.

In order to check your understanding in applying the Dix
method, use the information in Figure 4-15 to generate the
Dix results in Table 4-5. Preferably, obtain one set of solutions
by hand (or hand calculator) before using ReflectIVel.

If long spread lengths are a necessity, it is possible to gain more accuracy by incorporating additional, higher-order terms in the Dix approximation. Shah and Levin (1973)
develop this extension and include an error analysis.

Analyzing a Field Seismogram Containing Multiple Reflections

Let's apply a Dix analysis to a field seismogram recorded in an area where bedrock depths
tend to average 50 m to 100 m. The water table in the area is fairly close to the surface, and
there is a thick sequence of glacial lake clays overlain and underlain by sands and gravels.
Although there are several possible reflections on the seismogram, only two are selected for
analysis. These are indicated on Figure 4-17.

Figure 4-17 Field seismogram from the Connecticut Valley, Massachusetts. Geophone
traces are labeled 1–12. The first geophone is located 60 m from the energy source. The
geophone interval is 3 m. Reflections selected for analysis are marked by arrows. Timing
lines are at 10-ms intervals. The record encompasses 200 ms. Note that the record begins
at 50 ms.

After compiling time-distance values for these two reflections, we perform a Dix analysis. The most convenient approach is to use ReflectSolve for each group of data in order to obtain rms velocities. Then we use ReflectIVel to solve for interval velocities and thicknesses.

Solutions are shown in Figure 4-18. The deepest reflector likely is the bedrock surface, because its computed depth agrees well with depths to bedrock in nearby wells. Note that, unlike with the refraction method, we are not able to determine the bedrock velocity and cannot use that piece of information to identify the reflector as the bedrock surface. The best guide is, of course, well data, but, lacking that, we use the observation that in the local area the bedrock reflector usually produces the strongest reflection on the record. The shallower reflector is believed to be the bottom of a clay layer in contact with coarse gravels. This identification also is based on reference to well information.

DIPPING INTERFACE

If you consider for a moment how a dipping interface will affect reflection time-distance values, we hope that you will realize that, in general, data from a traverse in a single direction will not reveal that the interface is inclined. As in the case with refraction work, we always must make an adjustment in our field design to ensure that inclined interfaces are recognized. If we proceed in a similar fashion to refraction surveying, by collecting data from forward and reverse traverses, we will have data points that define different slopes and, therefore, different velocities. This would indicate an inclined interface for essentially the same reasoning as discussed in Chapter 3.

In reflection work it is convenient, for a number of reasons that should become more obvious as we work through the remainder of this chapter, to use a geophone spread dis-

Figure 4-18   An x^2-t^2 plot and results of Dix analysis for reflections identified in Figure 4-17.

tributed equally on either side of a center shot (a split-spread arrangement). This is the geometry we will use to develop a travel-time equation for a single dipping interface, which, of course, will hold also when the dip equals zero.

Derivation of Travel-Time Equation

Figure 4-19 diagrams a reflecting horizon that is inclined upward to the right. In order to maintain the same conventions as we elected when developing equations for the refraction method, the dip angle is taken to be positive when the slope is upward toward the right. The vertical distance EP to the interface beneath the shot point is thickness and is denoted by h. The perpendicular to the interface, EM, is denoted by j. Since we will be utilizing a center shot, x is taken to be positive to the right and negative to the left. Thus, for a positive angle, $+x$-values increase up-dip. The derivation which follows is based on a presentation by Slotnick (1959, p. 48–52).

In developing reflection equations a useful device is the concept of the image point. If we imagine extending the line EM, which is perpendicular to the reflecting horizon, an equal distance on the other side of the interface, this gives us the line MQ, which is also of length j. The image point is at Q (see Fig. 4-19). If we now draw a line from Q to G, which is the geophone position, we have a particularly useful geometric construct. Because triangles EMN and MQN are congruent, EN is equal to QN. Therefore, the path QNG is equal in length to the actual path traversed by the reflected wave, ENG.

This relationship makes it straightforward to develop our travel-time equation by using triangle EQG. Before we move on to this objective, however, we note another useful application of the image point. Once Q is located, if we draw a line QG to some point on the surface, we need only draw a line from the energy source to the point where QG crosses the reflecting horizon N to construct the ray which will be reflected from the interface to point G. This procedure makes the task of constructing a ray path to emerge at a selected surface position a trivial exercise. Try constructing such a ray path without using an image point!

Path distance is most easily derived by using the law of cosines:

$$a^2 = b^2 + c^2 - 2bc \, (\cos \, A) \tag{4-23}$$

where a, b, and c are the sides of the triangle and A is the angle opposite side a. Thus, for the triangle in Figure 4-19,

$$(QG)^2 = (EQ)^2 + (EG)^2 - 2(EQ)(EG) \cos(90 - \beta) \tag{4-24}$$

$$(QG)^2 = (2j)^2 + (x)^2 - 2(2j)(x) \cos(90 - \beta)$$

$$(QG)^2 = 4j^2 + x^2 - (4jx)\cos(90 - \beta)$$

and

$$(QG)^2 = 4j^2 + x^2 - (4jx) \sin \beta \tag{4-25}$$

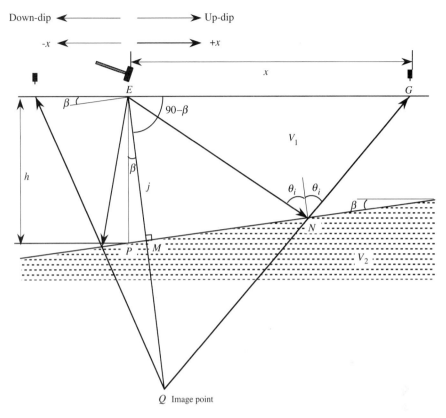

Figure 4-19 Diagram illustrating symbols used in derivation of a travel-time equation for a reflected ray. Dip convention is similar for that used in refraction analysis. β is positive as illustrated.

The travel-time equation thus becomes

$$t^2 = \frac{4j^2 + x^2 - 4jx \sin \beta}{V^2} \qquad (4\text{-}26)$$

or

$$t = \frac{\left(4j^2 + x^2 - 4jx \sin \beta\right)^{1/2}}{V} \qquad (4\text{-}27)$$

Equation 4-26 can be shown to have the form of a hyperbola, as was the case for our horizontal interface equation. However, one would think that there should be some difference in form or position of the hyperbola, due to the dipping horizon. One way to investigate the difference is to utilize dynamic Table 4-6 to create time-distance data and then to graph this data. Note that all values in Table 4-6 follow the conventions established at the beginning of this section. Thickness is the vertical distance from the energy source to the reflecting hori-

TABLE 4-6   Reflection Time-Distance Values for a Dipping Interface Using a Split-Spread Arrangement

Geophone	Distance (m)	Time (ms)
12	− 36	49.09
11	− 33	47.91
10	− 30	46.79
9	− 27	45.73
8	− 24	44.74
7	− 21	43.81
6	− 18	42.96
5	− 15	42.18
4	− 12	41.49
3	− 9	40.88
2	− 6	40.36
1	− 3	39.94
Source	0	39.61
1	3	39.38
2	6	39.25
3	9	39.23
4	12	39.30
5	15	39.48
6	18	39.76
7	21	40.13
8	24	40.60
9	27	41.17
10	30	41.82
11	33	42.55
12	36	43.36

Velocity (m/s)	**1500**
Thickness (m)	**30**
Dip (°)	**8**
Increment (x) (m)	**3**

t-zero (ms)	39.61
t-min (ms)	39.23
x-min (m)	8.27

zon. The positive x-values are oriented toward the right and increase up-dip, because the dip angle in the table is positive.

Investigate the assertion in the preceding text that velocities determined by forward and reverse traverses above a dipping interface using reflection data will be greater than and less than the true velocity, as was the case for refracted energy. Utilize the positive x portion of Table 4-6 only, copy time-distance values for a positive angle and then copy time-distance values for a negative angle of the same magnitude. Use these values to prepare an $x^2 - t^2$ plot.

Table 4-6 was used to produce time-distance values for a horizontal interface by setting β equal to zero and for an interface dipping at 8°. Travel-time curves for each set of values and a diagram of the model associated with each are illustrated in Figure 4-20. As you already know, the travel-time curve for the horizontal reflecting horizon is a hyperbola with an axis of symmetry that is parallel to the time axis and passes through the point $x = 0$. If we refer to the minimum time on this hyperbola as t_{min} and the x-position at which time is a minimum as x_{min}, then clearly $t_0 = t_{min}$ and $x_0 = x_{min}$.

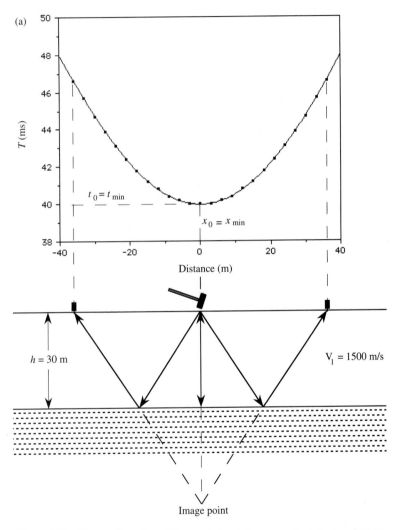

Figure 4-20 Diagram illustrating differences in travel-time curves for reflections from (a) a horizontal interface and (b) an inclined interface. All parameters are the same except for the dip of the interface. Model values are the same as in Table 4-6.

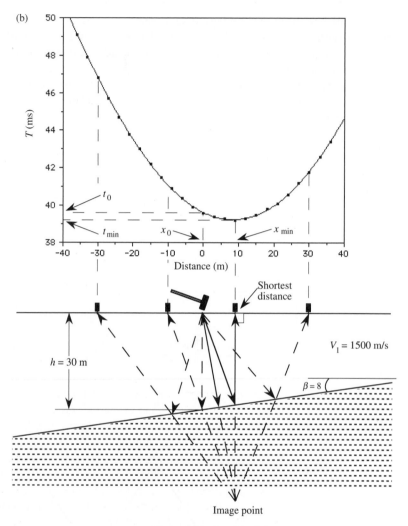

Figure 4-20   *(continued)*

If we now investigate Figure 4-20(b), it is obvious that these relationships no longer hold. The shortest path and, therefore, the one of least time at constant velocity, is no longer the vertical path located beneath the energy source. That this statement is correct can be proved by looking at the vertical line that passes through the image point. Because this line is perpendicular to the ground surface, it is the shortest line of all the lines emanating from the image point and contacting the ground surface. Therefore, the ray path constructed using this vertical line must be the shortest ray path. Based on the geometry of constructing the image point, this ray path always must be located up-dip from the energy source location, and, thus, the least time position (the position of the hyperbola symmetry line), which has coordinates x_{min}, t_{min}, is always offset up-dip from x_0. Figure 4-20 should help clarify these relationships.

Now the reason for a split-spread arrangement becomes less murky. If the reflection arrivals are relatively conspicuous, their hyperbolic form should immediately indicate if the reflections are from a dipping surface and, if so, the direction of dip. Our next task is to determine how to calculate dip amount, thickness, and velocity. First, let's develop a few more relationships which we then can use to arrive at these unknown values.

Determining Dip, Thickness, and Velocity

If we differentiate Eq. 4-26 and set the result equal to zero, we can find the value of x (x_{min}) for which the time term is a minimum. Accordingly,

$$\frac{dV^2 t^2}{dx} = 2tV^2 \frac{dt}{dx} = 2x - 4j \sin \beta$$

$$\frac{dt}{dx} = \frac{x - 2j \sin \beta}{tV^2} = 0 \tag{4-28}$$

and

$$x_{min} = 2j \sin \beta \tag{4-29}$$

We now can find an expression for t_{min} by substituting Eq. 4-29 into Eq. 4-26.

$$t_{min}^2 = \frac{4j^2 + (2j \sin \beta)^2 - 4j(2j \sin \beta) \sin \beta}{V^2}$$

$$t_{min}^2 = \frac{4j^2 \cos^2 \beta}{V^2} \tag{4-30}$$

$$t_{min} = \frac{2j \cos \beta}{V} \tag{4-31}$$

Next we derive an expression for t_0, which, since $x = 0$ at t_0, also is easily produced from Eq. 4-26.

$$t_0 = \frac{2j}{V} \tag{4-32}$$

If we have good data, we should be able to acquire values for t_0 and t_{min} from a travel-time curve. Realizing this and examining Eqs. 4-31 and 4-32, it follows that

$$\frac{t_{min}}{t_0} = \frac{\dfrac{2j \cos \beta}{V}}{\dfrac{2j}{V}} \tag{4-33}$$

$$\frac{t_{min}}{t_0} = \cos \beta \tag{4-34}$$

or

$$\beta = \cos^{-1}\left(\frac{t_{min}}{t_0}\right) \qquad (4\text{-}35)$$

Based on this derivation, the sign of β must be determined from the dip direction, which should be clearly indicated on the travel-time curve.

Because we now know β and can determine x_{min} in the same fashion we obtained t_{min}, Eq 4-29 supplies j.

$$j = \frac{x_{min}}{2 \; \sin \beta} \qquad (4\text{-}36)$$

Finally, we want to determine thickness, h. From Figure 4-19 it follows that

$$h = \frac{j}{\cos \beta} \qquad (4\text{-}37)$$

At this point we have values for h, j, β, t_{min}, t_0, and x_{min}. Any of several of the preceding equations can be used to determine V. The template in dynamic Table 4-7 labeled t-min/t-0 takes care of the computational tasks once the appropriate values are obtained from the travel-time curve.

However, it may not be this easy. You most likely can visualize seismograms on which reflection arrivals may not be especially distinct, so that time picks are somewhat uncertain. Moreover, a series of reflection arrivals may not exhibit much normal move-out, so that the exact position and value of t_{min} is not easy to extract from a travel-time curve. Might there be other ways to approach a solution for thickness and dip?

Determining Dip, Thickness, and Velocity—Another Approach

If we prepare an x^2-t^2 plot for the data in Table 4-6, we produce the pattern illustrated in Figure 4-21. The pattern is dominated by two sets of points. One set represents down-dip

TABLE 4-7 Templates for Calculating Thickness, Dip, and Velocity for An Inclined Reflector

t-min/t-0			
t-zero (ms)	39.6	Beta (°)	7.9
t-min (ms)	39.2	Thickness (m)	30.2
x-min (m)	8.3	Velocity (m/s)	1511.2

x2-t2 Average Data			
Velocity (m/s)	1500	j (m)	29.7
t-zero (ms)	39.6	Beta (°)	8.0
x (m)	27.0	Thickness (m)	30.0
t (ms)	41.2		

times and the other up-dip times. If we try to fit a straight line to these points, the result is a line that appears to bisect the data points. The equation for this line is labeled *actual data*. If we use the slope of this line to determine a velocity in the conventional manner, the result is 1499 m/s, compared with the 1500 m/s used in Table 4-6 to generate the time-distance values. If we solve for thickness as in the case of a horizontal layer, the result is 29.70 m. This is essentially the same as the perpendicular thickness of 29.71 computed using the dip of 8° and thickness of 30 m that generated the data. Clearly, there seems to be some merit in this approach, but why does it work? After all, the data are for an inclined reflecting horizon.

Because the line we produced in Figure 4-21 appears to bisect time values for equivalent distances from the source, let's investigate the result of taking an average by summing our travel-time Eq. 4-26 for an inclined interface and substituting +x- and -x-values.

$$t^2 + t^2 = \frac{4j^2 + (-x)^2 - 4j(-x)\sin\beta}{V^2} + \frac{4j^2 + (+x)^2 - 4j(+x)\sin\beta}{V^2} \quad (4\text{-}38)$$

$$2t^2 = \frac{x^2 + 4j^2 + x^2 + 4j^2}{V^2}$$

$$t^2 = \frac{x^2 + 4j^2}{V^2} \quad (4\text{-}39)$$

Equation 4-39 is the same as Eq. 4-6, with the exception that j replaces h, which tells us that an average of our time values at equivalent distances from the source can be used to determine velocity and the value of j. Once we have these values, we can substitute them and

Figure 4-21   An x^2-t^2 graph for the reflection data in Table 4-6. One line is fit to the data points. The other line is fit to the average of time values at corresponding distances from the energy source.

one of our time-distance data pairs into Eq. 4-26 to arrive at a value for β. Knowing β we solve for h using Eq. 4-37. The template in dynamic Table 4-7 labeled "x^2-t^2 Average Data" takes care of the computational task once the appropriate values are obtained from the x^2-t^2 plot.

A Return to Normal Move-Out

There is yet another approach to determining dip, thickness, and velocity when dealing with an inclined reflecting horizon. Before exploring this final method, we believe it is useful to return for a moment to the topic of normal move-out (NMO). When we first introduced NMO, we developed only the straightforward definition. However, especially when dealing with computer-based approaches to working with reflection data, it is more convenient to express NMO in terms of the time at $x = 0$, or t_0. Let's see how we might accomplish this.

Recall that the definition of NMO is the difference in reflection travel-times from a horizontal reflecting surface due to variations in the source-geophone distance, or

$$T_{NMO} = t_x - t_0$$

$$T_{NMO} = \frac{\left(x^2 + 4h_1^2\right)^{1/2}}{V_1} - \frac{2h_1}{V_1} \tag{4-5}$$

Now let's examine our basic travel-time equation once again so that we can express as much as possible in terms of t_0. For convenience in the development we will represent V_1 by V and h_1 by h.

$$t_x = \frac{\left(x^2 + 4h^2\right)^{1/2}}{V} \tag{4-1}$$

$$t_x^2 = \frac{x^2}{V^2} + \frac{4h^2}{V^2} = \frac{x^2}{V^2} + t_0^2 \tag{4-40}$$

$$t_x^2 = \frac{x^2 + V^2 t_0^2}{V^2} = \left(\frac{x^2}{V^2 t_0^2} + 1\right) t_0^2 \tag{4-41}$$

$$t_x = t_0 \left(1 + \frac{x^2}{V^2 t_0^2}\right)^{1/2} \tag{4-42}$$

Remembering that our goal is to derive an equation that expresses NMO in terms of t_0, we see that Eq. 4-42 presents a problem, because a t_0 term is part of an expression that is raised to the one-half power. Reaching backward in time, we will use a relationship learned in high-school mathematics in order to expand

$$\left(1 + \frac{x^2}{V^2 t_0^2}\right)^{1/2}$$

According to the generalized binomial theorem,

$$(1 + z)^a = 1 + az + \frac{a(a - 1)}{2 \cdot 1} z^2 + \frac{a(a - 1)\cdots(a - n + 1)}{n!} z^y + \cdots$$

So, if we set

$$a = \frac{1}{2} \quad \text{and} \quad z = \left(\frac{x^2}{V^2 t_0^2}\right)^{1/2}$$

Eq. 4-42 can be expressed as

$$t_x = t_0\left(1 + \frac{x^2}{2 V^2 t_0^2} - \frac{x^4}{8 V^4 t_0^4} + \frac{x^6}{16 V^6 t_0^6} + \cdots\right) \qquad (4\text{-}43)$$

Examining the variables within the parentheses, we see that we can ignore all but the first two terms, *if*

$$\frac{x}{V t_0} <<< 1$$

If we do so, then our expression reduces to

$$t_x = t_0 + \frac{x^2}{2 t_0 V^2} \qquad (4\text{-}44)$$

Recalling our objective to express T_{NMO} in terms of t_0, we see that since

$$T_{NMO} = t_x - t_0$$

then

$$T_{NMO} = \frac{x^2}{2 t_0 V^2} \qquad (4\text{-}45)$$

However, we must keep in mind the conditions under which Eq. 4-45 is valid. Equation 4-7 tells us that our former qualification

$$\frac{x}{V t_0} <<< 1$$

is equivalent to saying

$$\frac{x}{2 h} <<< 1$$

Thus, for Eq. 4-44 to be a useful approximation, horizontal distance divided by twice the thickness must be much less than one. In cases where this is not so, a much better approximation results from using the first three terms in Eq. 4-43 rather than just the first two. Therefore, we have

$$T_{\text{NMO}} = \frac{x^2}{2t_0V^2} - \frac{x^4}{8t_0^3V^4}$$ (4-46)

Table 4-8 lists values computed using Eq. 4-45 (column labeled "First 2 terms") and Eq. 4-46 (column labeled "First 3 terms"). These can be compared with actual NMO values calculated using Eq. 4-5. The column labeled "Condition" lists the value of x/t_0V. Because this is a dynamic table, you can investigate values for which Eq. 4-45 provides a good approximation for NMO values and for which the use of Eq. 4-46 is advised.

Figure 4-22(a) illustrates the error involved in using Eq. 4-45 for various thicknesses. Figure 4-22(b) is the same except that Eq. 4-46 was used. Note that the maximum vertical scale value in Figure 4-22(a) is 0.6 ms, whereas that in Figure 4-22(b) is only 0.1 ms. Of course, all the errors in Figure 4-22(b) are much less than those in Figure 4-22(a). With the exception of a thickness value of 30 m (the one used in Table 4-8), all the errors are less than 0.10 ms. If you reexamine some of the field seismograms you've worked with in previous sections, you will see that such errors are acceptable when we are fortunate to pick events with a precision of 0.50 ms. However, the error for a thickness of 30 m is 0.55 ms at a source-geophone distance of 36 m. This error increases to 6.32 ms at a distance of 72 m. Such an error clearly is unacceptable. This analysis emphasizes that Eq. 4-45 should not always be utilized when dealing with the depths of reflectors typically encountered in shallow exploration surveys. Once again we see the importance of attempting to keep geophone spreads relatively short when reflecting depths are not great.

TABLE 4-8 Comparison of Actual NMO Values with Values from Eqs. 4-45 and 4-46

Geophone	Distance (m)	Travel-Time (ms)	NMO (ms)	Condition	First 2 terms	First 3 terms
Source	0	40.00	0.00	0.00	0.00	0.00
1	3	40.05	0.05	0.05	0.05	0.05
2	6	40.20	0.20	0.10	0.20	0.20
3	9	40.45	0.45	0.15	0.45	0.45
4	12	40.79	0.79	0.20	0.80	0.79
5	15	41.23	1.23	0.25	1.25	1.23
6	18	41.76	1.76	0.30	1.80	1.76
7	21	42.38	2.38	0.35	2.45	2.37
8	24	43.08	3.08	0.40	3.20	3.07
9	27	43.86	3.86	0.45	4.05	3.84
10	30	44.72	4.72	0.50	5.00	4.69
11	33	45.65	5.65	0.55	6.05	5.59
12	36	46.65	6.65	0.60	7.20	6.55

Velocity (m/s)	**1500**
Thickness (m)	**30**
Increment (m)	**3**

(a)

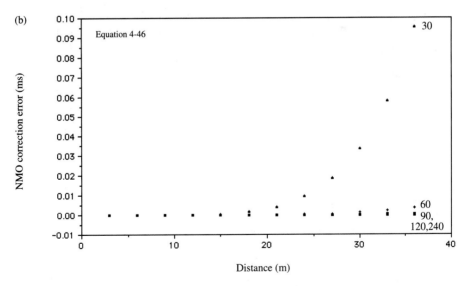

(b)

Figure 4-22 Plots of differences in NMO values using Eq. 4-5 and an approximation using Eqs. 4-45 and 4-46. Thicknesses used were 30, 60, 90, 120, and 240 m.

> The preceding paragraphs discussed the effects of source-geophone distance on the error involved in using Eq. 4-45 to calculate NMO. Determine how velocity affects this approximation and how these errors compare with those due to source-geophone distances.

The preceding discussion considered only NMO calculations for a single horizontal interface. What level of complexities will be introduced when multiple interfaces are introduced? If you consider the problem for a moment, it might occur to you that we will be able to use, in some fashion, V_{rms}. You certainly are correct. Taner and Koehler (1969) demonstrated that Eq. 4-45, or Eq. 4-46 for that matter, is valid for multiple horizontal interfaces if we substitute V_{rms} for the velocity of layer 1:

$$T_{NMO} = \frac{x^2}{2 t_0 V_{rms}^2} \tag{4-47}$$

In a following section when we discuss data-processing techniques, we will investigate the accuracy of this approximation using a dynamic table. Now, however, it is time to return to our initial objective in this major section—determining dip, thickness, and velocity when an inclined reflecting interface is present.

Determining Dip, Thickness, and Velocity—Yet Another Approach

This "yet another approach" essentially makes use of the displacement of the hyperbolic form of the travel-time curve when dealing with an inclined interface. Figure 4-23(a) graphically illustrates normal move-out, and Figure 4-23(b) illustrates a new relationship, *dip move-out*. Dip move-out (DMO) is the difference in travel-time to geophones at equal distances from the source position ($x = 0$) in a split-spread geometry, or

$$T_{DMO} = t_{+x} - t_{-x} \tag{4-48}$$

Because dip move-out is related to the dip of the reflecting horizon, it follows that there should exist a way to utilize this time to determine dip. We begin with the basic travel-time equation for reflections from a horizontal interface, rearrange this equation to incorporate t_0, and then expand the resulting equation using the generalized binomial theorem as we did in our recent discussion of normal move-out. In the following development, please recall the distinction we make between the quantities h and j. Thus,

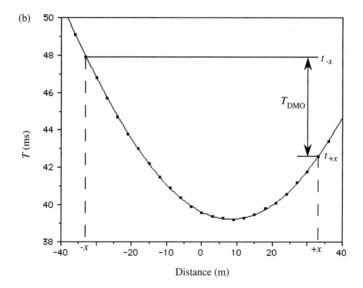

Figure 4-23 Comparison of normal move-out and dip move-out definitions on time-distance curves. Dip move-out (DMO) is the difference in travel-time to geophones at equal distances from the source position ($x = 0$) in a split-spread geometry.

$$t_x = \frac{\left(4j^2 + x^2 - 4jx \sin \beta\right)^{1/2}}{V} \qquad (4\text{-}27)$$

$$t_x{}^2 = \frac{x^2}{V^2} + \frac{4j^2}{V^2} - \frac{4jx \sin \beta}{V^2} \qquad (4\text{-}49)$$

$$t_x{}^2 = t_0{}^2 + \frac{x^2}{V^2} - \frac{4jx \sin \beta}{V^2} \qquad (4\text{-}50)$$

$$t_x{}^2 = t_0{}^2 \left(1 + \frac{x^2 - 4jx \sin \beta}{4j^2}\right) \qquad (4\text{-}51)$$

and

$$t_x = t_0 \left(1 + \frac{x^2 - 4jx \sin \beta}{4j^2}\right)^{1/2} \qquad (4\text{-}52)$$

Expanding Eq. 4-52, we have

$$t_x = t_0 \left[1 + \left(\frac{x^2 - 4jx \sin \beta}{8j^2}\right) - \left(\frac{x^2 - 4jx \sin \beta}{32j^2}\right)^2 + \cdots\right] \qquad (4\text{-}53)$$

Once again, we take only the first two terms in the expansion, and, remembering that $t_0 = 2j/V$, we arrive at

$$t_x = t_0 + \frac{x^2 - 4jx \sin \beta}{4jV} \qquad (4\text{-}54)$$

When we developed the x^2-t^2 *average data* method we summed reflection travel-time equations for $+x$ and $-x$ cases. However, recalling that we are developing a method to take advantage of the availability of dip move-out which we defined in Eq. 4-48, we see that

$$T_{\text{DMO}} = t_{+x} - t_{-x} = \left(t_0 + \frac{(+x)^2 - 4j(+x)\sin \beta}{4jV}\right) \qquad (4\text{-}55)$$
$$- \left(t_0 + \frac{(-x)^2 - 4j(-x)\sin \beta}{4jV}\right)$$

$$T_{\text{DMO}} = -\frac{2x \sin \beta}{V} \qquad (4\text{-}56)$$

and

$$\beta = \sin^{-1}\left(-\frac{VT_{\text{DMO}}}{2x}\right) \qquad (4\text{-}57)$$

We determine T_{DMO} for a selected x-value on our travel-time curve and V from an x^2 - t^2 plot. This gives us β from Eq. 4-57 and j, and then h, from one of several equations presented previously.

Use Table 4-6 to generate time-distance data for inclined reflection horizons where thicknesses are 30 m and 300 m. Determine β using Eq. 4-57. Are problems in accuracy present for cases in which thickness is not very much greater than offset distance? If so, must we abandon the dip move-out approach to determining β, or is there an alternative?

In concluding an analysis of methods to determine horizon dips, it is important to realize that the true dip of a plane cannot be determined from a seismic line unless one is fortunate to orient the survey perpendicular to the strike of the plane. Often, if true dips are desired, survey lines are arranged perpendicular to one another in order to provide the information necessary to determine the true dip. If dips are low (less than 10°), then the standard methods found in structural geology textbooks may be applied. These demonstrate how true dips can be determined from apparent dips acquired along two different orientations. If dips are not low, then more complicated calculations are needed, because the line along which reflection points occur is not parallel to the survey line. An excellent development of the problem with understandable solutions is presented in Slotnick (1959, p. 127–162).

ACQUIRING AND RECOGNIZING REFLECTIONS FROM SHALLOW INTERFACES

At the beginning of this chapter we mentioned the seemingly anomalous situation of reflection methods being dominant in petroleum exploration and refraction methods being dominant in shallow exploration. We suggested a few reasons why this might be so, among which was the problem of recognizing many reflection arrivals and the vast data processing required to solve this difficulty and to meet other requirements when working with reflections. As you work through this chapter, perhaps you also will appreciate that velocity data may not be as straightforward to acquire or as accurate as that produced by refraction work. In many situations, therefore, the requirements favor the information supplied by refraction work.

We also pointed out some of the limitations of refraction surveys and several advantages of working with reflections. Surely, the possibility of producing seismic sections that reveal great detail about the shallow subsurface is a powerful incentive for mastering reflection work in shallow exploration.

You should recall the two criteria for recognizing reflections on seismograms: (1) similarity of waveform and (2) exhibition of normal move-out. However, to apply these and other tests we first must produce a record on which reflections are sufficiently clearly developed that these tests are possible. Several strategies are available—equipment design, data processing, and survey design. In almost all cases all are used in varying proportions to pro-

duce meaningful records. As such, each will be discussed in sections of this chapter to follow. Here we concentrate on a fundamental concept of survey design that often results in
data suitable for interpretation with a minimum of computer processing.

The Optimum Window

Let's accept Figure 4-24(a) as a simulated seismogram from a typical shallow geologic section in the Connecticut Valley of Massachusetts. The model used is located in the upper
right corner of Figure 4-24(a). We assume that our instruments are configured to maximize
reflection arrivals. Waveforms for the several wave types illustrated are very simplified (see
Appendix B), but the locations are reasonably accurate. We ignore the reflection from the
water table, which is at a depth of 5 m. The water-table refraction is shown because it is
often clearly developed on shallow-reflection records. At the moment we especially are
concerned with the relative positions of the reflections from the interfaces at 30 m and 50 m
and the air wave and ground roll (surface wave).

In order to make this simulated seismogram more realistic, imagine a fair amount of
noise spread about the record so as to render the individual pulses less clearly defined. Also
extend the ground roll significantly toward increased time (see Fig. 4-26(a) for the extent of
ground roll interference on an actual record). It should be fairly obvious that the reflections

Figure 4-24 An illustration of the importance of source-receiver distance in reflection
surveying. (a) Typical reflection record showing positions of low-velocity waves (ground
roll and air wave), refraction from the water table, and reflections from two interfaces. (b)
An increased source-geophone distance moves low-velocity waves toward greater time at a
rate faster than reflections which become easier to discern. Position of optimum window
indicated. (c) Increased bedrock depth moves the bedrock reflector toward increased time
and changes the position of the optimum window for the bedrock reflection.

Figure 4-24   *(continued)*

will be all but impossible to discern. The source offset to the first geophone is 3 m in Figure 4-24(a). If we increase the source offset to 15 m and keep all other parameters constant, the result is illustrated in Figure 4-24(b). The two reflections now are much clearer because interference from the low-velocity waves is moved later in time relative to the reflections.

The reflections and refraction are moved later in time also, but not as much, since their average path velocities are much greater. The bedrock reflection begins to emerge from the air wave at a source-geophone distance of about 30 m. This is because as we increase the source offset to 30 m, we move the low-velocity interference still farther to the right and produce a record on which the bedrock reflection is visible on every trace.

This range of source-receiver distances within which reflections are most clearly seen is often referred to as the *optimum window* (Hunter et al., 1984). The "near side" of the optimum window is clearly defined by the coincidence of the least time limit of the low-velocity waves (noted by the shaded slanting line in Figure 4-24(b)) and by the reflections of interest. However, the upper limit or the "far side" of the optimum window is much less obvious. Several considerations influence the location of the maximum source-receiver distance. As source-receiver distances increase, reflections from shallow interfaces may begin to interfere with reflections from deeper interfaces, such as the bedrock reflection. Figure 4-25 illustrates such a case. The model used is slightly different from that in Figure 4-24. Ground roll is not included, although the limit of the low-velocity waves is indicated. The reflection from the 10-m-deep interface does not interfere with the bedrock reflection in Figure 4-25(a) for which the source offset is 15 m and the geophone interval is 3 m. However, in Figure 4-25(b), where the initial source offset is 30 m, this same reflection begins to merge with the bedrock reflection on the 63-m geophone trace. If we are primarily interested in the bedrock reflection, the optimum window extends from approximately 39 m to 63 m. However, if we hope to separate reflections from all three interfaces, the far-side limit shrinks drastically to a maximum, at most, of 45 m.

If the water-table refraction is particularly well developed, it may interfere with shallow reflections as the source-receiver distance is increased. Compare the position of this refraction with the first reflection in Figure 4-24(a) and (b). Also, as source-receiver distances increase, amplitude and phase changes in the incoming pulse may make it difficult to correlate from trace to trace on a given record. Pullan and Hunter (1985) document this effect quite vividly. They conclude that phase changes should not be a problem when tracing bedrock reflections as long as source-receiver distances do not exceed bedrock depth. We developed a restriction of this same order when investigating the Dix method and when developing an expression for NMO in terms of t_0. By now you should understand that appropriate source-receiver distances are quite sensitive to the depth of the horizon of interest. You will need to keep in mind what these limits are and the conditions under which they apply.

Finally, note that the position of the optimum window may vary in a survey if the depths of the interfaces being studied change. The optimum window selected in Figure 4-24(b) would no longer be applicable in Figure 4-24(c), because the bedrock surface has increased in depth from 50 m to 80 m.

After you feel you have a good grasp of the optimum window concept and the near-side controls, study Figure 4-26. Illustrated in this figure are two field seismograms from different sites in the Connecticut Valley of Massachusetts. The air wave and ground roll are present on both records, but the air wave is more clearly defined on Figure 4-26(a). The (a) and (b) parts of Figures 4-24 and 4-26 correlate reasonably well with each other, although in both cases the limit of the low-velocity waves is present at greater times on the actual

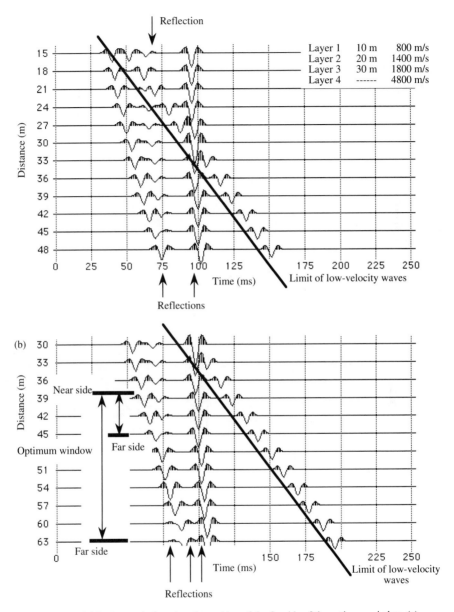

Figure 4-25 Factors influencing the position of the far side of the optimum window. (a) Three reflections from the designated model. (b) Source-receiver distance increased. Position of near side of optimum window clearly delimited; position of far side less clear. At the larger source-geophone distances, the shallow reflection interferes with the bedrock reflection.

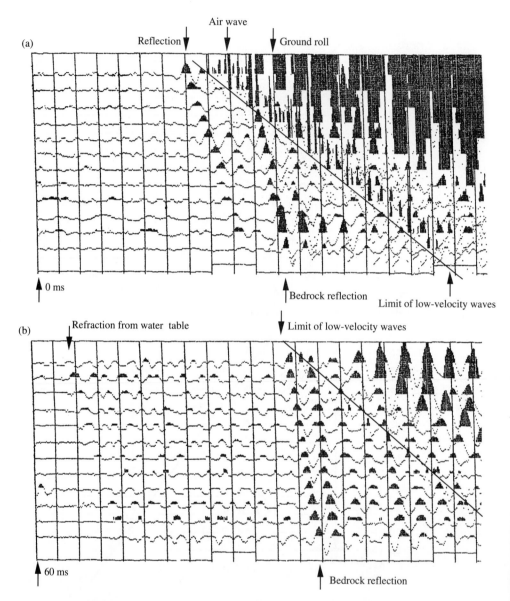

Figure 4-26   Field seismograms from the Connecticut Valley, Massachusetts. The geophone interval for both records is 3 m. (a) Source offset is 30 m. The record contains two prominent reflections, a well-defined air wave, and obvious ground roll. Compare with Figure 4-24(b). (b) This record is from a different site than (a). The source offset is 60 m. The air wave is present but is difficult to discern. The water-table refraction is just visible at about 75 ms on the first trace.

field records. This correlation should demonstrate that modeling programs such as ReflectModel can provide useful results when used with knowledge of their limitations.

How, then, is the optimum window determined? In the best case well logs exist for the exploration area. These should enable the production of useful models which can reduce initial field efforts considerably. In many cases, however, one simply employs a split spread and continues to expand its length by adding geophone cable to each end. Eventually, assuming conditions are satisfactory for recording reflections, reflections will emerge from the low-velocity waves. Spread length can be increased as time, funds, and conditions permit in order to develop distinct NMO and to watch for interfering reflections and phase changes. Care should be taken in acquiring this initial record in order to define the best possible optimum window.

> A 120-m-thick unit with a velocity of 2000 m/s overlies a 250-m-thick unit with a velocity of 4000 m/s. This latter unit in turn overlies a unit with a velocity of 6000 m/s. Determine the optimum window.

Multiple Reflections

The preceding discussion was directed toward a field procedure that attempts to separate reflections from other interfering wave types. Our primary goal, of course, is to produce a record that contains in recognizable form all reflections from the depth range of interest. One problem we have not as yet addressed is that of multiple reflections. We've considered simple reflection paths from a source to an interface and then to a receiver. However, many equally likely, but more complex, paths are followed by a portion of the incident energy.

Three relatively simple possibilities are illustrated in Figure 4-27(a). If we added another unit in the relatively thick sequence labeled "saturated sediments," the number of possible paths would be much more numerous. The reflections we previously considered that follow the direct path to an interface and back to a receiver (that consist of only one reflection at an interface) are referred to as *primary reflections*. Those reflections arriving at a receiver position that underwent more than one reflection during their travel are referred to as *multiple reflections*. Multiple reflections, or multiples, typically are divided into two main categories. *Short-path multiples*, as their name implies, are very similar in total path to a primary, but a short portion of their path consists of a second reflection event (Fig. 4-27(a)). *Long-path multiples* consist of a second reflection event with a path that is a substantial portion of a corresponding primary (Fig. 4-27(a)).

As you might guess, short-path multiples arrive very shortly after a corresponding primary and, as such, are not discerned as separate events. Their main effect is to change the shape of the overall primary pulse by adding additional undulations or a "tail." This prolonging of the primary arrival may itself tend to overwrite or obscure other primaries arriving close in time to our initial primary. However, long-path multiples stand a good chance of appearing as distinct events, which can easily be misinterpreted as primaries. If such a mistake occurs, subsurface reconstructions will be incorrect. How serious a problem are long-path multiples likely to be, and what strategy might we develop to deal with them?

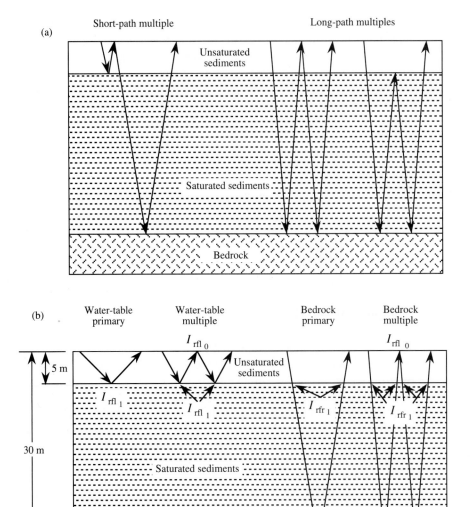

Figure 4-27 Various types of multiple reflections. (a) Long-path and short-path multiples. (b) Symbols used in text discussion of losses due to reflection and refraction at interfaces for primary and multiple reflections.

In attacking this problem, first let's return to our discussion of energy partitioning in Chapter 2 and apply this information to a rather simple subsurface model that is common in shallow surveying (Fig. 4-27(b)). Please remember that this exercise is intended to be mainly illustrative and does not pretend to cover the range of multiple arrivals that might occur in more complicated situations. Here we address only water-table and bedrock primaries and one possible long-path multiple for each interface. We recall that P-wave energy incident upon a velocity discontinuity is partitioned among reflected and refracted P- and S-waves. We compute this partitioning using Table 2-7 and, therefore, make the assumption of normal incidence, which clearly is not the case in our model. Since on normal incidence no S-waves are generated, our computed values will be on the high side, which is acceptable for the purposes of our calculations. Symbols in Figure 4-27(b) are the same as used in Eq. 2-31. For example, I_{rfl_1} refers to the energy fraction reflected at interface 1.

In order to compute the relative energy of the multiples and primaries in Figure 4-27(b), we use Table 2-7 to determine the energy fractions refracted and reflected at each interface by entering the following densities and velocities: air—$\rho = 1.0$ g/cm^3, velocity = 335 m/s; unsaturated sediments—$\rho = 1.8$ g/cm^3, velocity = 500 m/s; saturated sediments—$\rho = 2.0$ g/cm^3, velocity = 1800 m/s; and bedrock—$\rho = 2.6$ g/cm^3, velocity = 4800 m/s. For each wavepath we list all refractions and reflections that occur from source to receiver. The list for the bedrock multiple, for example, is $(I_{rfr_1})(I_{rfl_2})(I_{rfr_1})(I_{rfl_0})(I_{rfr_1})(I_{rfl_2})(I_{rfr_1})$. Substituting the values obtained from Table 2-7 into these lists results in energy partitioning values for the wavepaths in Figure 4-27(b). Normalizing these values to the water-table primary results in the values listed in the first row of Table 4-9. If we determine the path lengths for both primaries and multiples, we also can employ Tables 2-5 and 2-6 to determine energy losses due to absorption and spherical spreading. These results also are listed in Table 4-9.

The results of this analysis for the model of Figure 4-27 demonstrate that sufficient energy is present for the water-table multiple to constitute an identifiable pulse on a field record, whereas the bedrock multiple is too depleted to be noticed. Our analysis suggests that, in general, we need not be overly concerned with multiples unless we have quite large velocity contrasts and relatively simple paths with few boundaries to reduce energies through partitioning.

Some multiples may have sufficient amplitudes to appear on field records, however, so we must adopt approaches for identifying them and removing them from processed seismograms. In order to search for possibilities, once again let's develop time-distance data for a specific case. We assume a single layer 75 m thick with a velocity of 1800 m/s and primaries and multiples of the type illustrated in Figure 4-27(b). Time-distance data for this model are illustrated in Figure 4-28(a). We see that the multiple times are *approximately* but not exactly twice the primary times. An x^2 - t^2 analysis results in an identical velocity but twice the thickness for the multiple (Fig. 4-28(b)). This results from the fact that the path length of the multiple for a given source-receiver distance is identical to that which would be followed by an imaginary primary to an imaginary interface at exactly twice the

TABLE 4.9 Relative Energy Comparisons Among Primary Reflections and Multiples

	Water-table Primary	Water-table Multiple	Bedrock Primary	Bedrock Multiple
Energy partitioning	1.000	0.359	0.353	0.045
Plus absorption	1.000	0.323	0.189	0.013
Plus spherical spreading	1.000	0.314	0.160	0.009

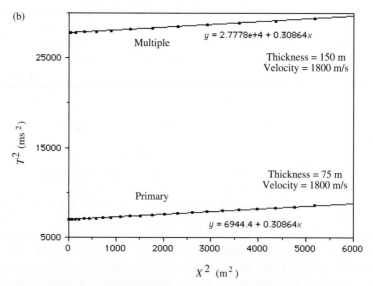

Figure 4-28 Comparison of primary reflection and multiple for a single layer with a velocity of 1800 m/s and a thickness of 75 m. (a) Time-distance plot for primary and multiple. (b) $x^2 - t^2$ plot for primary and multiple. (c) NMO plot for primary and multiple.

thickness of the actual interface. Finally, if we plot normal move-out values for both the primary and the multiple, it is obvious that the multiple possesses less NMO (Fig. 4-28(c)). This difference in NMO also is evident from Figure 4-28(a).

Thus, reflection events of smaller amplitude and twice the time of other reflection events are good candidates for multiples. If analysis produces closely similar velocities to

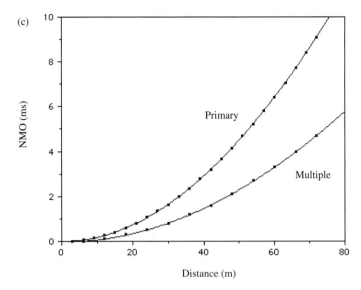

Figure 4-28 *(continued)*

the event of greater amplitude, and thickness values that are twice as great, a multiple event is indicated. Because of the difference in NMO values, the amplitudes of multiples can be further reduced during processing as described in a later section. Multiples constitute a serious problem in the deep-reflection surveys utilized by petroleum exploration firms, and much effort is expended in identifying and eliminating them. They are a less serious consideration in shallow work, but one nonetheless should be alert for such a possibility.

Diffractions

In Chapter 2 we noted that when a wave encounters a sudden change in curvature of an interface, diffraction occurs (see Figure 2-19). We recall that effects of diffraction are most pronounced when sharp changes in curvature of interfaces have radii that are similar to those of seismic wavelengths. In reflection work we attempt to utilize energy sources that produce a high frequency content in the source signal and utilize high-frequency geophones. Because 100-Hz frequencies are associated with 15-m wavelengths, we should expect to encounter diffracted energy in our work.

In Chapter 3 when we discussed refraction travel-time curves associated with discontinuous reflectors we learned that diffracted waves could occur as first arrivals in some situations but that, in general, it would not be an easy matter to confidently recognize these as diffractions on field records. In reflection work, however, as we continuously record geophone response, we stand a much better chance of recognizing diffraction arrivals. Therefore, let's proceed to determine what pattern diffractions should illustrate on field seismograms.

Figure 4-29(a) illustrates the general case of a step in a reflecting surface with the energy source offset from the vertical projection of the step. For the particular geometry

(a)

(b)

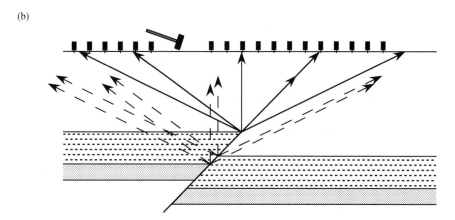

Figure 4-29 Diffraction at an edge. (a) Reflection and diffraction paths from a step in a horizontal interface. Symbols are explained in the text. (b) Some diffraction paths from three edges in a faulted horizontal unit.

illustrated, reflections will arrive in the normal manner with the exceptions that (1) there is a finite limit to the right of the step for reflections from the shallower portion of the interface, and (2) reflections from the deeper portion of the interface will begin to arrive at some distance after this limit. After energy arrives at the upper edge of the step, diffractions will begin to spread outward toward all geophones in the spread. Diffractions also will be generated from the lower edge of the step, but for simplicity we do not consider them in this analysis.

Our previous reflection equations obviously are valid for the two sets of reflections involved. It is a trivial matter to calculate the last point at which reflections from the upper interface arrive and the first point at which reflections from the lower interface arrive. Travel-times t_t for diffractions consist of two components: (1) the travel-time t_d of a diffracted wave from the edge of the step to a given horizontal distance x_g, and (2) the time t_s for the energy from the source to reach the step edge. The path distances are straightforward to determine, because right triangles can be applied in both cases. For diffraction travel-time

$$t_d = \frac{\left(x_g^2 + h_1^2\right)^{1/2}}{V_1} \tag{4-58}$$

Similarly, for travel-time from source to edge

$$t_s = \frac{\left(x_s^2 + h_1^2\right)^{1/2}}{V_1} \tag{4-59}$$

and, finally,

$$t_t = \frac{\left(x_s^2 + h_1^2\right)^{1/2}}{V_1} + \frac{\left(x_g^2 + h_1^2\right)^{1/2}}{V_1} \tag{4-60}$$

The diffraction times determined by Eq. 4-58 plot as a hyperbola, as do the total times determined by Eq. 4-60, since the first term in this equation simply adds a constant time to each diffraction time (compare diffraction times in Figure 4-30 (a) and (c). Figure 4-30(a) shows a typical travel-time plot for the model in Figure 4-29(a). The reflection arrivals produce their characteristic hyperbolic curve symmetrical about the source with the exception that part of the curve is displaced later in time by the presence of the step in the reflector. The diffractions also produce a hyperbolic curve, as mentioned previously. This curve illustrates greater NMO than the reflections even though both diffractions and reflections travel entirely in the V_1 medium. Note that the minimum diffraction time is located directly above the position of the step's edge.

If you are interested in working with diffraction arrival patterns, experiment with Table 4-10. Column 3 lists diffraction times from an edge to geophones located at the horizontal distances from the vertical projection of the edge listed in column 2. Column 5 lists total travel times. If the shot is not directly above the edge, the correct horizontal positions for geophones for the total travel times are listed in column 4. The shot position is determined relative to the edge position but is always at $x = 0$.

Figure 4-30(b) illustrates a situation similar to that in Figure 4-29(b) which illustrates beds displaced by a high-angle reverse fault. Diffractions from the uppermost three edges are plotted as well as reflections from the uppermost two horizons on the upthrown block. Reflections from the downthrown block are not shown in order to avoid clutter. Note that diffractions provide some information about the position and direction of inclination of the fault, as diffractions from each edge lie vertically above the edge. Also note the overlap of reflection and diffraction arrivals, especially the reflections from the second interface. On a field seismogram these overlaps would create a good deal of interference, thus obscuring reflections of interest.

Figure 4-30(c) illustrates the special case when the shot point is directly over the edge. This is included because of the symmetrical relation between reflections and diffractions. It perhaps is easier to observe the difference in normal move-out in this case. If we

(a)

(b)

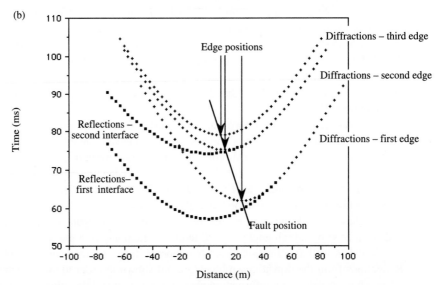

Figure 4-30   Diffraction and reflection patterns over a vertical step and an inclined fault. See Figure 4-29 for symbols and models. (a) Diffractions and reflections for the general case where the energy source is displaced from the diffracting edge: $h_1 = 40$ m, $h_2 = 5$ m, and $x_s = -24$ m. (b) Some diffractions and reflections from a model similar to that in Figure 4-29 (b); parameters similar to (a) except additional edges are present at 52 m and 55 m and are offset in the $-x$-direction as indicated. (c) Diffractions and reflections from a vertical step where the energy source is located directly over the step: $h_1 = 40$ m, $h_2 = 5$ m. (d) Arrival times in (c) corrected for NMO of reflection times. (e) Simulated seismic section based on times in (d).

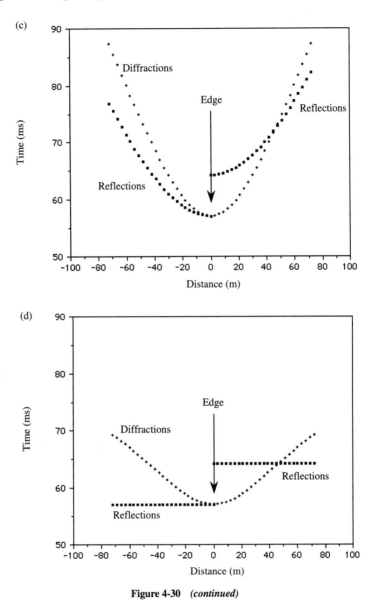

Figure 4-30   *(continued)*

correct all times in Figure 4-30(c) for normal move-out, the result is the pattern in Figure 4-30(d). Due to the NMO difference between reflections and diffractions, the correction flattens the reflections but only reduces the amplitude of the diffraction hyperbola. If we take Figure 4-30(d) and flip it so that time increases downward and draw a wavelet at each arrival point, we create the simulated seismic section in Figure 4-30(e).

(e)

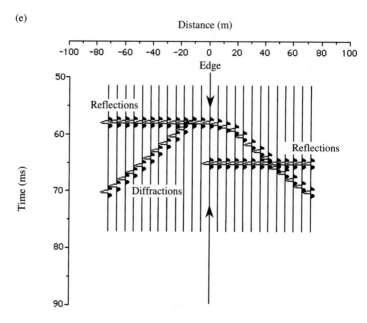

Figure 4-30 *(continued)*

TABLE 4-10 Diffraction Travel-Times

Geophone	Distance (m)	Diffraction Time (ms)	Offset Correction (m)	Total Time (ms)
Source	0	28.6	− 9	57.9
1	− 3	28.7	− 12	57.9
2	− 6	28.9	− 15	58.2
3	− 9	29.3	− 18	58.6
4	− 12	29.8	− 21	59.1
5	− 15	30.5	− 24	59.8
6	− 18	31.3	− 27	60.6
7	− 21	32.3	− 30	61.6
8	− 24	33.3	− 33	62.6
9	− 27	34.5	− 36	63.8
10	− 30	35.7	− 39	65.0
11	− 33	37.0	− 42	66.3
12	− 36	38.4	− 45	67.7
13	− 39	39.9	− 48	69.2
14	− 42	41.4	− 51	70.7
15	− 45	43.0	− 54	72.3
16	− 48	44.6	− 57	73.9
17	− 51	46.3	− 60	75.6
18	− 54	48.0	− 63	77.3
19	− 57	49.7	− 66	79.0
20	− 60	51.5	− 69	80.8
21	− 63	53.3	− 72	82.6
22	− 66	55.1	− 75	84.4
23	− 69	57.0	− 78	86.3
24	− 72	58.8	− 81	88.1

Velocity (m/s)	1400
Thickness (m)	40
Increment 1 (m)	−3
Increment 2 (m)	−3

Shot offset (m)	9
Extra time (ms)	29.3

Design a model similar to that in Figure 4-29(a). Place the step at somewhat greater depth and reduce its amount of relief. Place the shot point fairly far to the right of the step. Using Table 4-1 and Table 4-10 gather appropriate travel-times for reflections from both portions of the offset interface and for diffractions from the upper edge of the step. Be sure not to show reflections at positions where they cannot occur. Finally, graph the data.

This section emphasizes the misleading effect diffractions might have in suggesting the presence of structures or interfaces that simply do not exist. This is in addition to the other deleterious effects noted previously. Diffraction effects are especially common in seismic sections acquired during petroleum exploration efforts, and a great deal of effort and funds have been expended to deal constructively with this problem. Most of this occurs during data processing and will be mentioned briefly in a following section. Diffraction effects generally are not as obvious a problem in shallow work, largely due to the fact that short geophone spreads often are sufficient to solve the problem at hand. In these cases diffraction patterns are not sufficiently well developed to be recognized as such. However, many diffractions likely are present on records and function mainly as interference, the origin of which cannot be ascertained. When continuous profiling is employed for shallow investigations, diffractions may appear as clearly developed as on many deeper records. For example, in Figure 4-36 study carefully the wavelet patterns in and around the reflections produced by bedrock hills. Certainly, many patterns reminiscent of those in Figure 4-30 are in evidence. Other good illustrations of diffraction patterns on seismic sections can be found in Dobrin and Savit (1988, p. 310), Kearey and Brooks (1984, p. 90–93), and Robinson and Coruh (1988, p. 185). Now let's turn to a general discussion of how one goes about creating a section such as that in Figure 4-36.

COMMON FIELD PROCEDURES

In this section we primarily are interested in three fundamental field operations utilizing geophone spreads that are in common use for shallow-reflection surveys. Before beginning that discussion, however, a few comments are in order concerning equipment considerations. Much of this material was covered in Chapter 2, and you are urged to read once again the sections dealing with energy sources, geophones, and seismographs.

Equipment Considerations

In reflection work we constantly are concerned with enhancing the high-frequency component of the seismic signal. One reason is to remove lower-frequency noise in order to improve our chance to observe reflections. The other reason is to improve resolution so that we may observe detailed structures in the shallow subsurface.

As an example of the importance of frequency in vertical resolution, consider Figure 4-31. This diagram illustrates two thin reflectors spaced 7.5 m apart. We assume average velocity is 1500 m/s. Also illustrated are two seismic pulses with wavelengths λ of 10 m and 20 m, which have frequencies of 150 and 75 Hz, respectively. These are considered to be sine curves for simplicity. The signal received at a given geophone from such a reflection will be the composite waveform illustrated. Clearly, the 150-Hz signal delineates both thin beds, but the 75-Hz signal does not. Although Figure 4-31 is diagrammatic only and thus is simplified (for instance, reflections from the bottom of each bed are not considered), it does emphasize the role of signal frequency in vertical resolution. Many seismologists maintain that the limit of vertical resolution is $\lambda/4$ (Sheriff and Geldart, 1982, p. 119; Sheriff, 1989, p. 330–331), but in shallow work the more likely limit is, at best, $\lambda/2$.

Does frequency affect horizontal resolution, and, if so, in what way? In considering this question, we must recall that wavefronts actually are spherical and not the planar wavefronts we have assumed for our derivations. The radius R_1 of the area on the reflector primarily responsible for producing the signal (the first Fresnel zone) is determined by

$$R_1 = \tfrac{1}{2} V_1 \left(\frac{t_0}{f}\right)^{1/2} = \left(\tfrac{1}{2} \lambda h_1\right)^{1/2} \tag{4-61}$$

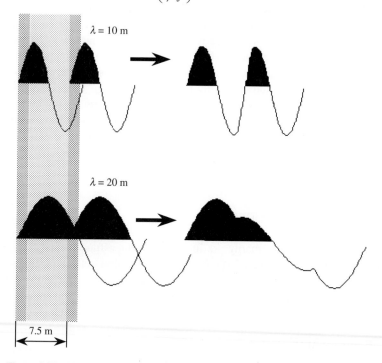

Figure 4-31 Diagrammatic explanation for increased vertical resolution achieved from higher-frequency (shorter-wavelength) pulse. The 10 m wavelength pulse delineates both reflectors, but the 20-m pulse effectively records only a single reflection event.

where V_1 is velocity, t_0 is travel-time, h_1 is thickness, and f is frequency (Sheriff and Geldart, 1982, p. 119–121). If we assume a velocity of 1500 m/s and a thickness of 30 m, then a frequency of 100 Hz has a Fresnel zone with radius equal to 15 m, and a 50-Hz signal leads to a radius of 21 m. We achieve better horizontal resolution with higher frequencies, as was the case in vertical resolution. Of course, geophone spacing also determines in what detail we sample the subsurface. For horizontal reflecting surfaces, the area of reflection is located halfway between receiver and source. If our geophone spread has a geophone interval of 10 m, we are sampling the subsurface at intervals of 5 m, whereas an interval of 3 m results in a 1.5-m spacing of samples (see Fig. 4-32).

In order to create and receive these high frequencies, we use energy sources that tend to produce a high-frequency spectrum with substantial energy values. As noted in Chapter 2, this often tends to be a shotgun source for shallow work. High-frequency geophones, typically 100 Hz, are desirable, as is a seismograph with filters that can remove lower frequencies. These typically are set at 100 or 200 Hz. In good conditions when one is interested only in mapping the bedrock surface, such equipment demands may not be necessary. We have had good results with bedrock reflections from an average depth of 50 m using a hammer source, 14-Hz geophones, and 100-Hz low-cut filters. However, we are able to obtain better quality reflections and more detail in the material above bedrock when using a 10-gauge shotgun source, 100-Hz geophones, and 200-Hz low-cut filters.

Although not directly related to frequency considerations, a signal-enhancement seismograph is essential. When using high-frequency geophones and low-cut filters, a substantial portion of the seismic energy is removed. It often is mandatory to fire two or more shots, thereby summing the incoming signal, in order to produce a record with discernible reflections. Digital recording capabilities also are necessary for all but the most basic reflection survey in order to process the recorded data by computer. In the following section, typical geophone spreads for reflection work are discussed. Two of the three rely on digital processing to greater or lesser degrees, and the remaining one can benefit from some processing.

Geophone Spreads

In this section we plan to cover three fundamental approaches to the acquisition of seismic reflection data. One is the standard split spread we already mentioned. The others, used primarily when seismic sections are desired, are referred to as the *common-offset* method and the *common-depth-point* method.

Split Spread.   When one utilizes reflection surveying primarily to determine subsurface characteristics at one location, split spreads with a center shot arrangement normally are employed. The first step is to deploy a geophone string and to record data beginning with a small shot offset. This serves two purposes. First, it supplies data that can be used for refraction determination of the near-surface velocities to be used in static corrections. Second, it supplies some information concerning the nature of the optimum window. Most likely the shot offset will have to be increased until the optimum window is clearly delineated.

This is a critical phase of the survey and must not be given short shrift. The major subsurface target of the exploration work needs to be clearly defined, as this affects the optimum-offset choice. Also, one needs to conduct an analysis of noise factors and to observe how low-cut filters affect the quality of the desired reflections. For example, a 200-Hz low-cut filter may remove a substantial portion of noise but reduce the amplitude of target reflections to an undetectable level. Once the correct shot offset is in hand and the best filter combination selected, the general survey commences.

Typically in shallow-reflection surveys, a group of 12 (or 24) geophones is strung along a preselected survey line (GL_1 (Group L_1) in Fig. 4-32). If possible the geophones are placed in augered holes in order to improve ground coupling and to reduce interference from the air wave. One or more shots are taken at the preselected shot offset (E_1 in Fig. 4-32). The geophone string then is moved to the right of the shot point to GR_1, and more shots are taken. This process and geometry results in records containing the best possible reflections of interest, permits dip determinations if dips are present, and hopefully produces good NMO to aid in delineating reflections and for velocity and depth determinations. If the survey purpose is to supplement refraction work or to obtain information that was not possible with refraction (too large an energy source required, etc.), the survey might end at this stage.

Typically more information is desired, so additional lines are deployed. The geophone string GR_1 is left in place, and the shot is moved to E_2. The string now functions as GL_2 and records information from additional points on the reflector. The next step is to move GL_2 to GR_2, and so forth. This process of accumulating data along a line is termed *profiling*. Note that in the case illustrated in Figure 4-32 not every point on the reflector is sampled. If this is unacceptable, then we can institute a slightly different procedure in order to assure complete coverage. A seismic section can be produced from this data by correcting for statics and removing NMO as discussed in the following section on data analysis. However, if a seismic section and continuous coverage are desired, we usually would select either the common-offset or common-depth-point method.

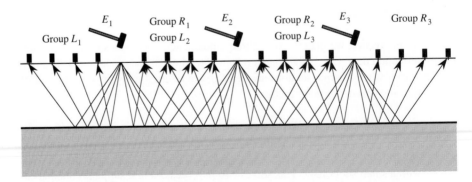

Figure 4-32 A simple form of seismic profiling with single-coverage, center-shot spreads. A 4-geophone spread is shown for simplicity, although shallow-exploration spreads typically contain 12 or 24 geophones.

Common Offset. The common-offset method (also termed optimum-offset) was developed by Hunter et al. (1984) and is remarkably simple to implement and actually quite efficient. As already described, one first must delineate the optimum window and determine the optimum offset. Once this value is known, the geophones are distributed, and the first shot (E_1 in Figure 4-33) is taken. The channel on the seismograph corresponding to geophone G_1 is then "frozen" so that it will not be affected by subsequent shots. The remaining channels then are cleared to remove all data, and the next shot is taken at E_2. Channel G_2 is frozen, the display is cleared, and shot E_3 is executed. This process continues until all 12 channels (or 24) contain data. The data then are transferred to disk or tape via digital recording capabilities, all channels are cleared, and the process continues.

Although only every fourth shot position is illustrated in Figure 4-33, note that common-offset procedures produce continuous subsurface coverage. Also, and most important, is the lack of normal move-out in the field seismogram, as all source-receiver distances are the same. Thus, NMO corrections are not required to produce a seismic section, and the raw data give a reasonable indication of the reflector geometry. Close examination of Figure 4-33 does, however, reveal two important relationships to keep in mind. First, the "relief" of the reflector as portrayed on the seismogram will be enhanced or depressed depending on the velocity of the overlying material. Of course, this can be corrected, as we will see when

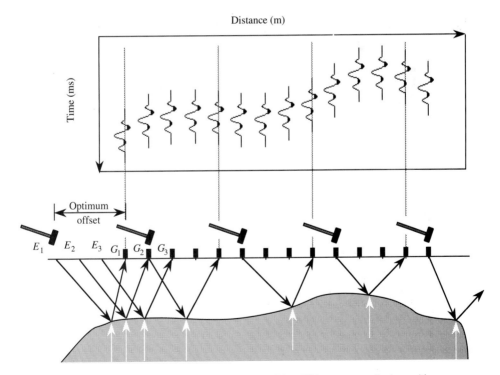

Figure 4-33 An example of common-offset profiling. White arrows emphasize position of reflecting points relative to position of receiving geophones.

discussing data processing operations. Second, (as shot points in the diagram are to the left of geophones) the undulations on the reflector are shifted to the right relative to the reflecting point but by unequal amounts depending on the local orientation of the reflecting surface. Therefore, there is a "stretching" distortion in the horizontal plane as well. In most cases where relief on reflectors is modest and/or interfaces have only slight dips and for most survey purposes, this horizontal distortion is not appreciable and is acceptable.

Although common-offset records normally should be digitally processed, the processing requirements can be quite simple, and the expense is not a major factor. The major consideration is removing effects of elevation differences in geophone locations and variations in the thickness and velocity of near-surface material.

Figure 4-34 contains a field seismogram taken to determine optimum offset and one developed using the common-offset technique. After analyzing the record in Figure 4-34(a), we selected an optimum offset of 42 m and then proceeded to obtain the common-offset record. This latter record brings a word or two of caution to mind concerning common-offset records. Note that the water-table refraction lines up very nicely and could easily be mistaken for a reflection. The same is true for the beginning of ground roll. When you are shooting at a constant source-receiver distance, assuming a uniform subsurface, all events large enough to be seen on the record are going to line up. You, therefore need to keep all possibilities in mind and check for them through simple computations, by shooting a refraction record first, and by careful observation of the split-spread records shot to determine the optimum offset.

Figure 4-35 contains the same record types as Figure 4-34 but from a different location. Note the position of the air wave in Figure 4-35(a). In Figure 4-35(b) (common offset) it could easily be mistaken for a reflection. It also interferes with a possible reflection in Figure 4-35(b), which suggests that the optimum offset chosen was not the best choice. Finally, note that in Figure 4-35(b) the reflectors appear to be curved, suggesting some structure or relief. However, because the water-table refraction exhibits the same curvature, this demonstrates that one needs to make corrections for elevation, thickness, or velocity variations. If we make this type of correction visually to Figure 4-35(b) by arranging the water-table refractions into a straight line, the other reflectors become more or less horizontal also.

As an example of what can be achieved with the common-offset method, let's examine Figure 4-36. This remarkable record of bedrock highs and lows was produced using a 7.25-kg hammer, 50-Hz geophones, 300-Hz low-cut filter, 12-channel signal-enhancement seismograph, and an Apple II computer for digital processing. Although this section was produced at what might be termed an "optimum site" in terms of available structure, bedrock depths, and good energy coupling, it does emphasize the capabilities of the common-offset reflection method at localities for which the reflection potential is moderate to good.

The seismic section illustrated in Figure 4-37 also exhibits several strong reflectors. As we might expect, the bedrock reflector is clearly delineated. Compact till also has a high velocity, and the contrast between till and overlying sands and gravels would be expected to produce a good reflection, which it does. Note also the strong reflections due to contacts between clay beds and sands and gravels.

(a)

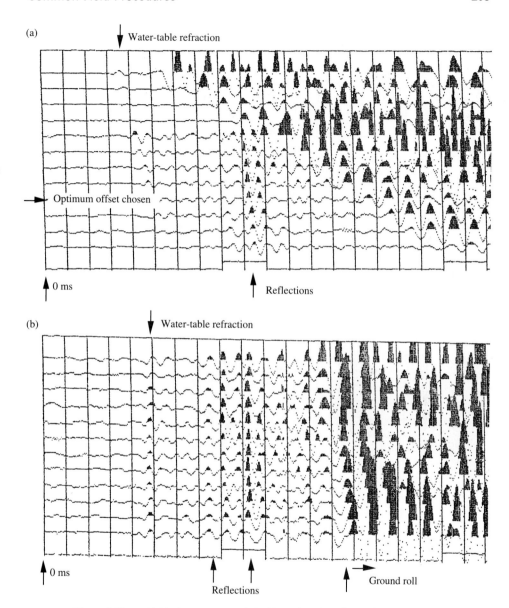

Figure 4-34 (a) Field seismogram recorded with an 18-m shot offset and a 3-m geophone interval. (b) Field seismogram recorded using the common-offset method. Offset to each geophone is 42 m. Note that the water-table refraction could easily be mistaken for a reflection.

Common Depth Point.   Much of what we have said about field and equipment considerations also apply to this method. The main objective of common-depth-point (CDP) investigations is to sample each subsurface point several times. If the information on

(a)

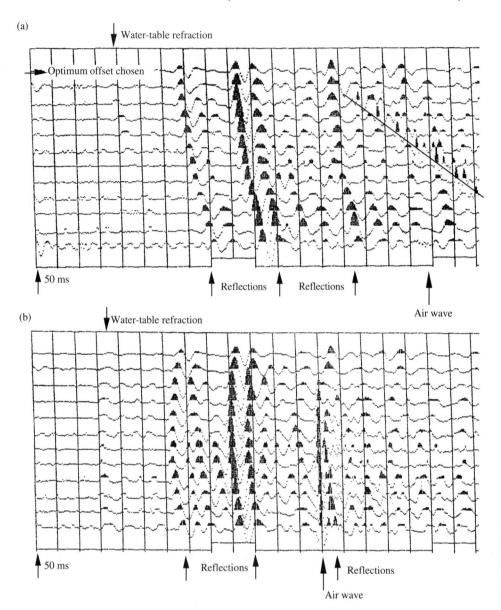

Figure 4-35 (a) Field seismogram recorded with a 60-m offset and a 3-m geophone interval. (b) Common-offset record shot with a 60-m offset. Note the way the air wave lines up and appears to be a reflection.

each record is then summed after suitable preprocessing, to be described in the next section, true reflection arrivals will be enhanced and various unwanted signals will tend to be reduced or eliminated, thereby producing superior records. A key component of CDP work

Time (ms)

Figure 4-36 Seismic section recorded near Shawville, Quebec. Shot offset 30.5 m. Geophone interval 3 m. Maximum relief of the bedrock hills is approximately 40 m. Notice good reflections within uncon-solidated glacial sediments deposited between the bedrock highs. (Modified from Hunter, J. A., Pullan, S. E., Burns, R. A., Gagne, R. M., and Good, R. L., 1984, Shallow seismic reflection mapping of the overburden-bedrock interface with the engineering seismograph—some simple techniques: Geophysics, v. 49, p. 1381–1385.)

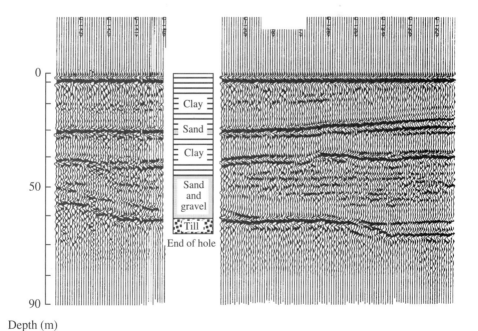

Depth (m)

Figure 4-37 Seismic section recorded near Val Gagné, Ontario. Notice the excellent reflections from bedrock, till, and clay-sand interfaces. Survey parameters: shot offset = 15 m, 300-Hz low-cut filters, 100-Hz geophones, 12-gauge Buffalo gun source. (Courtesy James Hunter and Susan Pullan, Geological Survey of Canada.)

is extensive computer processing of digital records for velocity analysis, removal of NMO, summing of traces from the same subsurface point, and various other crucial corrections.

In order to understand the basic design of the CDP method, let's examine Figure 4-38(b). For simplicity we'll assume we are operating with a string of 6 geophones, although 12 is the actual minimum we would use. The first shot point is E_1, and the geophone string for this shot is labeled $G_{11},\ldots,G_{61}$ (geophones 1,...,6 and shot 1). The field record corresponding to shot E_1 is so designated in Figure 4-38(a). The shot point then is moved to E_2, and the geophone string position becomes $G_{12},\ldots,G_{62}$. Another field record is recorded, the shot point is moved again, and so on. There are a number of procedures for speeding field work so that the entire string of six geophones is not moved for every shot. One obvious option is to move the geophone nearest the shot next to the most distant geophone after each shot. However, the main procedural point to grasp is the manner in which the shot is moved along the line.

Figure 4-38(b) clearly illustrates that such a procedure will result in several traces corresponding to each subsurface reflecting point. These points on the reflector are identified by white arrows, each of which notes the number of reflections corresponding to that point. For example, the point labeled "CDP for gather" is a reflecting point common to three reflection traces. If you carefully trace out the reflection paths to this point and correlate them with the correct shot and geophone position label, you should arrive at the three geophone identifiers that are surrounded by boxes in Figure 4-38(b). The corresponding traces are identified in like fashion in (a). Because of the myriad of reflection paths in the diagram, this might appear to be very complicated. Actually the field procedure is simple. All that is needed is an orderly plan for record keeping. Remember that all records are being stored digitally and that any can be accessed easily. Note that there is an orderly progression in the record traces corresponding to a given CDP. The records for the point we already discussed are G_{51}, G_{32}, and G_{13}. The records for the next point to the right are G_{61}, G_{42}, and G_{23}.

If we extract these CDP traces (*gather* them together), we then have a *CDP gather*. Such a gather is illustrated in Figure 4-38(c) as well as the reflection paths corresponding to these traces. In an actual survey, more geophones would be used, and we would have more traces/CDP. In our illustration we have threefold coverage. In most shallow exploration work the minimum coverage would be sixfold. As long as we keep the distance between shot points and the geophone interval the same, and if n represents the number of geophones, our coverage is equal to $n/2$-*fold*.

It is easy to see why a CDP gather is ideal for determining the NMO associated with a given reflection, since all traces on the record are from the same subsurface point. Moreover, after we correct for NMO, we can sum these traces to produce an enhanced reflection signal. In this simple case any other undulations on these traces are unlikely to undergo substantial enhancement when the traces are summed. Therefore, the CDP method is extremely effective in improving the amplitude of reflections while at the same time reducing unwanted noise.

Now it is time to turn to a discussion of the various digital processing techniques required to transform raw reflection data into a seismic section. Unlike split-spread and common-offset profiling, which provide only singlefold coverage, common-depth-point profiling demands digital processing from the onset. Because of the computer power

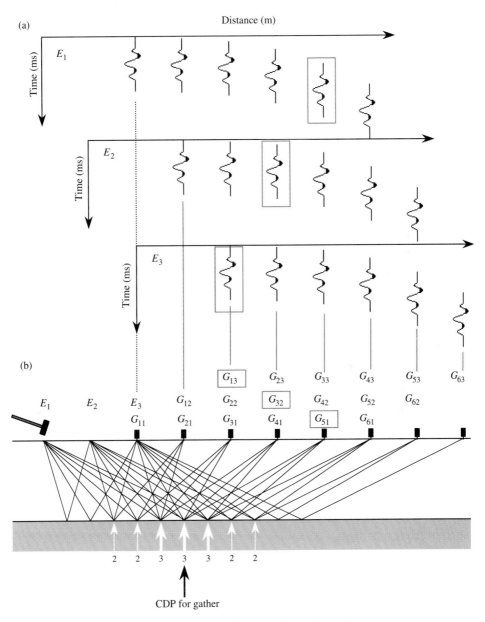

Figure 4-38 An example of common-depth-point (CDP) profiling. (a) Field seismograms from shot points E_1, E_2, and E_3. Traces from a common point on the reflector are outlined. (b) Geophone positions and shot points for CDP profiling. White arrows designate positions on reflector from which more than one trace is recorded. (c) All traces from the same reflecting point (labeled "CDP gather" in (b)) gathered together into one CDP record.

207

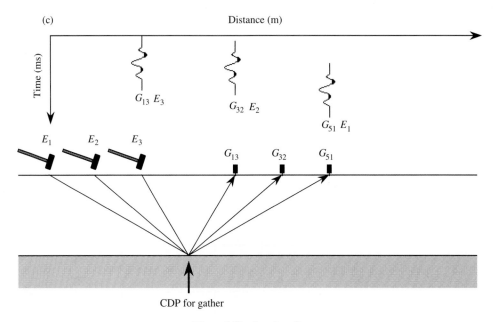

Figure 4-38 *(continued)*

required to perform this task, the CDP method has received less use in shallow-reflection work to date than the other methods. Personal computers now are capable of performing CDP processing, and as their power increases and cost decreases, it is likely that CDP will become the method of choice in shallow exploration, as it is in the exploration endeavors performed by petroleum firms.

COMPUTER PROCESSING OF REFLECTION DATA

As mentioned previously, enormous amounts of computer time on very high speed computers are regularly utilized by petroleum exploration companies to process seismic reflection data. Such time, expense, and expertise normally are not available to the small (sometimes one person) companies active in solving geotechnical problems using shallow-reflection surveys. Although this is rapidly changing, our goal in this section is to concentrate on processing that is essential to meet the normal objectives of the shallow survey while briefly introducing techniques that may be applied to shallow-reflection data and that normally are utilized when targets are located at greater depths.

The Static Correction

The first step in processing reflection data is to apply both a topographic correction and what is called the *weathering correction*. The reasons and strategy are similar to those used in refraction work and were discussed in Chapter 3. In what is perhaps the majority of cases, a low-velocity layer that varies in thickness is present at the surface. Because the velocity of this layer is substantially lower than the layer underlying it, travel-times are affected unequally to an extent that cannot be ignored.

A generalized situation is presented in Figure 4-39(a). In the case of shallow-reflection surveys the low-velocity layer often consists of unsaturated sediments which have low velocities relative to their saturated counterparts situated beneath the water table. Hence the term weathering correction is not entirely appropriate here but is retained because of its

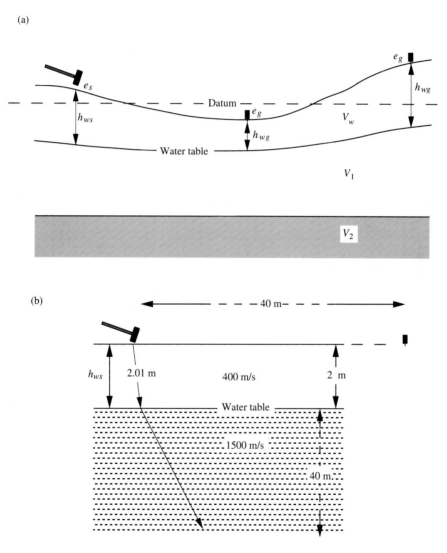

Figure 4-39 Static corrections. (a) Parameters used in elevation and weathering corrections. (b) A model demonstrating the validity of the vertical-path assumption.

widespread usage in the petroleum industry. The symbols h_{wg} and h_{ws} refer to the thickness of the low-velocity layer beneath the geophone and shot point, and e_g, e_s, and e_d refer to the elevations of the geophone, shot point, and datum, respectively.

In making these corrections we assume vertical paths through the low-velocity layer. Due to the velocity contrasts this is not a bad assumption, but in situations where relief is excessive or the velocity contrast is not large, it might be wise to sample the error introduced by assuming vertical paths. Is this assumption acceptable for the investigation diagramed in Figure 4-39(b)? Let's assume that our maximum source-geophone offset is 40 m and that the thickness of the 1500 m/s layer is 40 m. We use Snell's law and the given velocities to calculate the angles of incidence and refraction for the raypaths illustrated. Using these angles and the thickness of the low-velocity layer, we find that the actual ray path and the vertical path differ by 0.01 m. At 400 m/s this translates into a time difference of 0.025 ms. If you examine the time scale on any of the field seismographs illustrated in Chapters 3 or 4, you should see that this time difference is negligible.

The first step is to remove the effect of the low-velocity layer. Let t_w represent the weathering correction time. In the case illustrated in Figure 4-39(a) the correction represents the time of travel at 400 m/s less the time of travel at 1500 m/s (essentially we calculate the delay due to this material possessing a velocity of 400 m/s rather than a velocity of 1500 m/s). In mathematical form

$$t_w = \frac{h_w}{V_w} - \frac{h_w}{V_1} \tag{4-62}$$

where h_w represents either h_{wg} or h_{ws}, depending whether we are correcting for the geophone or the shot.

Once the weathering correction is applied, we effectively change the 400-m/s velocity to 1500 m/s. The topographic correction now is applied simply by dividing the difference in elevation between the geophone (or shot) and the datum by, in our example, 1500 m/s. Therefore, the elevation correction is

$$t_e = \frac{e_g - e_d}{V_1} \tag{4-63}$$

These corrections are simple to apply if the necessary information is available. If reflection spreads do not have source-geophone distances sufficiently small to sample refractions from the base of the low-velocity layer, then special refraction surveys are undertaken. If possible, the datum should be chosen as the lowest point in the survey, because statics always, therefore, reduce arrival times. Note that the static correction is applied to the entire trace. Every point on a trace from a given geophone is moved the same amount in time depending on static correction for that geophone position.

In some situations a simpler approach may suffice. Assume the base of our low-velocity layer is the water table. If we have common-offset data, we can typically identify the water-table refraction arrival. Also, if relief along our spreads is not great, then the water table is very close to horizontal. Simply shifting the water-table refractions so they all begin at the same time is an elegant but simple way to adjust later arrivals on all traces. Our datum in this case is the water table, so depth determinations, for instance, are relative to it, not the ground surface. The seismogram illustrated in Figure 4-35(b), as already mentioned, serves as an excellent example of why statics must be applied. This record also emphasizes how straightforward it is to apply the water-table refraction approach to static corrections. We set the first trace constant as a reference. Each succeeding trace is shifted so that the beginnings of all refraction arrivals are at the same time as our reference trace. You can visually apply this method and see the curved reflections straighten. Remember that this method works only given our assumptions concerning the water table and common-offset data, wherein all source-geophone distances are the same.

In contrast to static corrections *dynamic* corrections move different parts of a given trace different amounts in time. We now turn to these.

Correcting for Normal Move-Out

If our primary goal in a reflection survey is simply to identify reflectors and determine thicknesses and velocities within a very limited area, we might work entirely with split-spread data, identify reflections visually, and apply x^2 - t^2 analyses to arrive at a solution for subsurface structure. Usually, however, if we undertake reflection work, our goal is a view of the subsurface in time and distance coordinates (i.e, a seismic section). In this case we attempt to plot every reflection arrival at its correct t_0—as if it traveled a vertical path to the reflector and back. In other words we correct for NMO.

Of course, we could take split-spread data, determine thicknesses and rms velocities, and use these values in Eq. 4-5 to determine NMO values (see Eq. 4-64).

$$T_{\text{NMO}} = \frac{\left(x^2 + 4h^2\right)^{1/2}}{V_{\text{rms}}} - \frac{2h}{V_{\text{rms}}} \qquad (4\text{-}64)$$

We could then use these NMO values to correct various reflections on all traces. This would be done via computer but still would demand a tremendous amount of time and effort on the part of the person undertaking such an analysis. Imagine all the steps involved and the number of maneuvers in a relatively simple case where each 12-geophone spread contained six obvious reflections and where eight spreads were recorded. It wouldn't take much of that kind of labor to search for a better way.

One approach is to record only common-offset data, which do not require NMO corrections. In such a case our section does not display vertical path arrivals, but we still produce an acceptable picture of the subsurface if we keep our sampling interval small (e.g., geophone intervals of from 1 to 3 m). However, we still require velocity information to obtain depths for our sections, so a somewhat more automated approach remains desirable.

Once we begin to acquire CDP data in any quantity, some automation is mandatory. In addition, the preceding approaches work only with reflection arrivals that are clear and obvious. As long as we are automating, why not inquire if other advantages might also accrue through computer processing—such as identifying weak reflections? Although a number of strategies exist for automated velocity determinations, we concentrate here on outlining three separate designs which for convenience we label "Velocity Analysis A," "Velocity Analysis B," and "Velocity Analysis C." All involve computer processing and have elements in common, but B is somewhat more sophisticated than A, and C typically employs much more sophisticated algorithms and demands considerably more processing power than either A or B.

Velocity Analysis A.   A typical computer program used to facilitate this analysis uses an equation such as Eq. 4-65 to compute the NMO correction.

$$T_{NMO} = \frac{x^2}{2t_0 V_{st}^2} - \frac{x^4}{8t_0^3 V_{st}^4} \qquad (4\text{-}65)$$

Of course, we know the offsets x used in our field design, but we do not know velocities or t_0. Remembering that we do not want to employ the x^2 - t^2 method but wish to somehow automate this determination, we rely on the computer's computational speed to solve the problem essentially by a trial-and-error method. Here we refer to the velocities we select for trials as *stacking velocities* or V_{st}. These are similar in value to root-mean-square velocities (determined via Dix's method) but are arrived at through the trial-and-error process described next.

Let's examine a very simplified outline of this procedure using a dynamic table, Table 4-11, as an aid. Assume we have a 24-trace record that contains two identifiable reflections. Figure 4-40(a) illustrates a time-distance curve for two reflections (time is oriented so as to increase downward as in a seismic section). In order to reduce clutter we present only points on this curve rather than wavelets. Our goal is to find values of V_{st} and t_0 for each reflecting horizon such that NMO is completely removed for arrivals from that horizon—all points plot at the same time.

Our computer program will allow us to enter maximum and minimum values and increment sizes for both velocity and t_0. In its simplest implementation, the program takes these values, calculates NMO corrections for one combination, adjusts all digitized data points on all traces by the computed amount, and plots the results (Fig. 4-40(b)). Another pair of values is selected, corrections once again computed, and results plotted (Fig. 4-40(c)). The process continues until all combinations of t_0 and V_{st} that originally were designated are utilized and results plotted. You can gain a feeling for the process by entering trial values in Table 4-11 and observing the results of the corrections. Although you can enter only three pairs of trial values at a time, you can observe how quickly the computations proceed. Therefore, attempting a much larger number of trials would not be insignificant in processing time but would be much more efficient than attempting to employ x^2 - t^2 analyses.

Although Figure 4-40(d) illustrates a successful correction for the earlier of the two reflections on the record, the points at 60 m and more do not line up nearly as well as the points at distances less than 60 m. Can you explain this discrepancy knowing that the reflecting horizon is at 45 m and that the velocity to this first interface is 1400 m/s?

TABLE 4-11 Using V_{st} and t_0 to Automate Velocity Analysis

Distance (m)	Reflection Times (ms)	NMO Correction 1 (ms)	Corrected Time (ms)	NMO Correction 2 (ms)	Corrected Time (ms)	NMO Correction 3 (ms)	Corrected Time (ms)
0	64.3	00.0	64.3	00.0	64.3	00.0	64.3
3	64.3	00.0	64.3	00.1	64.3	00.0	64.3
6	64.4	00.1	64.3	00.3	64.1	00.1	64.3
9	64.6	00.3	64.3	00.6	64.0	00.2	64.4
12	64.9	00.6	64.3	01.1	63.7	00.3	64.5
15	65.2	00.9	64.3	01.7	63.4	00.5	64.6
18	65.6	01.3	64.3	02.5	63.1	00.8	64.8
21	66.0	01.7	64.3	03.3	62.7	01.0	65.0
24	66.5	02.2	64.3	04.3	62.2	01.4	65.2
27	67.1	02.8	64.3	05.4	61.7	01.7	65.4
30	67.8	03.5	64.3	06.6	61.1	02.1	65.6
33	68.5	04.2	64.3	07.9	60.6	02.6	65.9
36	69.2	04.9	64.3	09.3	59.9	03.0	66.2
39	70.1	05.8	64.3	10.7	59.3	03.5	66.5
42	70.9	06.6	64.3	12.3	58.7	04.1	66.8
45	71.9	07.5	64.3	13.8	58.1	04.7	67.2
48	72.9	08.5	64.4	15.4	57.4	05.3	67.6
51	73.9	09.5	64.4	17.0	56.8	05.9	68.0
54	75.0	10.5	64.4	18.7	56.3	06.6	68.4
57	76.1	11.6	64.5	20.3	55.8	07.3	68.8
60	77.3	12.7	64.6	21.9	55.4	08.1	69.2
63	78.5	13.8	64.7	23.5	55.0	08.8	69.7
66	79.7	15.0	64.8	25.0	54.8	09.6	70.1
69	81.0	16.1	64.9	26.4	54.6	10.4	70.6
72	82.3	17.3	65.0	27.7	54.7	11.2	71.1

Model Values:

Increment (m)	3
Thickness (m)	45
Velocity (m/s)	1400

Correction velocities (m/s):

Velocity 1	1400
Velocity 2	1000
Velocity 3	1800

Correction t at $x=0$ (ms):

t_{0_1}	64.3
t_{0_2}	64.3
t_{0_3}	64.3

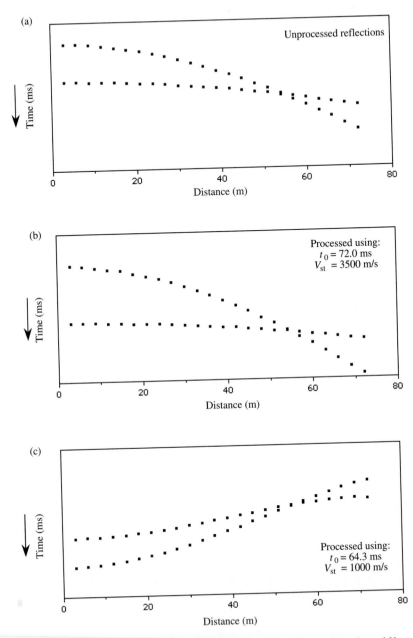

Figure 4-40 Velocity analysis of reflections processed using various values of t_0 and V_{st}. (a) Original reflections. (b) and (c) Unsuccessful corrections. (d) Early reflection successfully processed and stacking velocity determined. (e) Later reflection successfully processed and stacking velocity determined.

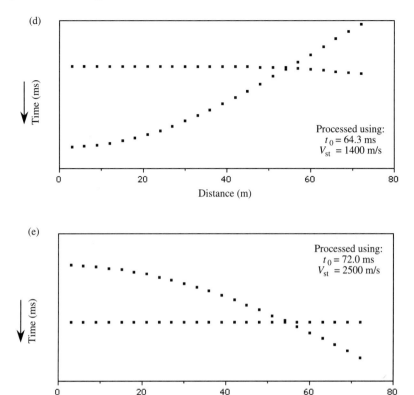

Figure 4-40 *(continued)*

If we entered reasonable ranges for t_0 and V_{st} before running our correction program, we should find some traces for which the corrections are successful (Fig. 4-40(d) and (e)). Using these values we can produce a final seismic section in which all reflections are corrected for NMO and also have the information necessary to determine interval velocities and thicknesses.

Assume that the digital sampling interval on our seismic traces is 0.5 ms. Further assume that a reflection arrives at 64.0 ms at a shot-geophone offset of 3 m. Velocities must be between 500 and 3000 m/s. Use Table 4-11 to determine realistic values for the minimum value, maximum value, and trial increment for the zero-offset time.

Although this approach is not elegant, it works reasonably well and does not require more than modern personal computer capabilities. It is adequate for much shallow exploration work, is not difficult nor expensive to implement, and is equally useful for routine data or for CDP data. One drawback is the potentially large number of trial results that need to be scanned to pick the trial values that successfully remove NMO from a reflection. Its major drawback, however, is that one works only with reflections that can be identified in order to be checked for NMO removal. Methods B and C do not possess this drawback, but as you might surmise, impose much more extensive computational demands.

Velocity Analysis B.   A portion of this approach is similar to A in that many trials are executed using combinations of t_0 and V_{st}. For each set of trial values all traces are shifted the correct amount to correct for NMO. After this adjustment the computational algorithm moves through all traces at small time increments. At each time increment the amplitudes of all traces are added together. If the trial values are correct for a given reflection, after the NMO correction, all troughs and peaks for that reflection will line up on the record. When added together, these will result in a signal that is much greater than for other trial values that are not correct for this reflection. If we plot amplitudes against all trial values for a record on which several reflections are present, we should see several large amplitude peaks. The t_0 and V_{st} associated with these peaks give us the information we need to produce a seismic section.

Velocity Analysis C.   In this approach we assume that each point on a selected trace is the onset of a possible reflection and then test this assumption on all other traces. Of course, most tests will fail, but some will succeed. In this brief account we concentrate on how such tests are made and how success or failure is determined.

First we utilize Eq. 4-42 and substitute V_{st}.

$$t_x = t_0 \left(1 + \frac{x^2}{V_{st}^2 t_0^2} \right)^{1/2}$$
(4-66)

for the velocity term. If we select a t_0 and a V_{st}, we can calculate reflection times for the source-geophone offsets that we used in our field survey. As we know, these times will plot along a hyperbolic curve. Figure 4-41(a) illustrates a time-distance curve for an actual reflection and two curves determined using Eq. 4-66 and the t_0 and V_{st} values identified in the diagram. Clearly, these are not the correct values, because correct values would produce a curve that plots on the actual reflection curve. Figure 4-41(b) illustrates two curves for which V_{st} is correct but for which t_0 is not. This trial and error process is the same as in our A and B methods, but here we are not correcting traces and looking for reflections with no NMO (either visually or by computing amplitudes), but are comparing actual reflection patterns with computed curves. How do we make these comparisons?

First we compute a curve using a trial value of t_0 and V_{st}. This curves begins at t_0 and crosses all traces on which we will correct for NMO. We then move forward a few milliseconds in time on the first trace in order to "capture" the form of whatever exists within our time window. Once the form of any displacements on the trace is defined, we move across

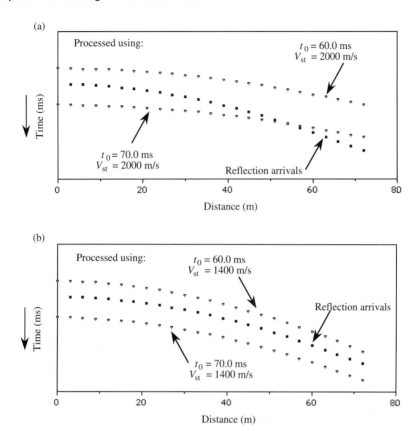

Figure 4-41   Hyperbolic curves computed for various values of t_0 and V_{st} compared with the time-distance curve for an actual reflection.

other traces, following the hyperbolic curve computed using the initial t_0 and V_{st}, remaining within our time window, and comparing curve forms from trace to trace. If an actual reflection lies within the window and follows the computed curve, the comparison algorithm will be successful and will return a high value. If not, a low value will be returned. In Figure 4-42(a) our time window cuts across the actual reflection, so comparison of waveforms yields a low value. In Figure 4-42(b) two combinations using a constant V_{st} and two values of t_0 fail, but one set of trial values, as indicated on the diagram, define a window that parallels the NMO of the true reflection arrivals. This set of trial values would yield a very high comparison success value. After all initially selected trial values of t_0 and V_{st} are used, a table of comparison success values is supplied. These comparison values are analyzed to determine successful t_0 and V_{st} trials which most likely indicate reflections (or multiples).

Stacking CDP Gathers

At this point we have applied static corrections, undertaken velocity analysis, and computed NMO corrections for our traces. Although we noted in our discussions of these two correc-

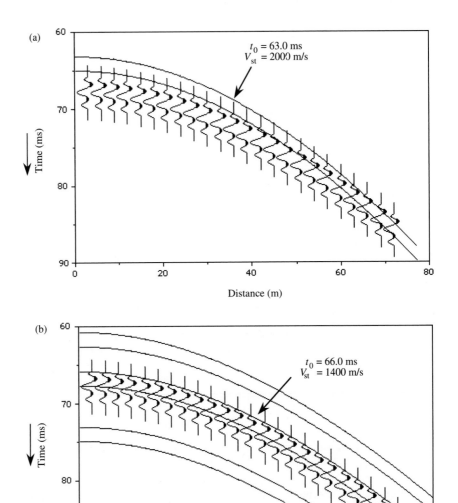

(a)

$t_0 = 63.0$ ms
$V_{st} = 2000$ m/s

Time (ms)

Distance (m)

(b)

$t_0 = 66.0$ ms
$V_{st} = 1400$ m/s

Time (ms)

Distance (m)

Figure 4-42 Velocity analysis method of comparing waveforms within a hyperbolic curve window defined by trial values of t_0 and V_{st}. (a) Trial values define a curve which cuts across reflection arrivals and for which comparison of waveform within the window would return a low value. (b) Windows defined using the correct V_{st} and three values of t_0, one of which is correct. Comparison of waveform within the correct window clearly would meet with success and result in a high value.

tions that they are applied to field data collected using a variety of field designs, most modern processing strategies utilize common-depth-point data. Recall that we gathered together traces representing reflections from the same depth point. Now that static and NMO corrections have been applied to these gathers, waveforms representing reflection arrivals should line up across the record if the reflecting surface is horizontal (Fig. 4-43). If these traces are summed together (or *stacked*), reflection arrivals will be enhanced, whereas noise and other arrivals (such as refractions) in general should be degraded.

The stacked traces can now be presented in their correct position (for vertical travel paths), each at the position of their common reflecting point, in order to create a seismic section. If diffractions are minimal and interfaces have only low dips or are horizontal, for many routine shallow investigations processing activities may end at this stage. After all, we have a reasonable representation of the subsurface in time and have the values required to compute interval velocities and thicknesses. However, there are a number of additional processing stages available to improve the data still further. The time and cost of applying these additional techniques often is not justified for shallow work of limited extent and duration, but usually is justified when deeper targets with greater rewards are sought.

Migration

In our discussion of the production of seismic sections we assumed that reflecting points were directly beneath the zero-offset position. However, if the reflecting horizon is inclined, this will not be the case. Consider Figure 4-44(a). Here the reflector dips to the right, and the actual zero-offset path is indicated by the solid black line with arrowheads at each end. When plotted in the conventional manner, the true reflecting point assumes the position identified as "apparent reflecting point." Although this clearly is not a trivial matter, its seriousness is underscored by Figure 4-44(b).

First consider the reflecting surface inclined to the right. The apparent reflecting points related to this reflector define a line (dashed in the diagram) that is displaced upward and to the right compared to the real reflector and is inclined at a different angle. If we add

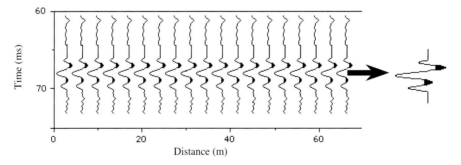

Figure 4-43   CDP gathers corrected for NMO and then stacked to yield waveform of enhanced amplitude.

(a)

(b)

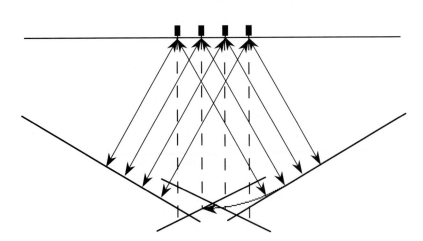

Figure 4-44 Displacement of true reflecting points when zero-offset traces are plotted directly beneath receiver. (a) Diagram illustrating that displacement of reflecting point is about a circular arc and equal in degrees to dip of reflecting horizon. (b) Diagram illustrating same displacements as in (a) but demonstrating how false images can be produced when reflecting points are not plotted in their correct position.

a reflector dipping to the left, as would be the case if a syncline was present, this surface also is displaced. The displacement of one point is illustrated by an arc with an arrowhead attached. The combination of the two apparent displacements creates a pattern of two crossing lines that possibly could be mistaken for an anticline and, at the very least, obscures the true structure.

We need to discover a way to reposition these apparent reflecting points to their true positions. This process is referred to as *migration*. Modern migration techniques are computer based and are extremely sophisticated. Because of this and because migration is not normally attempted in much shallow work, we do not pursue the theory and operation of migration in this text. Several of the references in the Suggested Reading at the end of this chapter contain excellent summaries of this topic. Our goal in the next few paragraphs is to demonstrate two straightforward approaches to migration that formed the basis of techniques used to migrate sections before computers assumed this task.

Consult Figure 4-44(a) once again. Both lines with arrowheads at each end are the same length. Given a surface receiving point and a reflection time, we know that the reflecting point must be located somewhere on a circular arc with radius of length $V_1(t_0/2)$. If we know the dip β of the reflecting surface, simple geometry demonstrates that the displacement is through the angle β.

Consider Figure 4-45(a). If we take each apparent reflecting point and move it along a circular arc, we see that a line tangent to all these arcs defines the true position of the reflector. While such a procedure is applicable only in the simplest of cases, it does emphasize that true reflecting surfaces can be reconstructed.

Figure 4-45(b) illustrates how the apparent dip of a reflector can be determined using the apparent reflecting points. Once the dip is known, the position of the actual point is determined by calculating values for m and y. In this illustration lines with arrowheads at

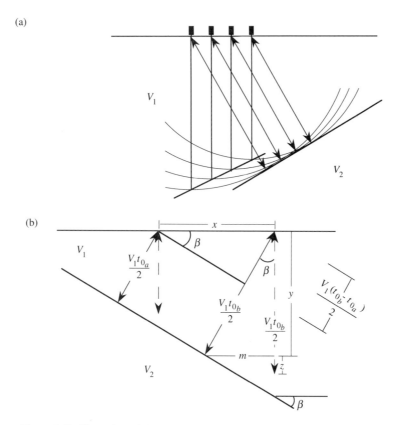

Figure 4-45 Illustrations of simple manual (noncomputer) methods for migrating reflecting points. (a) Migrating about circular arcs. (b) Geometric method to compute horizontal and vertical displacement of reflecting point.

each end follow the same convention as in Figure 4-44. As the vertical zero-offset path and the actual zero-offset path both have the same length, we know the lengths of both the reflection paths illustrated. We can determine from the diagram that the difference in path lengths is related to the dip of the reflector by Eq. 4-67.

$$\sin \beta \; = \; \frac{V_1 \left(t_{0_b} - t_{0_a} \right) \big/ 2}{x} \qquad\qquad (4\text{-}67)$$

Given β, it is trivial to compute y, m, and z. Thus, knowing the position of the apparent reflecting point and having calculated m and z, we establish the position of the actual reflecting point.

In addition to placing inclined reflectors in their correct spatial positions and orientations, migration typically attempts to remove diffractions as well, We have already examined diffractions and realize how they often obscure reflections. They are relatively easy to identify visual, because of their typical hyperbolic form on seismic sections. Diffraction migration takes advantage of the special geometry of diffractions by employing computer algorithms that search for diffraction patterns and then remove them in a way that affects the surrounding waveforms as little as possible.

We are confident you realize by now that migrating a seismic section is possible, although exceedingly complex in practice, and that a migrated section is much superior to an unmigrated one for interpreting subsurface relationships. It also likely is evident that migration is costly in terms of computer processing time and in developing programs to perform the task. Thus, it only is attempted where the value to be gained outweighs the cost of the process. For those of you inclined to pursue migration in greater detail, a particularly readable summary with examples is presented by Waters (1981).

Waveform Adjustments

A number of additional operations typically are performed on seismic traces prior to final presentation as a seismic section. Most of these may be grouped into three categories which we refer to as *amplitude adjustments, frequency adjustments*, and *transmission adjustments*.

In the shallow subsurface especially, variations in transmission efficiencies due to pockets of heterogeneous materials and variations in ground coupling of geophones can result in marked differences in amplitudes of reflections from the same horizon. This is especially true for the common-offset method, as the shot occupies a different position for each trace. These variations may make reflections more difficult to identify or the seismic section less easy to interpret, because it becomes more difficult to trace individual reflections. It is a relatively simple matter to construct a computer program that compares waveforms within a narrow time window from trace-to-trace and then normalizes these trace-to-trace amplitudes. Also, arrivals from deep reflectors or interfaces with moderate reflection coefficients tend to have reduced amplitudes. It is again a straightforward matter to specify time windows in which amplitudes on all traces should be increased or decreased by a specified amount. Although a relatively crude approach, it often is remarkably effective in increasing the readability of shallow sections.

Frequency adjustments involve filtering of the digital seismic record. You likely recall our discussion of filtering of the incoming seismic signal to remove unwanted frequencies so that reflections would not be obscured as amplifiers became saturated with high-amplitude, low-frequency signals. Much of this effort was directed at removing or at least reducing ground roll with a large low-frequency component in order to preserve higher-frequency reflection signals. Of course, once we employ filters prior to recording data, the part of the signal removed is lost forever. As seismographs typically used for shallow exploration become more sophisticated and as digital memory limitations vanish, virtually all of the incoming signal can be preserved. We then apply digital filters to remove unwanted frequencies and to thereby enhance signals of interest. The beauty of this approach is that no information is lost in the filtering process, because the original digital record always is available. Thus, various filter combinations can be applied in order to achieve the best possible results for a particular geologic situation.

Transmission adjustments represent attempts to remove certain effects that altered the shape of wavelets as they moved through the subsurface. Two simple examples serve to emphasize the goals of this processing task. Recall that higher-frequency components of a seismic signal are removed at a greater rate than low-frequency components as the signal is transmitted through the earth. One type of transmission adjustment, therefore, is to attempt to restore the lost higher-frequency components. Because in frequency filtering we attempted to remove components, this attempt to restore components typically is referred to as *inverse filtering*. Another example involves short-path multiples. In our discussion of multiple reflections we noted that the main effect of short-path multiples is to lengthen the seismic wavelet. If lengthened wavelets appear at a number of positions on a trace, weaker reflections could be hidden or difficult to identify with confidence. Filtering these extended wavelets to remove all but the true reflection pulse clearly is a valuable operation. Yet another adjustment attempts to remove long-path multiples which, when strong, also obscure reflections occurring late in time on the trace.

Seismic Sections: Time Sections and Depth Sections

After some or all of the previous corrections and adjustments to seismic data occur, the final seismic section is plotted and interpreted. The section can be presented with time increasing downward, which is referred to as a time section (Fig. 4-46). In shallow exploration work, time sections are the norm, as the goals of the survey often do not warrant developing a *depth section*. A time section provides a general view of subsurface relationships and when combined with a depth scale based on a velocity analysis, may suffice. In surveys where the main interest lies within the saturated sediment column above bedrock, the processing required to convert time sections to depth sections may not be warranted, because velocities within the saturated sediments tend to be quite similar, and, hence, the spacing between reflectors on the seismic sections remains constantly related to actual thicknesses.

In situations where velocities vary with depth or may vary laterally, depth sections should always be prepared. This, of course, requires detailed knowledge of velocity values, both vertically and along the bearing of the seismic line. However, since computer-based

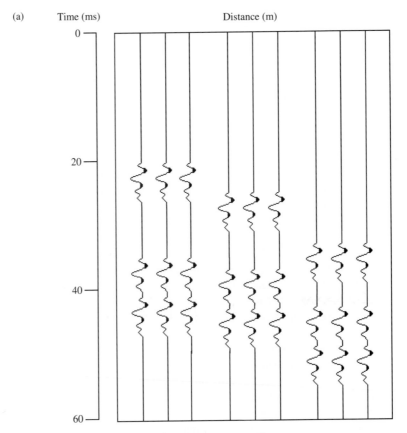

Figure 4-46   An illustration of time versus depth sections. (a) A time section in which thicknesses of all units remain constant but velocities of the upper two units vary laterally. (b) A depth section with all traces corrected according to well-defined velocity functions.

velocity analysis is used to prepare the seismic section by removing NMO effects in order to stack CDP gathers, there is little excuse for not applying the next step and preparing depth sections. Figure 4-46 attempts to demonstrate the value of depth sections compared to time sections. Figure 4-46(a) is a schematic of several traces with well-defined reflection arrivals. There appear to be three strong reflectors. If one makes an interpretation of these patterns based on the positions of the reflectors in time, it appears that the upper unit thickens to the right, because the first reflections arrive progressively later moving from left to right. It also appears that the second unit thins to the right, because the time interval between the earliest and next-earliest decreases to the right. Based on the difference in time between the first and second reflection arrivals and the second and third arrivals, the second unit appears considerably thicker than the third unit. Reflector patterns also suggest that the base of the third reflector is significantly deeper to the right.

None of this is true. In fact, unit boundaries are horizontal and unit thicknesses are constant across the entire section. Unit two and unit three are equal in thickness. Depth to

(b)   Depth (m)   Distance (m)

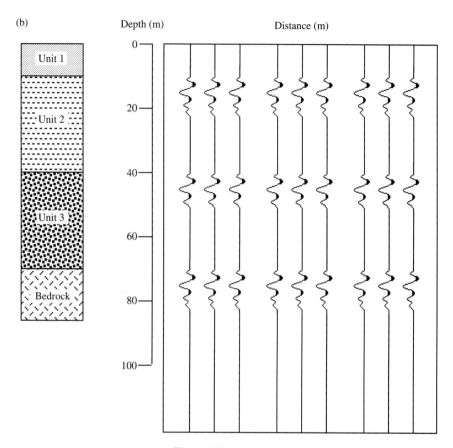

Figure 4-46 *(continued)*

bedrock remains constant. Each unit possesses a different velocity, which is greater than that of the overlying unit. The reason for the displacements of the reflections in time from left to right on the time section is a variation in the velocities of the upper two units. Velocities in the first unit range from 500 m/s for the first three traces to 400 m/s for the next three traces to 300 m/s for the three traces farthest to the right. Velocities in the second unit increase from left to right from 2000 m/s to 2500 m/s to 3000 m/s. Although these are simple simulated seismograms, the effects portrayed occur all too frequently in the field. When one needs to make decisions based on depths, unit thicknesses, and the patterns of thicknesses, it is critical to have a depth section with which to work.

APPLYING THE SEISMIC REFLECTION METHOD

Although *routine* shallow reflection work dates only from the latter half of the 1980s, a number of important applications already exist in the literature. Examples already examined in this chapter emphasize the applicability of the method to detailed mapping of bedrock topography and, when conditions are good, the stratigraphy of unconsolidated materials overlying bedrock. These relatively straightforward applications clearly contain a great deal of valuable information for engineering and groundwater concerns. Another example of this type of information is presented next.

Shallow-reflection work also possesses considerable value in terms of hazard assessment. Detailed information from the shallow subsurface in zones of recently active faults is very valuable, as are seismic sections from areas with potential of collapse due to sinkhole formation or mining activities. Studies concerned with each of these hazards also are examined in this section.

Whately, Massachusetts

In Chapter 3 we outlined the aquifer contamination problem that Whately faced and noted that refraction data were used to secure alternative groundwater sources primarily through depth-to-bedrock determinations. When the Whately study commenced, shallow-reflection techniques were not in general use, so the method did not contribute to the main study's conclusions. Since that time, however, we have run several reflection lines within Whately. These lines, although not nearly as extensive as those in Figure 4-37, are very similar in appearance. They nicely map bedrock topography and contain distinct reflections in the sediment column. One of these reflections correlates with the base of the clay that delineates the top of the confined aquifer. A study now is underway to map the aquifer geometry on a regional basis in order to ensure a well-planned development of this resource.

Meers Fault, Oklahoma

The Meers fault, located in southwestern Oklahoma, has produced significant surface displacements within the last 2000 years. Although a recognizable scarp marks the trend of the fault, no historical record of activity exists (Lawson et al., 1979). Detailed information from the shallow subsurface might contribute information useful in establishing the degree of hazard the Meers fault poses. For this reason and in order to investigate the potential of shallow reflection surveys in detailing shallow structures, a reflection survey was undertaken by Miller, Steeples, and Myers (1990, p. 18–25). The summary that follows is taken from the results of this work.

The survey used the common-depth-point approach with a 24-channel seismograph, 100-Hz geophones, and analog low-cut filters. A typical geophone spread used geophone intervals of 2 m with a shot offset of 20 m. Data reduction consisted dominantly of static corrections, muting ground roll and the air wave, velocity analysis, NMO corrections, frequency filtering, CDP stacking, and automatic gain control. Figure 4-47 illustrates a typical final record and structural interpretation.

In addition to providing detailed information about the deformation zone surrounding the Meers fault, this survey demonstrated that shallow reflection work can reveal a great deal of structural complexity (see structural interpretation in Figure 4-47). The survey recorded reflections from depths up to 200 m, achieved a minimum bed resolution of 6 m,

Figure 4-47 CDP stack and a line-drawing interpretation of the stacked section for a seismic line across the Meers fault. (From Miller, Richard D., Steeples, Don W., and Myers, Paul B., 1990, Shallow seismic reflection survey across the Meers fault, Oklahoma: Geological Society of America Bulletin, v. 102, p. 18–25.)

and demonstrated that the detection limit for faulting is approximately 2 m. Structural conclusions included demonstrating that the major fault is a high-angle reverse fault, that secondary faulting is present within a depth range of 30 m to 200 m, and that the majority of secondary deformation is within 100 m of the Meers fault scarp. One obvious application of such detailed shallow structural information is to aid in selection of sites for trenching in hopes of determining the displacement history of the fault and the present hazard associated with it.

Cavity Detection

The ability to locate and outline subsurface cavities resulting from dissolution or mining activities promises substantial mitigation of these hazards. Shallow seismic reflection surveys possess this ability, assuming that prominent reflectors are present.

Steeples, Knapp, and McElwee (1986, p. 295–301) were able to recognize sinkholes

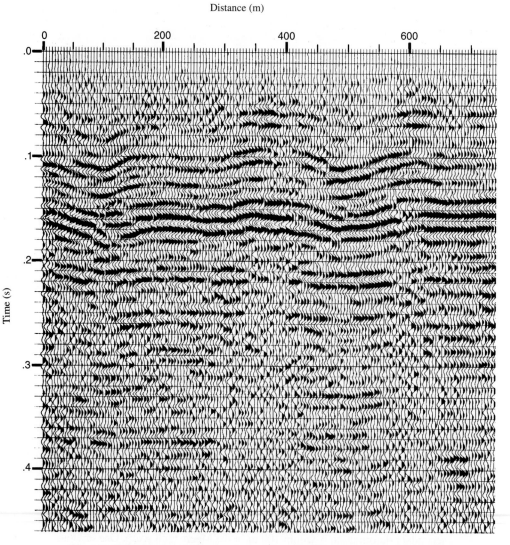

Figure 4-48   Seismic reflection record section illustrating disruptic of strata due to sinkhole formation, Reno County, Kansas. Disturbed stratification is evident for section times less than 170 ms. Sinkhole formation is due to salt dissolution. The base of the salt unit is at 210 ms. (Courtesy Richard D. Miller and Don W. Steeples, Kansas Geological Survey.)

produced by dissolution of the Hutchinson Salt along Interstate Highway 70 in Russell County, Kansas. Subsidence associated with sinkhole formation tended to be gradual and caused the highway to be twice rebuilt at sinkhole sites. However, a catastrophic collapse at a salt-water disposal well associated with a nearby oil field heightened public concern. In order to better define this hazard a seismic reflection survey was conducted in the area. The survey demonstrated that it was possible to recognize sinkholes by displaced strata. Because reflections were secured from the top and bottom of the salt unit, changes in thickness due to solution could be determined. This determination delineated areas where continued sinkhole development and/or growth could be expected. These areas can now be monitored more closely in order to reduce the degree of hazard. Another area of sinkhole formation in Kansas due to the same process is illustrated in Figure 4-48. Because dissolution at this site also is beneath a major highway, the reflection record is invaluable in defining areas of greatest hazard.

Collapse of roofs in shallow coal mines constitutes an acute hazard in many coal-mining regions. Unfortunately, the location of many abandoned mines is unknown, and, therefore, the degree of the hazard cannot be ascertained nor hazard-reduction efforts undertaken. Branham and Steeples (1988) used shallow reflection techniques to locate abandoned mines by tracing reflections from a coal unit. Areas where reflections from the coal are absent represent voids formed from removal of the coal by mining activities. The capability

Figure 4-49 CDP seismic section illustrating locations of faults cutting coal seams. (From Gochioco, Lawrence M., and Cotten, Steven A., 1989, Locating faults in underground coal mines using high-resolution seismic reflection techniques: Geophysics, v. 54, p. 1521–1527.)

to record coal reflections from depths as shallow as 10 m promises a relatively inexpensive approach to quantifying the mine hazard.

In addition to cavities, faults cutting coal seams obviously constitute a serious hazard. Gochioco and Cotten (1989) demonstrated that shallow reflection work is capable of finding such faults. Knowledge of the location of these faults before mining commences permits an entry system to be planned that increases safety. Figure 4-49 illustrates how these faults appear on a CDP seismic section.

Other Applications

Other informative studies include those by Slaine et al. (1990), who mapped overburden stratigraphy and bedrock relief in the vicinity of a proposed hazardous waste facility; Miller et al. (1990), who delineated an intra-alluvial aquifer's geometry, offset relations of the aquifer and surrounding sediments, and bedrock structural relationships; and Pullan and Hunter (1990) who illustrated the utility of the shallow seismic reflection technique in mapping a number of buried bedrock valleys.

PROBLEMS

4-1 Refer to Figure 4-50(a). Determine the thickness and velocity of material above the first reflecting horizon.

4-2 A prominent reflection is indicated in Figure 4-50(b). What is the velocity and thickness of material above this reflecting horizon?

4-3 Figure 4-51 illustrates a center-shot, wiggle-trace seismogram. The shot offset is 34 m, and the geophone interval is 48 m. Three prominent reflections are indicated by arrows. Assuming these are the only reflections present, determine rms velocities, interval velocities, and unit thicknesses.

4-4 Using the following time-distance values, interpret the subsurface geology as completely as possible.

Distance from shot (m)	Reflection 1 (ms)	Reflection 2 (ms)	Reflection 3 (ms)
3	21.4	62.3	79.4
6	25.0	62.4	79.5
9	30.1	62.6	79.6
12	36.1	62.9	79.9
15	42.5	63.2	80.1
18	49.2	63.6	80.5
21	56.2	64.1	80.9
24	63.3	64.7	81.3
27	70.4	65.4	81.8
30	77.6	66.1	82.4
33	84.9	66.9	83.0
36	92.2	67.7	83.7

(a)

Reflections

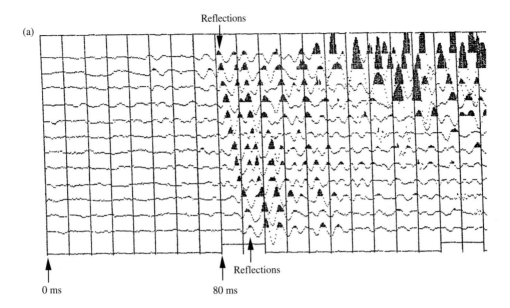

Reflections

0 ms   80 ms

(b)

Water-table refraction   Reflections

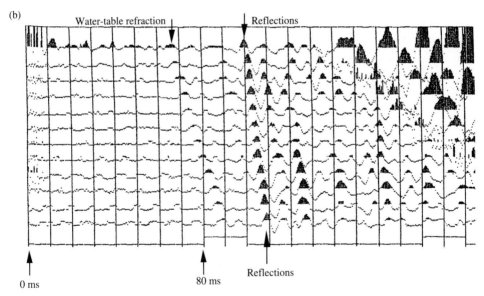

0 ms   80 ms   Reflections

Figure 4-50 (a) Field seismogram recorded with a 40-m offset and a 3-m geophone interval. (b) Field seismogram recorded with a 50-m offset and a 3-m geophone interval.

(a)

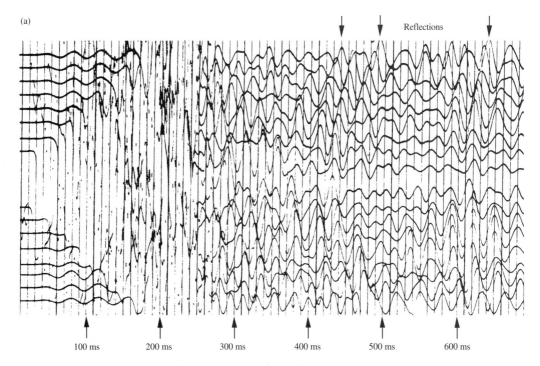

Figure 4-51   (a) Center-shot, wiggle-trace seismogram. Shot offset is 34 m; geophone interval is 48 m. Reflections for analysis indicated by downward-pointing arrows. (b) Enlarged view of (a) to aid in picking reflection times.

4-5   Using the following time-distance values, interpret the subsurface geology as completely as possible.

Distance from shot (m)	Reflection 1 (ms)	Reflection 2 (ms)	Reflection 3 (ms)
3	40.5	102.5	116.8
13	47.7	103.1	117.2
23	61.0	104.4	118.0
33	77.2	106.3	119.2
43	94.9	108.9	120.9
53	113.3	112.1	123.0
63	132.2	115.8	125.5
73	151.4	120.0	128.4
83	170.8	124.6	131.6
93	190.3	129.7	135.1
103	209.9	135.1	138.9
113	229.5	140.8	143.0

(b)

Figure 4-51   *(continued)*

4-6   Interpret the time-distance data illustrated in Figure 4-52(a).

4-7   A target horizon in a reflection survey is believed to be at an approximate depth of 35 m. Explain why the maximum shot-geophone distance should not exceed 35 m. Include two different reasons in your explanation.

(a)

(b)

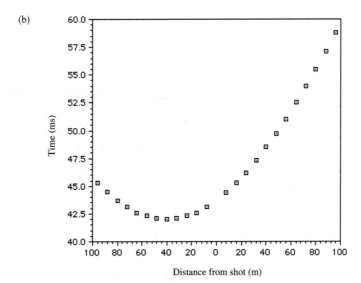

Figure 4-52   Reflection time-distance data. (a) Data for Problem 6. (b) Data for Problem 10.

4-8 Use both the t_{min}/t_0 and the $x^2 - t^2$ methods to interpret the following time-distance data from a dipping reflector.

Distance from Shot (m)	Reflection Time (ms)
36	22.10
33	21.44
30	20.80
27	20.21
24	19.66
21	19.15
18	18.68
15	18.27
12	17.92
9	17.62
6	17.38
3	17.21
0	shot
3	17.06
6	17.09
9	17.18
12	17.34
15	17.56
18	17.84
21	18.19
24	18.59
27	19.04
30	19.54
33	20.08
36	20.67

4-9 Although the record illustrated in Figure 4-53 was recorded under noisy conditions, the prominent reflection is reasonably obvious. Determine the thickness and velocity of the material above the reflecting horizon. Develop a method to assess the degree of uncertainty in your results due to the difficulty in determining the correct arrival time for the reflection at each geophone.

4-10 Using dip move-out, interpret Figure 4-52(b).

4-11 An 80-m thick sequence of saturated sands with a velocity of 1350 m/s overlies a 40-m thick sequence of till with velocity of 2800 m/s. The till is in contact with bedrock which has a velocity of 5500 m/s. Determine the optimum window when the target is the bedrock reflection.

4-12 Calculate the depth to bedrock and the velocity of the sediments above bedrock using the field record in Figure 4-26(b).

Reflections ↓

Reflections

0 ms   80 ms

Figure 4-53   Field seismogram recorded with a 36-m offset and a 3-m geophone interval.

4-13   Construct a travel-time curve for the accompanying model. Show reflection times for all geophones and diffractions from the upper edge of the step.

4-14 Calculate elevation and weathering corrections for the three geophone locations in the accompanying sketch.

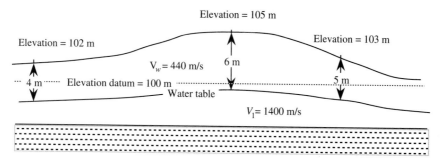

Elevation = 105 m

Elevation = 102 m

Elevation = 103 m

$V_w = 440$ m/s

6 m

... 4 m ...... Elevation datum = 100 m

5 m

Water table

$V_1 = 1400$ m/s

4-15 Determine V_{st} and t_0 for each of the reflections in the simulated seismogram. The correct arrival times are listed below the seismogram. Use Table 4-11.

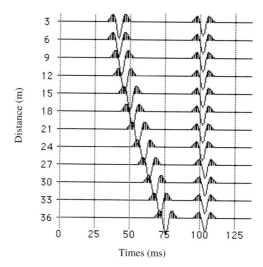

Distance (m)

3
6
9
12
15
18
21
24
27
30
33
36

0   25   50   75   100   125

Times (ms)

Distance (m)	Time (ms)	Distance (m)	Time (ms)
3	33.71	3	93.35
6	34.80	6	93.42
9	36.55	9	93.53
12	38.87	12	93.68
15	41.67	15	93.87
18	44.85	18	94.10
21	48.33	21	94.38
24	52.07	24	94.69
27	56.00	27	95.05
30	60.09	30	95.45
33	64.31	33	95.89
36	68.64	36	96.37

4-16 Perform migration on the traces illustrated in the following seismic section in order to portray subsurface relationships in their true form.

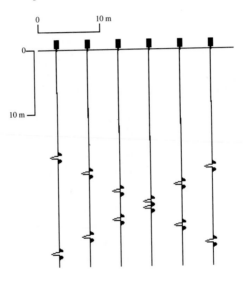

REFERENCES CITED

Branham, K. L., and Steeples, D. W., 1988, Cavity detection using high-resolution seismic reflection methods: Mining Engineering, v. 40, p. 115–119.

Dix, C. Hewitt, 1955, Seismic velocities from surface measurements: Geophysics, v. 20, p. 68–86.

Dobrin, Milton B., and Savit, Carl H., 1988, Introduction to geophysical prospecting, 4th ed.: New York, McGraw-Hill, 867 p.

Green, C. H., 1938, Velocity determinations by means of reflection profiles: Geophysics, v. 3, p. 295–305.

Gochioco, Lawrence M., and Cotten, Steven A., 1989, Locating faults in underground coal mines using high-resolution seismic reflection techniques: Geophysics, v. 54, p. 1521–1527.

Hunter, J. A., Pullan, S. E., Burns, R. A., Gagne, R. M., and Good, R. L., 1984, Shallow seismic reflection mapping of the overburden-bedrock interface with the engineering seismograph—some simple techniques: Geophysics, v. 49, p. 1381–1385.

Kearey, P., and Brooks, M., 1984, An introduction to geophysical exploration: London, Blackwell Scientific Publications, 296 p.

Lawson, J. E., Jr., DuBois, R. L., Foster, P. H., and Luza, K. V., 1979, Earthquake map of Oklahoma: Oklahoma Geological Survey Map GM-19, scale 1/750,000.

Miller, Richard D., Steeples, Don W., Hill, Robert W., Jr., and Gaddis, Bobby L., 1990, Identifying intra-alluvial and bedrock structures shallower than 30 meters using seismic reflection techniques *in* Ward, Stanley H., ed., Geotechnical and Environmental Geophysics; Volume 3: Geotechnical: Society of Exploration Geophysicists Investigations in Geophysics, v. 5, p. 89–97.

Miller, Richard D., Steeples, Don W., and Myers, Paul B., 1990, Shallow seismic reflection survey across the Meers fault, Oklahoma: Geological Society of America Bulletin, v. 102, p. 18–25.

Pullan, S. E., and Hunter, J. A., 1985, Seismic model studies of the overburden-bedrock reflection: Geophysics, v. 50, p. 1684–1688.

Pullan, S. E., and Hunter, J. A., 1990, Delineation of buried bedrock valleys using the optimum offset shallow seismic reflection technique *in* Ward, Stanley H., ed., Geotechnical and Environmental Geophysics; Volume 3: Geotechnical: Society of Exploration Geophysicists Investigations in Geophysics, v. 5, p. 75–87.

Robinson, Edwin S., and Coruh, Cahit, 1988, Basic exploration geophysics: New York, John Wiley, 562 p.

Shah, Pravin M., and Levin, F. K., 1973, Gross properties of time-distance curves: Geophysics, v. 38, p. 643–656.

Sheriff, Robert E., 1989, Geophysical methods: Englewood Cliffs, New Jersey, Prentice Hall, 605 p.

Sheriff, R. E., and Geldart, L. P., 1982, Exploration seismology; Volume 1: History, theory, and data acquisition: Cambridge, Cambridge University Press, 253 p.

Sheriff, R.E., and Geldart, L.P., 1983, Exploration seismology; Volume 2: Data processing and interpretation: Cambridge, Cambridge University Press, 221 p.

Slaine, David D., Pehme, Peeter E., Hunter, James A., Pullan, Susan E., and Greenhouse, John P., 1990, Mapping overburden stratigraphy at a proposed hazardous waste facility using shallow seismic reflection methods *in* Ward, Stanley H., ed., Geotechnical and Environmental Geophysics; Volume 2: Environmental and groundwater: Society of Exploration Geophysicists Investigations in Geophysics, v. 5, p. 273–280.

Slotnick, M. M., 1959, Lessons in seismic computing: Tulsa, Oklahoma, Society of Exploration Geophysicists, 268 p.

Steeples, Don W., Knapp, Ralph W., and McElwee, Carl D., 1986, Seismic reflection investigations of sinkholes beneath Interstate Highway 70 in Kansas: Geophysics, v. 51, p. 295–301.

Taner, M. Turhan, and Koehler, Fulton, 1969, Velocity spectra—digital computer derivation and applications of velocity functions: Geophysics, v. 34, p. 859–881.

Waters, Kenneth H., 1981, Reflection seismology, 2nd ed.: New York, John Wiley, 453 p.

SUGGESTED READING

Hunter, J. A., Pullan, S. E., Burns, R. A., Gagne, R. M., and Good, R. L., 1984, Shallow seismic reflection mapping of the overburden-bedrock interface with the engineering seismograph—some simple techniques: Geophysics, v. 49, p. 1381–1385.

Knapp, R. W., and Steeples, D. W., 1986a, High-resolution common-depth-point seismic reflection profiling: Instrumentation: Geophysics, v. 51, p. 276–282.

Knapp, R. W., and Steeples, D. W., 1986b, High-resolution common-depth-point seismic reflection profiling: Field acquisition parameter design: Geophysics, v. 51, p. 283–294.

Sheriff, R. E., and Geldart, L. P., 1982, Exploration seismology, Volume 1: History, theory, and data acquisition: Cambridge, Cambridge University Press, 253 p.

Sheriff, R. E., and Geldart, L. P., 1983, Exploration seismology, Volume 2: Data-processing and interpretation: Cambridge, Cambridge University Press, 221 p.

Steeples, Don W., and Miller, Richard D., 1990, Seismic reflection methods applied to engineering, environmental, and groundwater problems *in* Ward, Stanley H., ed., Geotechnical and environmental geophysics; Volume 1: Review and tutorial: Society of Exploration Geophysicists Investigations in Geophysics, v. 5, p. 1–30.

Waters, Kenneth H., 1981, Reflection seismology, 2nd ed.: New York, John Wiley, 453 p.

5

Electrical Resistivity

INTRODUCTION

As we move from seismic topics to a discussion of electrical methods for exploring the shallow subsurface, two striking differences are the diversity of electrical methods and their relative complexity in deriving quantitative interpretations from field data as compared to seismic methods.

Although there exists a wide range of methods that can be included under the heading of this chapter, these methods group naturally into two categories: (1) those in which current is applied to the earth and (2) natural energy sources.

Applied Currents

In *electrical-resistivity* (ER) methods direct current or low-frequency alternating current is applied at the ground surface, and the potential difference is measured between two points. Variations in resistance to current flow at depth cause distinctive variations in the potential-difference measurements which provide information on subsurface structure and materials.

When the current used in electrical-resistivity surveying is switched off, the voltage between the two points used to determine potential difference does not decline immediately to zero. Rather, a small residual voltage slowly decreases. A similar effect occurs when the current is switched on, suggesting that the current flow has electrically polarized the ground. The character of this *induced polarization* (IP) also reveals information on the subsurface.

A third method utilizing applied currents is *electromagnetic* (EM) surveying. In EM surveying a primary electromagnetic field is produced by passing alternating current through a coil. Conducting bodies beneath the surface generate secondary electromagnetic fields which are detected by a receiver coil. Differences between the primary and secondary fields once again provide information on the presence and characteristics of subsurface conducting bodies.

Natural Currents

The flow of charged particles in the ionosphere due to solar emissions is responsible for alternating currents that flow through the upper regions of the earth. This natural current flow (referred to as *telluric* currents) is altered by the varying conductivity properties of rocks. The telluric method takes advantage of these natural current variations by measuring potential differences at the surface and interprets these differences in terms of subsurface materials. The *magnetotelluric* method is similar to telluric surveying but measures the magnetic field as well as the electrical field.

The self-potential or spontaneous potential (SP) approach utilizes natural electrochemical activity. If an ore body is in contact with solutions possessing different compositions, resultant chemical reactions result in a flow of ions. This leads to potential differences and, thus, to a flow of current. A simplified view is a system behaving as a battery. As current is naturally produced, we need only employ two probes to measure potential differences at points on the surface.

A Brief History

The history of electrical prospecting essentially dates from the 1830s, when Robert W. Fox experimented with natural currents associated with sulfide ore deposits at Cornwall, England. Most of the work during the 1800s dealt only with these types of currents. In the early part of the second decade of the 1900s Conrad Schlumberger in France and Frank Wenner in the United States applied current to the ground and measured the resulting potential difference. These investigations established the direct-current resistivity method.

In 1914 Schlumberger found a rich ore deposit in Serbia using the self-potential method, and in 1917 the electromagnetic method was introduced by H. R. Conklin. Telluric currents were first studied by O. H. Gish and W. J. Rooney in the United States in the early 1920s. This work was continued and expanded upon by the Schlumberger group beginning in 1934. Virtually all this early work was quite naturally associated with ore deposits and their exploration. One of the earliest "noncommercial" applications was mapping of high-resistivity bedrock by I. B. Crosby and E. G. Leonardon in 1928 during an investigation of a proposed dam site.

With the exception of the magnetotelluric method, essentially all electrical methods had been investigated in some fashion by 1930. Since that time, progress has consisted of instrument refinement, development of a solid theoretical basis, and improvement in interpretative methods, primarily, if not entirely, due to enhanced computer capabilities. An especially complete review of the history of electrical prospecting is presented by Van

Nostrand and Cook (1966), from which the preceding comments were selected. As we suggested for a similar history of seismic reflection, reading such a complete history of a particular geophysical method, especially after completing the chapter dealing with the method, is not only satisfying but provides a useful perspective and, in a sense, is a useful review of concepts.

Chapter Goals

When introducing electrical methods in a text with the objectives of ours, we confront a number of problems. First, as you are now aware, several electrical methods exist, and portions of each contain unique theoretical aspects as well as operational procedures. Second, all methods are complex theoretically, and quantitative interpretations are not nearly as easy to achieve compared to the seismic methods with which you are familiar.

To cover electrical methods thoroughly, therefore, would take more room and time than we devoted to seismic methods even though electrical methods are not as widely utilized. A superficial coverage of each method results primarily in confusion, not mastery. In view of these concerns and constraints we will concentrate on one method only—electrical resistivity. Electrical resistivity is probably the most common method applied to shallow subsurface investigations, especially in groundwater studies. At the end of this chapter we provide an outline of the major additional methods in terms of procedure and applications. We hope that, mastery of one method and knowledge of the major applications of the others provide a reasonable solution in discussing electrical methods in an introductory text.

BASIC ELECTRICITY

The electrical-resistivity method works through the measurement of potential differences at points on the earth's surface that are produced by directing current flow through the subsurface. This leads to the determination of resistivity distribution in the subsurface and thence to an interpretation of Earth materials. Because we are working with resistance, current flow, and potentials, a brief review of basic electrical concepts seems warranted.

Figure 5-1 illustrates a basic electrical circuit containing a battery, connecting wires, and a resistor. The battery maintains a potential difference between two points—its positive

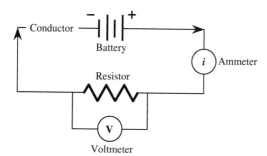

Figure 5-1 A simple electrical circuit illustrating standard symbols for common components.

terminal and its negative terminal. The battery thereby functions as a power source in moving charges through the circuit, much as a pump moves water through pipes. The convention adopted in this text is to define current flow as the movement of positive charges. To accomplish this flow the battery must move positive charges from a low potential at the negative terminal to high potential at the positive terminal. The work done in this potential change requires that a force be applied. This force is known as electromotive force or *emf*. The unit of emf is the *volt* (V).

A 9-V battery maintains a potential difference of 9 V between its terminals and thus has a certain potential for doing work. As noted, the movement of charges through the conducting wire is termed *current*. Specifically,

$$i = \frac{q}{t} \tag{5-1}$$

where i is current in amperes, q is charge in coulombs, and t is time in seconds.

Another and very important concept in electrical resistivity surveying is the current density j. Current density is defined as the current divided by the cross-sectional area of the material through which it is flowing,

$$j = \frac{i}{A} \tag{5-2}$$

Because current is defined as the movement of charges across a given cross-sectional area in a unit of time, maintaining a constant current and reducing the cross-sectional area through which it flows must result in reduced spacing of charges and, therefore, an increased current density. Figure 5-2 attempts to illustrate this relationship diagrammatically using arrows to represent movement of charges.

You recognize, most likely, that copper wire, wood, aluminum, and rubber possess varying resistances to the flow of current. Copper has very low resistance whereas rubber has an extremely high resistance. Resistance is quantified in the following way: 1 *ohm* (Ω) of resistance allows a current of 1 ampere to flow when 1 V of emf is applied.

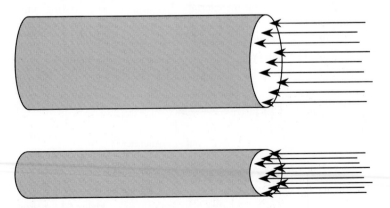

Figure 5-2 Diagram illustrating concept of current density j in wires with different cross-sectional areas. Current flow is represented by arrows.

Ohm's law, first presented by German physicist Georg Simon Ohm, states that current is directly proportional to voltage V and inversely proportional to resistance R, or

$$i = \frac{V}{R} \qquad (5\text{-}3)$$

Consider Figure 5-1. If the battery supplies 9 V, and the resistor has a value of 10 Ω, the current measured by the ammeter will be 0.9 amperes. Or, if resistance is increasing, it will take an increasing voltage to maintain the same current.

Because various geologic materials can be expected to have different resistances to current flow, it might seem fairly straightforward to measure current and voltage in order to measure resistance and to determine the material in the subsurface. One immediate complication is that resistance depends not only on the material but also its dimensions. Consider Figure 5-3. The sketches illustrate two resistors with different lengths l and cross-sectional areas A. If these resistors are constructed from the same material, it seems intuitively obvious that they will not have the same resistance to current flow. Remembering that current flow is the movement of charged particles in a given unit of time, we can use a water analogy. Consider an open pipe in which one section is filled with gravel. A pump creates a pressure difference in the pipe, and water flows. The gravel causes a resistance to flow relative to the open portion of the pipe. If we keep everything the same, but increase the length filled with the gravel, the resistance to flow increases and the flow rate of the water is reduced. If, however, we increase the diameter, the resistance to flow is reduced and more water flows.

This behavior suggests that the resistances of the resistors in Figure 5-3 depend on their length and cross-sectional areas and also to a fundamental property of the material used in their construction, which we term *resistivity* and denote by ρ. Based on our discussion, we can say that

$$R = \rho \frac{l}{A} \qquad (5\text{-}4)$$

or

$$\rho = R \frac{A}{l} \qquad (5\text{-}5)$$

Figure 5-3 Two resistors of different lengths l and different cross-sectional areas A.

The resistivity unit is resistance · length, which is commonly denoted by $\Omega \cdot$ m. Conductance is the inverse of resistance, and conductivity is the inverse of resistivity.

> Copper has a resistivity or $1.7 \times 10^{-8} \, \Omega \cdot$ m. What is the resistance of 20 m of copper wire with a cross-sectional radius of 0.005 m? Quartz has a resistivity of $1 \times 10^{16} \, \Omega \cdot$ m. What is the resistance of a quartz wire of the same dimensions?

CURRENT FLOW IN A HOMOGENEOUS, ISOTROPIC EARTH

Point Current Source

Because the resistivity method consists of applying current and measuring potentials, we begin by considering the potential at a point P_1 when current is applied at a point source C_1. We place the return current electrode at a very great distance and assume material of uniform resistivity ρ. Because air has infinite resistivity, no current flows upward. Thus, current flows radially outward through the earth equally in all directions so as to define a hemispherical surface (Fig. 5-4). Because current distribution is equal everywhere on this surface, which is at a distance r from the current electrode C_1, the potential also is equal. These surfaces are known as *equipotential* surfaces. If we define a very thin shell of thickness dr and employ Eqs. 5-3 and 5-4, we can define the potential difference across the shell to be

$$dV = i(R) = i\left(\rho \frac{l}{A}\right) = i\left(\rho \frac{dr}{2\pi r^2}\right) \qquad (5\text{-}6)$$

We now use Eq. 5-6 to determine the potential at P_1. In determining the potential at a point, we compare it to the potential at a point infinitely far away, which by convention is arbitrarily defined to equal zero. The most direct way to determine V is to integrate Eq. 5-6 over its distance D to the current electrode to infinity, or

$$V = \int_D^\infty dV = \frac{i\rho}{2\pi} \int_D^\infty \frac{dr}{r^2} = \frac{i\rho}{2\pi D} \qquad (5\text{-}7)$$

(Van Nostrand and Cook, 1966, p. 28). Equation 5-7 is the *fundamental* equation in our electrical prospecting discussions. Before using it to develop more practical relationships, let's examine what information we can glean from it, taken by itself. Assuming a resistivity and current, we can map the potential at any point in our section view of Figure 5-4. If we plot a sufficient number of points, we can draw lines through those points possessing the same potential. This defines the equipotential surfaces. Because current flow must be perpendicular to the equipotential surfaces, we can determine the direction of current flow. Of

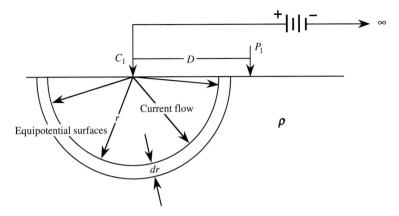

Figure 5-4 Diagram illustrating symbols and configuration used to determine potential at P_1 for a single point source of current C_1. The current sink, C_2, is at infinity. The two equipotential surfaces shown are separated by the distance dr.

course, in this case, we already know what the result will be, but in the next step the pattern will not be so evident.

Two Current Electrodes

Our next step is an attempt to determine the current flow in a homogeneous, isotropic Earth when we have two current electrodes. In this case the current must flow from the positive current electrode (the source) to the negative current electrode (the sink). The path of the current is not as obvious as in our previous discussion, so we once again proceed to determine the potential at a point. From this information we will determine the equipotential surfaces and then the current flow. For simplicity we present the derivation only for the case when the current electrodes and our potential points lie in the same plane (Fig. 5-5). The more general case is quite similar and can be examined in Van Nostrand and Cook (1966, p. 30–31).

The potential at point P_1 is determined by using Eq. 5-7. The effect of the source at $C_1 (+)$ and the sink at $C_2 (-)$ are both considered, and, therefore,

$$V_{P_1} = \frac{i\rho}{2\pi r_1} + \left(-\frac{i\rho}{2\pi r_2} \right) \tag{5-8}$$

Expressing r_1 and r_2 in terms of the x-z-coordinate system illustrated in Figure 5-5, we rewrite Eq. 5-8 as

$$V_{P_1} = \frac{i\rho}{2\pi} \left\{ \frac{1}{\left[\left(\frac{d}{2} + x \right)^2 + z^2 \right]^{1/2}} - \frac{1}{\left[\left(\frac{d}{2} - x \right)^2 + z^2 \right]^{1/2}} \right\} \tag{5-9}$$

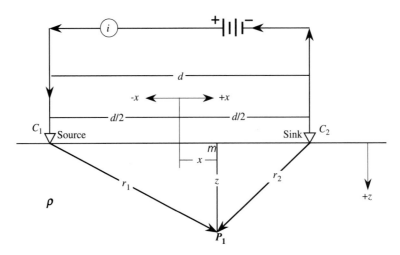

Figure 5-5 Diagram illustrating symbols and configuration used to determine potential at P_1 for a current source C_1 and sink C_2.

Now we need only calculate V_{P_1} at many points in the x-z-plane, draw contours through points of equal potential in order to define the equipotential surfaces, and, finally, draw current flow lines perpendicular to the equipotential surfaces.

Table 5-1 is a dynamic table based on Eq. 5-9. Parameters in the table are related to Figure 5-5. If you examine the values of the potential at the points in the table, you should be able to obtain a sense of the distribution of the equipotential surfaces. Figure 5-6 is a simple contour plot of the values in Table 5-1 with symbols also keyed to Figure 5-5. If you have a computer contouring package available to you, the spreadsheet template on disk named "Table 5-1 contour" is exactly the same as Table 5-1 but is configured to make it simple to transfer this information to a contouring program. Experiment!

The patterns of equipotential surfaces likely are fairly similar to what you would predict. Note that the current electrodes are displaced slightly from the centers of the hemispheres. Many contour lines close to the current electrodes were omitted for clarity, as you can tell if you examine contour line values and the potential values in Table 5-1.

We know that the current flow lines are perpendicular to the equipotential lines, but we do not know the manner in which the current is distributed. The mathematical analysis to determine the current distribution is fairly complicated but results in a simple equation that provides current distribution as a fraction of the total current (Van Nostrand and Cook, 1966, p. 30–34). Along a vertical plane midway between the two current electrodes, the fraction of the total current i_f penetrating to depth z for an electrode separation of d is given by

$$i_f = \frac{2}{\pi} \tan^{-1}\left(\frac{2z}{d}\right) \tag{5-10}$$

Dynamic Table 5-2 utilizes Eq. 5-10, so you can investigate current distribution for various current electrode separations. If you try several values for current electrode separa-

TABLE 5.1 Potential at a Point with Two Current Electrodes

Depth (m)	Horizontal Position (m)																				
	−10	−9	−8	−7	−6	−5	−4	−3	−2	−1	0	1	2	3	4	5	6	7	8	9	10
0	1.06	1.42	2.04	3.32	7.23	##	7.07	2.98	1.52	0.66	0.00	−0.66	−1.52	−2.98	−7.07	##	−7.23	−3.32	−2.04	−1.42	−1.06
1	1.03	1.36	1.91	2.90	4.91	7.17	4.75	2.57	1.39	0.62	0.00	−0.62	−1.39	−2.57	−4.75	−7.17	−4.91	−2.90	−1.91	−1.36	−1.03
2	0.95	1.22	1.60	2.16	2.85	3.20	2.70	1.85	1.11	0.52	0.00	−0.52	−1.11	−1.85	−2.70	−3.20	−2.85	−2.16	−1.60	−1.22	−0.95
3	0.84	1.04	1.28	1.56	1.82	1.89	1.68	1.28	0.83	0.41	0.00	−0.41	−0.83	−1.28	−1.68	−1.89	−1.82	−1.56	−1.28	−1.04	−0.84
4	0.73	0.86	1.01	1.15	1.25	1.25	1.12	0.89	0.60	0.30	0.00	−0.30	−0.60	−0.89	−1.12	−1.25	−1.25	−1.15	−1.01	−0.86	−0.73
5	0.62	0.71	0.79	0.87	0.90	0.88	0.79	0.63	0.44	0.22	0.00	−0.22	−0.44	−0.63	−0.79	−0.88	−0.90	−0.87	−0.79	−0.71	−0.62
6	0.53	0.58	0.63	0.67	0.67	0.64	0.57	0.46	0.32	0.17	0.00	−0.17	−0.32	−0.46	−0.57	−0.64	−0.67	−0.67	−0.63	−0.58	−0.53
7	0.44	0.48	0.51	0.52	0.52	0.48	0.43	0.34	0.24	0.12	0.00	−0.12	−0.24	−0.34	−0.43	−0.48	−0.52	−0.52	−0.51	−0.48	−0.44
8	0.38	0.40	0.41	0.41	0.40	0.37	0.33	0.26	0.18	0.09	0.00	−0.09	−0.18	−0.26	−0.33	−0.37	−0.40	−0.41	−0.41	−0.40	−0.38
9	0.32	0.33	0.34	0.33	0.32	0.29	0.25	0.20	0.14	0.07	0.00	−0.07	−0.14	−0.20	−0.25	−0.29	−0.32	−0.33	−0.34	−0.33	−0.32
10	0.27	0.28	0.28	0.27	0.26	0.23	0.20	0.16	0.11	0.06	0.00	−0.06	−0.11	−0.16	−0.20	−0.23	−0.26	−0.27	−0.28	−0.28	−0.27
11	0.23	0.23	0.23	0.22	0.21	0.19	0.16	0.13	0.09	0.04	0.00	−0.04	−0.09	−0.13	−0.16	−0.19	−0.21	−0.22	−0.23	−0.23	−0.23
12	0.20	0.20	0.19	0.19	0.17	0.15	0.13	0.10	0.07	0.04	0.00	−0.04	−0.07	−0.10	−0.13	−0.15	−0.17	−0.19	−0.19	−0.20	−0.20
13	0.17	0.17	0.16	0.16	0.14	0.13	0.11	0.08	0.06	0.03	0.00	−0.03	−0.06	−0.08	−0.11	−0.13	−0.14	−0.16	−0.16	−0.17	−0.17
14	0.15	0.14	0.14	0.13	0.12	0.11	0.09	0.07	0.05	0.02	0.00	−0.02	−0.05	−0.07	−0.09	−0.11	−0.12	−0.13	−0.14	−0.14	−0.15
15	0.13	0.12	0.12	0.11	0.10	0.09	0.07	0.06	0.04	0.02	0.00	−0.02	−0.04	−0.06	−0.07	−0.09	−0.10	−0.11	−0.12	−0.12	−0.13
16	0.11	0.11	0.10	0.10	0.09	0.08	0.06	0.05	0.03	0.02	0.00	−0.02	−0.03	−0.05	−0.06	−0.08	−0.09	−0.10	−0.10	−0.11	−0.11
17	0.10	0.10	0.09	0.08	0.07	0.06	0.05	0.04	0.03	0.01	0.00	−0.01	−0.03	−0.04	−0.05	−0.06	−0.07	−0.08	−0.09	−0.09	−0.10
18	0.09	0.08	0.08	0.07	0.06	0.06	0.05	0.04	0.02	0.01	0.00	−0.01	−0.02	−0.04	−0.05	−0.06	−0.07	−0.08	−0.08	−0.08	−0.09
19	0.08	0.07	0.07	0.06	0.06	0.05	0.04	0.03	0.02	0.01	0.00	−0.01	−0.02	−0.03	−0.04	−0.05	−0.06	−0.06	−0.07	−0.07	−0.08
20	0.07	0.06	0.06	0.05	0.05	0.04	0.03	0.03	0.02	0.01	0.00	−0.01	−0.02	−0.03	−0.03	−0.04	−0.05	−0.05	−0.06	−0.06	−0.07

Resistivity (ohm-m)	**50**
Current (amperes)	**1**
Electrode separation (m)	**10**
x-increment (m)	**1**
z-increment (m)	**1**
(Potential values are in volts)	

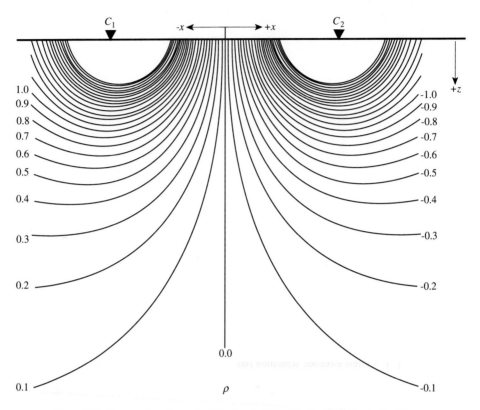

Figure 5-6 Contours based on potential values in Table 5-1. Contours represent positions of equipotential surfaces about a current source C_1 and sink C_2. Many contours near the current electrodes are not shown due to the close spacing of contours in this region.

tion, you should observe that the values in columns two and three in Table 5-2 do not change. A graph of these relationships is illustrated in Figure 5-7. Examining these relationships allows us to arrive at certain conclusions. Fifty percent of the current is confined above a horizontal plane with a depth of one-half the current electrode separation. Seventy percent of the current is confined above a depth equal to the electrode separation (see shaded lines in Fig. 5-7). Clearly then, the greater the electrode separation, the greater the depth to which a given percentage of current penetrates. Make a firm mental note, however, that this relationship is for a homogeneous, isotropic subsurface. Once we introduce a discontinuity, we must reexamine the current distribution question.

Now we can take the information that Eq. 5-10 supplies and combine it with Figure 5-6 in order to produce the current flow lines that are diagramed in Figure 5-8. We will return several times to these current flow lines, their spacing, and their distribution, so be sure to grasp what they represent and how they were derived.

TABLE 5-2 Percent Current Penetrating a Homogeneous, Isotropic Earth

Depth (m)	Depth/Electrode Separation	% of Total Current
1	0.1	13
2	0.2	24
3	0.3	34
4	0.4	43
5	0.5	50
6	0.6	56
7	0.7	61
8	0.8	64
9	0.9	68
10	1	70
11	1.1	73
12	1.2	75
13	1.3	77
14	1.4	78
15	1.5	80
16	1.6	81
17	1.7	82
18	1.8	83
19	1.9	84
20	2	84

Current electrode separation (m)	**10**

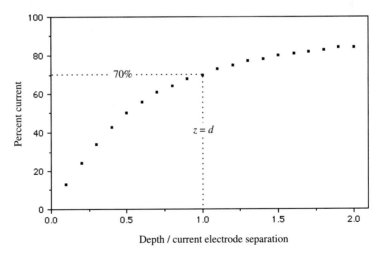

Figure 5-7 Plot of results in Table 5-2. Depth is z and current electrode separation is d. The data points illustrate the extent to which current penetrates into a homogeneous, isotropic Earth.

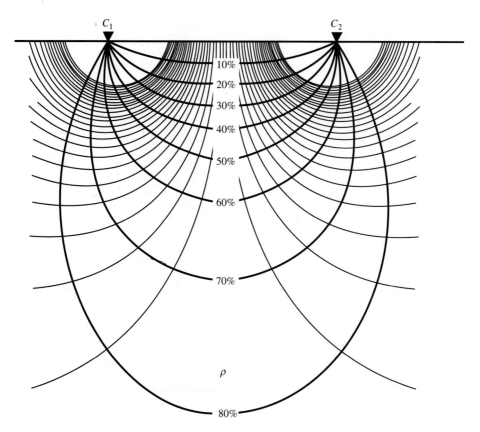

Figure 5-8 Equipotential surfaces and current lines of flow. Labels indicate percent of total current that penetrates to the depth of the line.

Two Potential Electrodes

In electrical-resistivity surveying our goal is to measure the potential difference between two points just as we often make this measurement in electrical circuits. Examine Figure 5-9 and compare it to Figure 5-5. Figure 5-9 illustrates two potential electrodes P_1 and P_2 that are located on the surface as are the current electrodes. Using the equation we have already derived to determine the potential at a point due to a source and a sink, we obtain the potential difference by determining the potential at one potential electrode P_1 and subtracting from it the potential at P_2. Using Eq. 5-8 we determine that

$$V_{P_1} = \frac{i\rho}{2\pi r_1} - \frac{i\rho}{2\pi r_2} \qquad (5\text{-}11)$$

and

$$V_{P_2} = \frac{i\rho}{2\pi r_3} - \frac{i\rho}{2\pi r_4} \qquad (5\text{-}12)$$

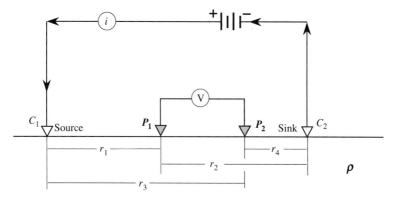

Figure 5-9   Diagram used to determine potential difference at two potential electrodes P_1 and P_2.

Therefore, the potential difference ΔV equals

$$\Delta V = V_{P_1} - V_{P_2} = \left(\frac{i\rho}{2\pi r_1} - \frac{i\rho}{2\pi r_2} \right) - \left(\frac{i\rho}{2\pi r_3} - \frac{i\rho}{2\pi r_4} \right) \tag{5-13}$$

or

$$\Delta V = \frac{i\rho}{2\pi} \left(\frac{1}{r_1} - \frac{1}{r_2} - \frac{1}{r_3} + \frac{1}{r_4} \right) \tag{5-14}$$

In the resistivity method, current is entered into the ground, potential difference is measured, and resistivity determined. Because resistivity is the unknown quantity we normally hope to determine, we solve Eq. 5-14 for ρ and obtain

$$\rho = \frac{2\pi\Delta V}{i} \left(\frac{1}{\dfrac{1}{r_1} - \dfrac{1}{r_2} - \dfrac{1}{r_3} + \dfrac{1}{r_4}} \right) \tag{5-15}$$

Perhaps we should test our understanding of Eq. 5-15 by applying it to a known situation. Let's assume we can place potential electrodes anywhere along the surface, as illustrated in Figure 5-9. Further, we will use the values in Table 5-1 for our test. Figure 5-10(a) presents one possible measurement, and Figure 5-10(b) another. Substituting the values in Figure 5-10(a) produces Eq. 5-16, which results in a resistivity value of 50 $\Omega \cdot$ m. A glance at the model values used to produce Table 5-1 confirms that the resistivity is 50 $\Omega \cdot$ m.

$$\rho = \frac{2\pi\,(4.5\,\text{V})}{1\;\text{ampere}} \left(\frac{1}{\dfrac{1}{3\;\text{m}} - \dfrac{1}{7\;\text{m}} - \dfrac{1}{8\;\text{m}} + \dfrac{1}{2\;\text{m}}} \right) = 50\;\Omega \cdot \text{m} \tag{5-16}$$

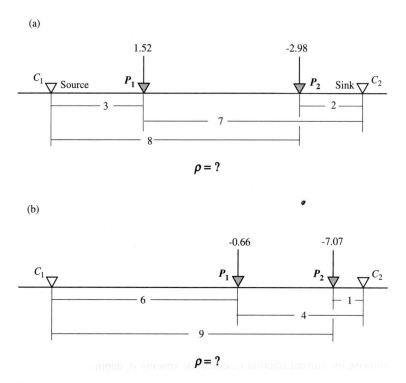

Figure 5-10   Measuring potential difference to determine ρ. Potential values for the indicated spacings are taken from Table 5-1. Solutions are presented in the text (Eqs. 5-16 and 5-17).

Equation 5-17 uses the values illustrated in Figure 5-10(b). As expected, evaluation of the equation gives $50 \ \Omega \cdot m$ as the resistivity.

$$\rho = \frac{2 \pi \left(6.41 \text{V}\right)}{1 \text{ ampere}} \left(\frac{1}{\dfrac{1}{6 \text{ m}} - \dfrac{1}{4 \text{ m}} - \dfrac{1}{9 \text{ m}} + \dfrac{1}{1 \text{ m}}} \right) = 50 \ \Omega \cdot m \qquad (5\text{-}17)$$

These calculations confirm that if we produce a current, measure the electrode spacings, and determine the potential difference, we can arrive at a value for the resistivity of the subsurface materials. However, up until this point we have assumed that the subsurface is a constant resistivity. Just as in our development of the seismic method, it is now time to introduce a horizontal surface that separates materials with different resistivities.

A SINGLE HORIZONTAL INTERFACE

We have already noted that derivation of equations for many electrical resistivity problems is mathematically quite complex and will not be attempted in this text. Our strategy in presenting the necessary equations varies. In some cases we simply introduce an important equation and note a reference which you may consult if you are interested and have the mathematical background to follow the derivation. In other cases we present a simplified case or present only part of the derivation if we feel this will help understand the concept. And finally, we may present only a qualitative explanation or couple such an explanation with an equation or with a partial derivation.

Current Distribution

An important goal in this section is to gain a qualitative understanding for the pattern of current distribution in the subsurface when a single horizontal interface separates materials of different resistivities. Our first step toward this goal is to employ an equation that tells us the fraction of the current that penetrates below the interface. This current fraction is given by

$$i_F = \frac{2\rho_1}{\pi\rho_2}(1 + k)\sum_{n=0}^{\infty} k^n \left\{ \frac{\pi}{2} - \tan^{-1}\left[\frac{2(2n + 1)z}{3a} \right] \right\} \qquad (5\text{-}18)$$

where

$$k = \frac{\rho_2 - \rho_1}{\rho_2 + \rho_1}$$

(Van Nostrand and Cook, 1986, p.89).

Although this equation is relatively complex in form, none of the quantities should surprise us: current fraction i_F, electrode spacing a, depth of interface z, and resistivities of the materials above and below the interface ρ_1 and ρ_2. Note that in this equation electrode spacing a refers to the spacing between each electrode in the array (Fig. 5-11(a)), whereas previously we usually referred to the spacing between current electrodes. The relationships that we can deduce from this equation form the basis for a qualitative interpretation of field measurements, so let's take some time to gather information.

Equation 5-18 is sufficiently straightforward to permit entry into a spreadsheet for evaluation, and Table 5-3 is the result. The appearance of this table is slightly different from those we've seen previously, as the equation being evaluated is broken down into several components due to its complexity. The major variables in the equation are placed in the upper left-hand corner. The result is given as percent current and is located in the upper right. The components are

$$k = \frac{\rho_2 - \rho_1}{\rho_2 + \rho_1}$$

$$C = \frac{2\rho_1}{\pi\rho_2}(1 + k)$$

$$k\text{-factor} = k^n$$

$$\text{remainder} = \left\{ \frac{\pi}{2} - \tan^{-1}\left[\frac{2(2n + 1)z}{3a} \right] \right\}$$

and

$$\text{result} = k\text{-factor times remainder}$$

(a)

(b)

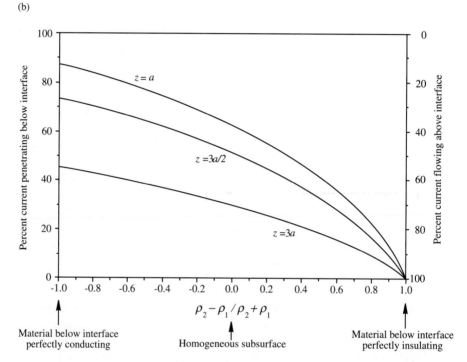

Figure 5-11 Current distribution when a single horizontal interface is present. (a) Electrode configuration. (b) Graph of current distribution. Current distribution is controlled by the relationship between interface depth z, electrode separation a, and the two resistivities ρ_1 and ρ_2.

The numbers in the "Results" column are added together (see the summation symbol in Eq. 5-18) and multiplied by C. This gives the current fraction, which is then presented as the percent current.

    If we select a depth for the interface and for electrode spacing (perhaps in the ratio $z = a$), and vary values for ρ_1 and ρ_2 to produce k-values ranging from -1.0 to 1.0, we can create a curve such as one of the three illustrated in Figure 5-11. All three of these curves were produced using values calculated by Table 5-3. If ρ_1 equals ρ_2, then we have the previous case of the homogeneous subsurface. The results should be the same as predicted by Eq. 5-10 and illustrated in Figure 5-7. If we insert equal resistivities into Table 5-3, place the depth of the interface at 24 m, and elect an electrode spacing of 8 m (which is a current electrode spacing of 24 m), the result in the table tells us that 30 percent of the current penetrates beneath the level of a plane located at a depth of 24 m (see curve $z = 3a$ in Figure 5-11(b) for $k = 0$). Figure 5-7 confirms this by illustrating that 70 percent of current flow is above a plane at a depth equal to the *current* electrode spacing.

Now let's examine the situation when $\rho_2 > \rho_1$. The material below the interface has a higher resistance to current flow (it is a poor conductor relative to the layer above). K-values will be positive for this case. The three curves in Figure 5-11(b) all have the same form, so we can generalize our comments. If we select any positive k-value, we see that substantially *less* current penetrates below the level of the interface compared to the homogeneous case (k-value = 0). In other words, current flow tends to avoid a poor conductor in favor of a good conductor. Perhaps a water analogy will make this phenomenon clearer. Visualize a homogeneous sequence of coarse sand with water flowing through. If at some depth z we then introduce a relatively low permeability layer, less water will flow below that depth level than in the homogeneous case. Of course, the extreme case is when the layer below

TABLE 5-3 Percent Current Penetrating Below a Single Horizontal Interface

Electrode spacing a in m	8		
Depth of interface (m)	8		
ρ_1	70		
ρ_2	30	Percent current	74
(Resistivities are in $\Omega \cdot m$)			

	k	-0.40	
	C	0.89	
n	k-factor	Remainder	Result
0.00	1.00	0.98	0.98
1.00	-0.40	0.46	-0.19
2.00	0.16	0.29	0.05
3.00	-0.06	0.21	-0.01
4.00	0.03	0.17	0.00
5.00	-0.01	0.14	0.00
6.00	0.00	0.11	0.00
7.00	0.00	0.10	0.00
8.00	0.00	0.09	0.00
9.00	0.00	0.08	0.00
10.00	0.00	0.07	0.00
		Sum of results	0.83
		Current fraction	0.74

the interface is perfectly insulating (or completely impermeable), and no current (or water) penetrates below that level.

If $\rho_2 < \rho_1$, the material below the horizontal interface is less resistant to current flow than the material above the interface and is, therefore, a better conductor. As Figure 5-11(b) demonstrates, more current will penetrate below the interface than in the homogeneous case. Using the water analogy once again, we replace our homogeneous material with a coarse gravel that has very high permeability. More water will now flow below the level of the interface and through the gravel. Note that a substantial percentage of the current flows through the material above the interface even if the material below the interface possesses very low resistivities (is almost perfectly conducting).

In our previous discussion of current flow in a homogeneous, isotropic subsurface we learned that as current electrode spacing is increased, the percentage of current flowing at depth is increased. For example, only 30 percent of the current penetrates below a depth z when the electrode spacing is equal to one-third the depth ($z = 3a$). However, 50 percent of the current penetrates deeper than z when the electrode separation is increased to two-thirds the depth ($z = 3a/2$), and 63 percent of the current penetrates below z when the electrode spacing equals the depth ($z = a$). Remember that we are using electrode spacing a as diagrammed in Figure 5-11(a). Thus, when the electrode spacing is equal to the depth of our plane of reference ($z = a$), the current electrode separation is three times the depth ($3z = 3a = d$).

Figure 5-11(b) indicates, as expected, that current penetration increases as electrode separation increases. However, the percentage of current penetrating below an interface is controlled by the relative magnitudes of ρ_2 and ρ_1 as well as by electrode separation. This relationship plays an important role in data acquisition design, so it is a good idea to understand thoroughly the implications of Figure 5-11.

Current Flow Lines and Current Density

The preceding discussion presents us with sufficient information to make a qualitative assessment of current flow lines and, more importantly, current density distribution when a horizontal interface is present. As a first step in this process, we must investigate what happens to the orientation of flow lines and equipotentials when crossing a boundary separating regions of differing conductivities or resistivities. Hubbert (1940, p. 844–846) demonstrated that the flow lines follow a tangent relationship such that

$$\frac{\tan \theta_1}{\tan \theta_2} = \frac{\rho_2}{\rho_1} \tag{5-19}$$

where θ and ρ are as defined in Figure 5-12(a). If the resistivity ρ_2 of the deeper material is greater, then the flow lines bend in toward the normal to the interface (Fig. 5-12(b)) and, as a consequence, are more widely spaced. However, if the reverse is true, as in Figure 5-12(c), the flow lines bend away from the normal, become oriented more parallel to the interface, and are closer together.

If we take this relationship and the information we developed to produce Figure 5-11, we can assess the effect of the horizontal interface. Figure 5-13(a) illustrates the pattern of

(a)

(b)

(c)

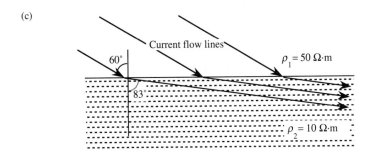

Figure 5-12   Refraction of current flow lines at a boundary separating materials of differing resistivities. (a) Symbols used in Eq. 5-19. (b) Refraction when $\rho_1 < \rho_2$. (c) Refraction
when $\rho_1 > \rho_2$.

current flow lines for a homogeneous subsurface ($\rho_2 = \rho_1$). If we increase the value of ρ_2
(Fig. 5-13(b)), more current flows above the interface, the current flow lines are spaced
more closely, and the current density is greater in the region above the interface relative to
the case of the homogeneous subsurface. If $\rho_2 < \rho_1$, the opposite effect takes place. A
greater percentage of current flows beneath the interface (Fig. 5-13(c)), the current flow
lines are spaced more widely in the material above the interface, and the current density
there is reduced.

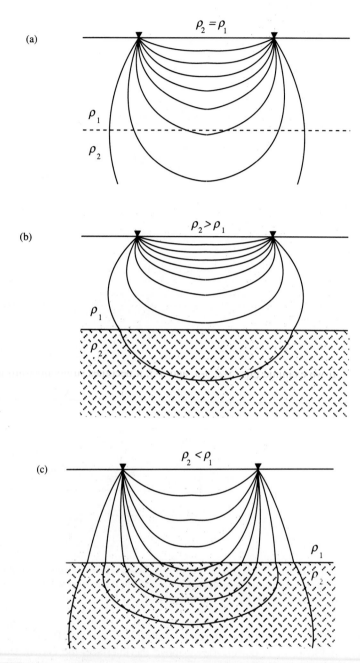

Figure 5-13 Qualitative distribution of current flow lines when a horizontal interface separates materials of differing resistivities. (a) Homogeneous subsurface. Dashed line marks position of horizontal interface in (b) and (c). (b) When material with greater resistivity is located beneath the interface, the current flow lines are more closely spaced than in (a). (c) When material with greater resistivity is located above the interface, the current flow lines are less closely spaced than in (a).

Our real purpose is to determine, qualitatively at the moment, how the presence of the interface affects ΔV readings at the surface. We are just about ready to take this step, but first we need to introduce the notion of *apparent resistivity*.

Apparent Resistivity

When we derived Eq. 5-15, we assumed a homogeneous, isotropic subsurface. As demonstrated previously, any combination of electrode spacings and current results in a potential difference that provides the correct value for the resistivity of the subsurface (as of course should be the case if our equation is correct). Once the subsurface is nonhomogeneous, the value determined for the resistivity is extremely unlikely to equal the resistivity of the material in which the electrodes are inserted. Equation 5-15 thus defines a different quantity, which is termed the *apparent resistivity* ρ_a. Inasmuch as nonhomogeneity is the rule, we write

$$\rho_a = \frac{2\,\pi\Delta V}{i} \left(\frac{1}{\dfrac{1}{r_1} - \dfrac{1}{r_2} - \dfrac{1}{r_3} + \dfrac{1}{r_4}} \right) \tag{5-20}$$

The question we now face is, What does this equation tell us? How do we interpret apparent resistivity values in terms of the subsurface geology?

If we examine Eqs. 5-2, 5-3, and 5-5, it becomes straightforward to demonstrate that the potential is proportional to current density:

$$V \propto j \tag{5-21}$$

When we measure the potential difference between the potential electrodes, the values are proportional to the current density in the cylinder of material near the surface that extends between these two electrodes. If we examine Eq. 5-20 and remember the relationship expressed in Eq. 5-21, we conclude that, all else being equal, variations in current density near the surface will result in variations in apparent resistivity (Mooney, 1958). Let's once again return to Figure 5-13.

In the case of Figure 5-13(a) we already know that Eq. 5-20 will yield a value of $\rho_a = \rho_1$. In the case represented by Figure 5-13(b) the current density is increased in the upper layer relative to the homogeneous case, so that ρ_a will be greater than ρ_1. Conversely, in Figure 5-13(c) the current density is decreased due to the lower resistivity material at depth, and therefore the value of ρ_a will be less than ρ_1. Notice that the electrode spacing remains constant in this example.

Qualitative Development of the Resistivity Pattern over a Horizontal Interface

By now you should have a good grasp of how the horizontal interface affects current flow, current density, and, hence, the resistivity that we measure at the surface (the apparent resistivity). But how do we use this information to explore the subsurface as we did using seismic methods—to determine the depth to the interface and the nature of the geologic materials above and below the interface? As a first step we continue our qualitative approach, but shortly we will develop quantitative methods.

If a horizontal interface is deep relative to electrode spacing a or current electrode spacing d as defined in Figure 5-11(a), it is unlikely that the presence of the interface will have much effect on the current density between the potential electrodes. Therefore, ρ_a will equal or be very close to ρ_1. This relationship is diagramed in Figure 5-14, where dashed lines represent current flow if no interface is present, and solid lines represent current flow due to the interface. In Figure 5-14(a) both sets of lines coincide. If we simply move this electrode array about on the surface, all readings will be the same, assuming the ρ_1-layer is laterally homogeneous. However, if we expand the electrode spacing, more current will flow at depth, and the current density will be affected by the ρ_2-layer. The apparent resistivity value will be greater than ρ_1 but still will be substantially less than ρ_2 (Fig. 5-14(b)). If electrode spacing is increased still more, the effect of the higher resistivity material below the interface will be even greater, as illustrated in Figure 5-14(c). The value of apparent resistivity will continue to increase and will gradually approach the ρ_2-value while becoming increasingly greater than the ρ_1-value.

In many resistivity investigations electrode spacing is plotted against apparent resistivity. If we sketch a curve based on our previous analysis, it should have a form similar to that in Figure 5-15(a). Of course, the actual form will depend on the resistivity contrast, the depth of the interface, and the electrode spacings. At the moment it is important only that you understand why the general shape of the curve is as illustrated. If we reverse the situation so that the material below the interface has a lower resistivity than the material above, the curve assumes the form illustrated in Figure 5-15(b). Various curves for other subsurface situations are illustrated in Mooney (1958), and we encourage you to study them.

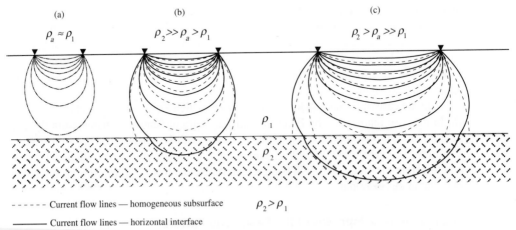

(a)   (b)   (c)

$\rho_a \approx \rho_1$ $\rho_2 \gg \rho_a > \rho_1$ $\rho_2 > \rho_a \gg \rho_1$

ρ_1

ρ_2

- - - - - Current flow lines — homogeneous subsurface $\rho_2 > \rho_1$

———— Current flow lines — horizontal interface

Figure 5-14   Effect on apparent resistivity as electrode spacing is increased. Dashed black lines represent current flow line distribution for homogeneous subsurface. Solid lines represent actual current flow lines due to horizontal interface. As the distance between current electrodes is increased, ρ_a approaches the value of ρ_2.

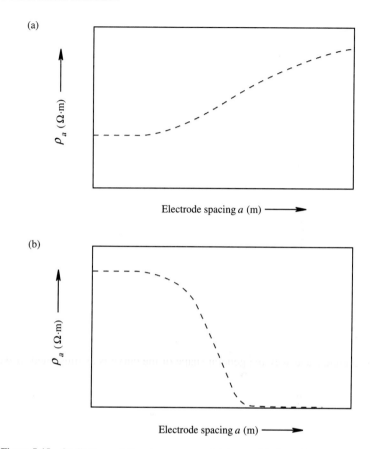

Figure 5-15   Qualitative variations in apparent resistivity ρ_a with electrode spacing a for a single horizontal interface. (a) Material with the greater resistivity lies below the interface. (b) Material with the greater resistivity lies above the interface.

Following the form of the previous discussion, describe why the curve in Figure 5-15(b) appears as it does.

Quantitative Development of the Resistivity Pattern over a Horizontal Interface

There are a number of approaches to solving the quantitative problem of the single horizontal interface. The simplest approach is one using an optical analogue which is suitable for solving a number of the problems we are interested in. We will partially develop the optical analogue method so you can gain some insight into one quantitative approach.

As a first step in the optical analogue approach, we assume a planar boundary which behaves as a semitransparent mirror. If we place a light source at point C_1 and an observation point at P_1, part of the light will travel directly to P_1 and part will be reflected from the mirror (Fig. 5-16). The total light reaching P_1 will be the sum of the two. The amount reflected depends on a property of the mirror referred to as the *reflection coefficient* (k) and is equal to the reflection coefficient times the intensity of the light source. In subsequent development of this approach it is more direct to construct an image of point C_1 on the other side of the mirror. This is the same approach we used in deriving equations for reflected rays in Chapter 4. Recall that C_1 and its image are equidistant from the boundary (mirror) and lie on a line that is perpendicular to the boundary. Recall also that the distance r_2 is equal to the reflected path. If our observation point is at P_2 rather than P_1, the amount of light reaching P_2 is only that which is transmitted through the mirror. This amount is due to

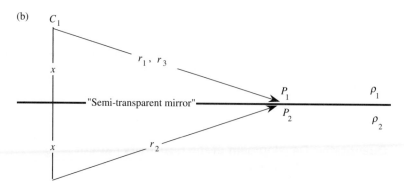

Figure 5-16   Determination of potential at points P_1 and P_2 using an optical analogue for a case when a single interface separates two regions of different resistivities. (a) General case. (b) Case when P_1 and P_2 are at the same position on the boundary.

the intensity of the light source, the *transmission coefficient* (1-k), and the distance of P_2 from C_1.

In order to develop the electrical case we place a current source at C_1 with intensity i and develop equations for the potential at P_1 and P_2 following the optical analogue as just outlined. The planar boundary separates materials with resistivities ρ_1 and ρ_2. We use a modified form of Eq. 5-7 to determine the potential at P_1 and P_2. Equation 5-7 is valid for a current source at the surface, as the equipotential surfaces are hemispheres. In Figure 5-16 the current source is within the ρ_1-material, and the equipotential surfaces are spheres. For such a case the equation takes the form

$$V = \frac{i\rho}{4\pi D} \tag{5-22}$$

where D is the distance from the current electrode to the point where the potential is being measured.

Remembering that the potential at P_1 is due to a direct and a reflected component, we write

$$V_{P_1} = \frac{i\rho_1}{4\pi r_1} + \frac{ik\rho_1}{4\pi r_2} \tag{5-23}$$

The potential at P_2 (due to transmission) is

$$V_{P_2} = \frac{i(1-k)\rho_2}{4\pi r_3} \tag{5-24}$$

The presence of the boundary imposes constraints on current flow across it. If P_1 and P_2 are both located on the boundary (Fig. 5-16(b)) at the same point, then the two potentials must be equal. In this case $r_1 = r_2 = r_3$, so we set Eqs. 5-23 and 5-24 equal and simplify to obtain

$$V_{P_1} = V_{P_2} = \frac{i\rho_1}{4\pi}\left(\frac{1}{r_1} + \frac{k}{r_2}\right) = \frac{i\rho_2}{4\pi}\left(\frac{(1-k)}{r_3}\right)$$

$$\rho_1\left(\frac{1+k}{r_1}\right) = \rho_2\left(\frac{1-k}{r_1}\right)$$

$$k = \frac{\rho_2 - \rho_1}{\rho_2 + \rho_1} \tag{5-25}$$

This derivation is useful for two reasons. First, we now have sufficient information to map the equipotentials for this two-medium boundary problem. Also, we know how to determine the reflection coefficient. You may recall that this same factor appeared in Eq. 5-18 when we first began our qualitative consideration of the horizontal interface problem. Although we accepted Eq. 5-18 without explaining its derivation, our present inquiry is beginning to shed light on various components of that equation.

We will not pursue this case further. Instead we now turn to the actual problem at hand—that of the horizontal interface. If we inspect Figure 5-17 and compare it to Figure 5-16, we see that we must add another surface in order to consider both the effect of the

(a)

(b)

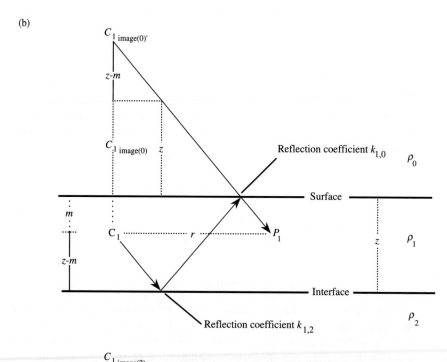

Figure 5-17 Determination of potential at point P_1 using an optical analogue for the case of a horizontal interface beneath the surface. (a) First images in upper and lower materials. (b) Second image in upper material.

earth's surface and the buried interface. In order to derive the potential equations we must consider both surfaces, but we proceed essentially as before.

A current source C_1 of intensity i is placed within the ρ_1-layer. The point P_1 at which we desire to know the potential is placed a distance r from C_1 at the same depth. The placement of C_1 and P_1 is for convenience; they will be placed on the surface at a later time in the derivation. The upper boundary (the earth's surface) separates ρ_0 from ρ_1. Our object of interest, the horizontal interface, serves as the lower boundary and separates ρ_2 from ρ_1. The reflection coefficient for the upper boundary is designated by $k_{1,0}$ and follows the form of Eq. 5-25 (using ρ_1 and ρ_0), as does the reflection coefficient for the lower boundary $k_{1,2}$. Image sources of C_1 are referred to as such with a number in parentheses referring to the layer in which they reside.

Referring to Figure 5-17(a) we proceed as in the previous example. The potential at P_1 due to C_1 is simply

$$^{\text{primary}}V_{P_1} \;=\; \frac{\rho_1 i}{4\,\pi\;r} \tag{5-26}$$

If we consider a single reflection from the upper boundary (equivalent to placing an image of C_1 in the upper material), the intensity of the current is i times the reflection coefficient $k_{1,0}$, and the contribution to the potential at P_1 is

$$^{\text{upper}}_{\text{first}}V_{P_1} \;=\; \frac{\rho_1 i k_{0,1}}{4\,\pi\left(r^2 + 4\,m^2\right)^{1/2}} \tag{5-27}$$

The qualifiers associated with the potential symbol V signify that the image is in the upper material and is the first image constructed there. It is demonstrated in Figure 5-17 that we can also have a single reflection from the lower boundary. This contribution to the potential is

$$^{\text{lower}}_{\text{first}}V_{P_1} \;=\; \frac{\rho_1 i k_{1,2}}{4\,\pi\left(r^2 + 4\,(z - m)^2\right)^{1/2}} \tag{5-28}$$

Reverting to the optical analogue for a moment, we can consider a ray emanating from C_1 reflecting from the lower boundary, reflecting from the upper boundary, and then arriving at P_1. As illustrated in Figure 5-17(b), this is equivalent to placing a second image in the upper layer ($C_{1_{\text{image}(0)}}$) which is located at a distance $2z$ from C_1. As this path involves reflections from both boundaries, both reflections contribute to a reduction in intensity. Current density, therefore, is multiplied by the product of both reflection coefficients. The equation for the potential due to the second image in the upper material is

$$^{\text{upper}}_{\text{second}}V_{P_1} \;=\; \frac{\rho_1 i k_{0,1} k_{1,2}}{4\,\pi\left(r^2 + 4\,z^2\right)^{1/2}} \tag{5-29}$$

We can continue this process and consider as a next step the path that is similar to that in Figure 5-17(b) but that reflects twice off the lower boundary and twice off the upper boundary. The equation for this path is

$$\text{upper} \atop \text{third} V_{P_1} = \frac{\rho_1 i k_{1,0}^2 k_{1,2}^2}{4\pi \left(r^2 + 16 z^2\right)^{1/2}} \tag{5-30}$$

Note that now each reflection coefficient is squared. We can continue this process, producing more and more images in the upper medium that contribute less and less to the potential at P_1. If you examine Eqs. 5-29 and 5-30, you should note the similarity. Equations incorporating more and more of this group of images would continue to have this form. This suggests that these equations can be expressed as the infinite series

$$\text{upper} \atop \text{even} V_{P_1} = \sum_{n=1}^{\infty} \frac{\rho_1 i k_{1,0}^n k_{1,2}^n}{4\pi \left[r^2 + \left(2nz\right)^2\right]^{1/2}} \tag{5-31}$$

(Keller and Frischknecht, 1966, p. 109) Up to this point we have four equations that represent contributions to the potential at P_1 (5-26, 5-27, 5-28, and 5-31). In addition to these, three other infinite series must be included. In arriving at Eq. 5-31 we considered only an even number of reflections which could be related to an image in the upper medium. Other combinations are: odd number of reflections—images in upper medium; even number of reflections—images in lower medium; and odd number of reflections—images in lower medium. The total potential at P_1 is the sum of all these.

The next step is to place C_1 and P_1 on the surface. In this case, since air has infinite resistivity, $k_{1,0} = 1$ and $m = 0$. These substitutions are made in all the equations we added together to determine the potential at P_1. These then are rearranged to arrive at

$$V_{P_1} = \frac{\rho_1 i}{2\pi r} \left\{ 1 + 2 \sum_{n=1}^{\infty} \frac{k_{1,2}^n}{\left[1 + \left(2nz/r\right)^2\right]^{1/2}} \right\} \tag{5-32}$$

(Keller and Frischknecht, p. 112). However, we are ultimately interested in determining the potential difference due to a source and a sink in a configuration such as that illustrated in Figure 5-11(a). Equation 5-32 provides the potential at P_1 due to the source at C_1, which are a distance a apart. Deriving an equation for the potential at P_2 due to C_1 is trivial, as the only difference is the separation of the two points, which is now $2a$. Thus, the equation for the potential difference is

$$\Delta V = V_{P_1} - V_{P_2} = \frac{\rho_1 i}{4\pi a} \left\{ 1 + 4 \sum_{n=1}^{\infty} \frac{k_{1,2}^n}{\left[1 + \left(2nz/a\right)^2\right]^{1/2}} - 2 \sum_{n=1}^{\infty} \frac{k_{1,2}^n}{\left[1 + \left(nz/a\right)^2\right]^{1/2}} \right\} \tag{5-33}$$

Remember that our ultimate goal is to develop an equation similar to Eq. 5-20 that provides values of apparent resistivity for various electrode spacings in the case of the horizontal interface. If we use the electrode spacings in Figure 5-11(a), we can rewrite Eq. 5-20 as

$$\rho_a = \frac{2\pi\Delta V}{i} \left(\frac{1}{\dfrac{1}{a} - \dfrac{1}{2a} - \dfrac{1}{2a} + \dfrac{1}{a}} \right) \tag{5-34}$$

and

$$\rho_a = \frac{2\pi a\Delta V}{i} \tag{5-35}$$

Equation 5-33 does not consider the contribution of the electrode at C_2, but this sim- ply doubles the potential difference ΔV. We make this adjustment and then rewrite Eq. 5-35 using the expression for ΔV given in Eq. 5-33 to obtain

$$\rho_a = \left(\frac{2\pi a}{i}\right)\left\{\frac{\rho_1 i}{2\pi a}\left[1 + 4\sum_{n=1}^{\infty}\frac{k_{1,2}^n}{\left[1 + \left(2nz/a\right)^2\right]^{1/2}} - 2\sum_{n=1}^{\infty}\frac{k_{1,2}^n}{\left[1 + \left(nz/a\right)^2\right]^{1/2}}\right]\right\}$$

and

$$\rho_a = \rho_1\left\{1 + 4\sum_{n=1}^{\infty}\frac{k_{1,2}^n}{\left[1 + \left(2nz/a\right)^2\right]^{1/2}} - 2\sum_{n=1}^{\infty}\frac{k_{1,2}^n}{\left[1 + \left(nz/a\right)^2\right]^{1/2}}\right\} \tag{5-36}$$

The form of Eq. 5-36 is valid only for the electrode configuration in Figure 5-11(a). Other electrode configurations require modifications of Eq. 5-36. These other configurations will be considered in a later section. It's been a long path to arrive at this point, and a number of steps have been neglected in order to keep the development to a minimum. We hope this provides you with sufficient insight into the optical analogue approach to obtaining a quantitative solution to the single-interface problem. If you are interested in a complete and thorough analysis, consult Keller and Frischknecht (1966).

Let's place Eq. 5-36 into a dynamic table (Table 5-4) in order to investigate the variation of apparent resistivity with electrode spacing. This table is designed much like Table 5-3. The major variables are in the upper left-hand corner. The result is apparent resistivity and is located in the upper right. The components are

$$k = \frac{\rho_2 - \rho_1}{\rho_2 + \rho_1}$$

$$k\text{-factor} = k^n$$

$$\text{divisor } 1 = \left[1 + \left(2nz/a\right)^2\right]^{1/2}$$

$$\text{divisor } 2 = \left[1 + \left(nz/a\right)^2\right]^{1/2}$$

$$\text{sum } 1 = k\text{-factor}/\text{divisor } 1$$

$$\text{sum } 2 = k\text{-factor}/\text{divisor } 2$$

$$\rho_a = \rho_1\left[1 + 4(\text{sum } 1 \text{ column total}) - 2(\text{sum } 2 \text{ column total})\right]$$

Finally, we have a quantitative approach to determining the variation of apparent resistivity with electrode spacing as defined in Figure 5-11(a) (all electrodes spaced at equal intervals). If we select the subsurface model as it appears in Table 5-4 and vary the electrode spacing, we can generate the curve in Figure 5-18(a). The general form of the curve should come as no surprise as it is very similar to our qualitative prediction (Fig. 5-15(a)). What may come

TABLE 5.4 Variation of Apparent Resistivity with Electrode Spacing for a Single Horizontal Interface

Electrode spacing a (m)	**1**
Depth of interface (m)	**20**
ρ_1	**10**
ρ_2	**100**
(Resistivities are in $\Omega \cdot$ m)	

Apparent resistivity ($\Omega \cdot$ m)   10

k	0.82				
n	k-factor	Divisor 1	Divisor 2	Sum 1	Sum 2
1.00	0.82	40.01	20.02	0.02	0.04
2.00	0.67	80.01	40.01	0.01	0.02
3.00	0.55	120.00	60.01	0.00	0.01
4.00	0.45	160.00	80.01	0.00	0.01
5.00	0.37	200.00	100.00	0.00	0.00
6.00	0.30	240.00	120.00	0.00	0.00
7.00	0.25	280.00	140.00	0.00	0.00
8.00	0.20	320.00	160.00	0.00	0.00
9.00	0.16	360.00	180.00	0.00	0.00
10.00	0.13	400.00	200.00	0.00	0.00
			Column totals	0.04	0.08

as a surprise, however, is the magnitude of the electrode spacing a necessary for the apparent resistivity to approach the resistivity of the second layer. As electrode spacing is increased, more current penetrates to a greater depth, so, as you already know, we "sample" material at increasingly greater depths. However, the shallow material always is included in the measurement. Therefore, in order for the measured apparent resistivity to approach the resistivity of the second layer, a great volume of the subsurface must be sampled. As you might imagine, this constitutes a serious limitation of the resistivity method, because it requires sufficient space in which to place electrodes.

Using ERModel

Although we could continue to use Table 5-4 and a graphing application to construct apparent resistivity curves, it certainly is easier and more elegant to develop a computer program written especially for this purpose. ERModel is such a program. It accepts resistivities and thicknesses for up to 10 layers, calculates apparent resistivities, and plots apparent resistivity against electrode spacing. Henceforth, for convenience, whenever we mention electrode spacing (or a-spacing, as is sometimes the practice), we mean the spacing between individual electrodes which is identified by the letter a in Figure 5-11(a). Later in this chapter other variations will be discussed and labeled differently.

(a)

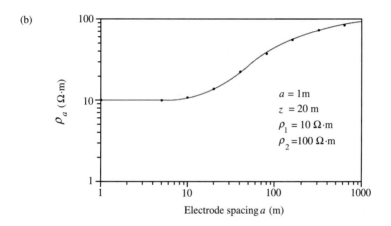

(b)

$a = 1\,$m
$z = 20\,$m
$\rho_1 = 10\ \Omega\cdot$m
$\rho_2 = 100\ \Omega\cdot$m

(c)

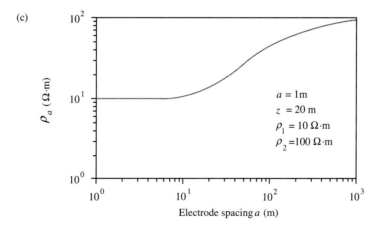

$a = 1\,$m
$z = 20\,$m
$\rho_1 = 10\ \Omega\cdot$m
$\rho_2 = 100\ \Omega\cdot$m

Figure 5-18 Apparent resistivity plotted against electrode spacing for geometry and parameters indicated. (a) Points calculated with Table 5-4. (b) Curve computed using ERModel.

Figure 5-18(b) illustrates a curve drawn by ERModel with the same input parameters as appear in Table 5-4. ERModel utilizes a different mathematical approach for solving the single horizontal interface problem, but the results are the same as the optical analogue produces. Actually, the results of Eq. 5-36 (which is used in Table 5-4) are plotted as small, filled circles in Figure 5-18(b). The line passing through these points is the ERModel curve from Figure 5-18(c) which is overlaid on these data points in order to emphasize that both approaches produce identical results.

Although procedures to interpret apparent resistivity curves in terms of subsurface geology have yet to be discussed, it is not too early to begin to assemble some basic properties of such curves. Apparent resistivity curves for the single horizontal interface have one of the two forms presented in Figure 5-15. Clearly, the exact form of the curve will depend on the resistivity values of the layers and the depth of the interface. Before proceeding further (and without examining Figure 5-19), try to sketch curves on the same set of axes for the following three subsurface models:

ρ_1 ($\Omega \cdot$ m)	ρ_2 ($\Omega \cdot$ m)	Depth to Interface
10	200	5
10	200	10
10	200	20

The important aspect of these curves is their similarities and differences, not the absolute values of the apparent resistivities and a-spacings.

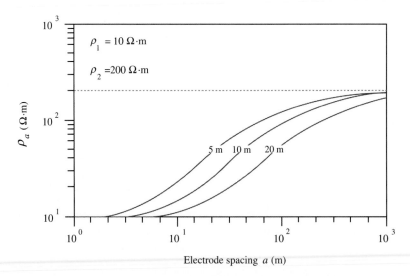

Figure 5-19   Apparent resistivity curves drawn by ERModel. Resistivity values constant for each curve. Depth of horizontal interface increases from left to right. The horizontal scale is logarithmic with six equal intervals for each decade. Because these are the preferred sampling intervals for field measurements, they are displayed on all curves calculated by ERModel and ERSolve. These values are 1.0, 1.47, 2.15, 3.16, 4.64, 6.81, 10.0, etc.

Now examine Figure 5-19, which was constructed by overlaying the three curves computed by ERModel for the three sets of values. You should be able to explain the general shape and relationship of the curves from the basic principles we developed in our discussion of the qualitative approach. All three curves have very similar shapes, but curves for the deeper interfaces are progressively offset to the right. This makes perfectly good sense if you remember the principle of current penetration, electrode spacing, and current density as affected by poor conductors. The 5-m curve begins to rise earlier because the effect of the shallow interface is felt even at small electrode spacings. In the case of the 20-m curve an electrode spacing of almost 10 m (current electrode spacing of 30 m) is necessary to reveal the presence of a higher-resistivity layer at depth. The 5-m curve begins to flatten noticeably at an a-spacing in the vicinity of 100 m, but the 20-m curve is still rising steeply at this point. It should be obvious that a 5-m-thick layer constitutes only a small fraction of the volume of material included in a measurement with an a-spacing of 200 m.

Therefore, the apparent resistivity for this a-spacing is quite close to the 200-$\Omega \cdot$m true resistivity of the deeper layer. This is not true when the interface is at 20 m, and so the ρ_a-value is only about 80 $\Omega \cdot$m.

Sketch a set of three curves following the layout of Figure 5-19. Use the same interface depths but reverse the resistivities so that the resistivity of the shallow layer is 200 $\Omega \cdot$ m and that of the deeper layer is 10 $\Omega \cdot$m. Be sure you can defend the relative positions of the curves.

Before we proceed to the next step, try to sketch three curves for which the interface depth remains the same, the resistivity of the shallow layer is constant at 10 $\Omega \cdot$m, but the resistivity of the deeper layer takes on values of 100 $\Omega \cdot$ m, 200 $\Omega \cdot$ m, and 500 $\Omega \cdot$m.

As a final exercise before we continue our general analysis by considering the effect of multiple interfaces, let's investigate one further set of relationships for the single interface case. In the preceding example resistivities remained constant and only the depth of the interface varied. Now let's hold constant the resistivity of the layer above the interface and the interface depth and vary the resistivity of the layer below the interface. Figure 5-20 provides a comparison for five apparent resistivity curves drawn under the stated constraints. Dashed lines are included in the figure in order to facilitate a visual comparison of the latter portion of each curve and the true resistivity of the deeper layer for which the curve was computed. The form and relationship of the curves should come as no surprise. As the resis-

Figure 5-20   Apparent resistivity curves for constant depth to an interface. Resistivity of the layer above the interface also is constant. Dashed lines are drawn at resistivity values assigned to the layer beneath the interface. Pay special attention to the different curve forms for the various resistivities assigned to the deeper layer.

tivity of the deeper layer increases, larger and larger electrode spacings are required before the slope of the curve begins to decrease and approach the resistivity value of that deeper layer.

The form of the curves in Figures 5-19 and 5-20 suggest that we should be able to devise a scheme to interpret apparent resistivity data in terms of resistivities and interface depths. We soon will see that this indeed is the case, but before considering interpretations of field data, it seems prudent to consider more complicated models.

MULTIPLE HORIZONTAL INTERFACES

Based on our development of a quantitative solution to the single-interface problem, it should be evident that we could continue in this direction for investigating the effect of more than one interface. However, the complexity of the image theory approach becomes extreme, so that in practice more advanced mathematical solutions usually are utilized. For our purposes it is sufficient to develop apparent resistivity curves using ERModel and to qualitatively analyze these curves based on the principles we discussed previously. In order to keep this discussion relatively brief, we consider only two horizontal interfaces (three layers). You are encouraged to use ERModel to conduct your own analysis of the three-interface (four layers) case.

Because we are dealing with three-layer resistivities in the two-interface case, four possible curve types exist: $\rho_1 > \rho_2 > \rho_3$ (Q), $\rho_1 > \rho_2 < \rho_3$(H), $\rho_1 < \rho_2 > \rho_3$ (K), and $\rho_1 < \rho_2 < \rho_3$ (A). The letters used to refer to the various curves are those typically used in electrical resistivity literature. We use them here only for convenience in referring to specific curve types. The four curves are illustrated in Figure 5-21. Interface depths are the same for all

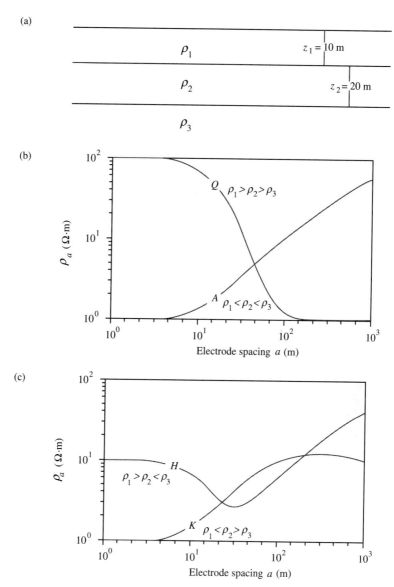

Figure 5-21 The four basic apparent resistivity curve types for two horizontal interfaces. Resistivities used to compute the curves are 1, 10, and 100 $\Omega \cdot$ m. Thicknesses for each case are as defined in (a).

curves and are defined in Figure 5-21(a). Resistivities used are 1, 10, and 100 $\Omega \cdot$ m. The ρ_1-, ρ_2-, and ρ_3- values for each curve are self-evident. Curves for $\rho_1 > \rho_2 > \rho_3$ (Q) and $\rho_1 < \rho_2 < \rho_3$ (A) are contained in Figure 5-21(b). These appear very similar to two-layer curves and could easily be misinterpreted as such, so it seems sensible to inquire further about this similarity. Figure 5-22 contains one of these three-layer curves (Q) (with $\rho_1 = 100 \, \Omega \cdot$m, $\rho_2 =$

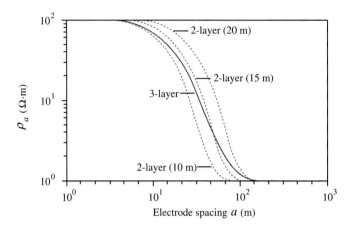

Figure 5-22   Comparison of two- and three-layer curves for similar depths and resistivities. The three-layer curve is a solid line (thickness of the first layer is 10 m; thickness of the second layer is 20 m; resistivities are 100, 10, and 1 $\Omega \cdot$ m). Resistivities for the two-layer cases are 100 and 1 $\Omega \cdot$ m. Depth of the interface is indicated for each curve.

10 $\Omega \cdot$ m, $\rho_3 = 1$ $\Omega \cdot$ m, $z_1 = 10$ m, and $z_2 = 20$ m) which is drawn with the solid line. The three dashed-line curves were computed using values of 100 $\Omega \cdot$ m for ρ_1 and 1 $\Omega \cdot$ m for ρ_2 and horizontal interface depths of 10 m, 15 m, and 20 m. Although the two- and three-layer curves are not the same, it could virtually be impossible to distinguish between them with field data which is acquired at discrete intervals and almost always contains some noise.

Figure 5-21(c) illustrates the remaining two curve types (H and K). These clearly indicate the presence of a low-resistivity layer within a sequence of higher resistivities or the presence of a high-resistivity layer lying between layers with lower resistivities. The form of curve K can be explained using the principles we developed previously. At small electrode spacings current density is affected only by the low resistivity (1 $\Omega \cdot$ m) of the shallow layer, which leads to small values of apparent resistivity. As electrode spacings increase, the higher resistivity (100 $\Omega \cdot$ m) of layer 2 affects current density, and apparent resistivities rise. As electrode spacings continue to increase, a greater percentage of current flows in the deepest layer, which has a lower resistivity (10 $\Omega \cdot$ m) than the intermediate layer. This also affects current density distributions, so that apparent resistivities begin to decrease. Note that even though the thickness of the second layer is substantial (20 m), apparent resistivities never exceed 13 $\Omega \cdot$ m even though the resistivity of the layer is 100 $\Omega \cdot$ m. Be sure you understand why curve H in Figure 5-21(c) has the particular form that it does.

How many basic curve types exist for the four-layer case? How many of these are unique in the sense that they clearly arise from four layers?

Although complexities and, therefore, identification and interpretation difficulties increase as the number of interfaces increases, the basic principles determining apparent resistivity values versus electrode spacing remain consistent. We will return to this subject during our discussion of interpreting field values but turn now to a different subsurface model—that of the vertical contact.

VERTICAL CONTACT

A rapid lateral change in resistivity can be modeled as a vertical contact. Such geometry often is taken to represent a vertical fault, but sudden changes in composition due to depositional processes or even a pronounced elevation gradient (vertical cliff) also fit into this model. The equations for this situation are no more complicated than for multiple horizontal interfaces. The method of images works well for the vertical contact. If you return to the section where we first considered the quantitative approach to a single horizontal interface and rotate Figure 5-16 by 90°, you should appreciate the similarity of the vertical contact problem to the single horizontal interface problem.

Complications arise, however, as electrode spacing is increased (or the entire spread is moved), for eventually one or more electrodes cross the contact (which is not a consideration when the interface is horizontal). This changes the boundary conditions, and a different potential equation must be applied. In all it requires five equations to satisfy all possible variations in electrode placements as illustrated in Figure 5-23(b). We've chosen not to derive or present the relevant equations here but rather to concentrate on the apparent resistivity patterns associated with measurements taken during traverses that cross a vertical contact. These equations and their derivation are present in Van Nostrand and Cook (1966, p. 52–54) and Telford et al. (1990, p. 556–557).

When discussing apparent resistivity patterns associated with horizontal interfaces, we increased electrode spacings in order to achieve increased current penetration and, therefore, changes in the observed apparent resistivity. However, two different approaches are feasible when a vertical contact is present. When dealing with such an abrupt lateral change, we are able to detect the contact either by fixing the center of the electrode spread and systematically increasing electrode spacing (*expanding-spread traversing*) or by holding electrode spacing constant and systematically moving the entire spread along a traverse line (*constant-spread traversing*). Expanding-spread traversing also is referred to as depth profiling, electrical drilling or resistivity sounding, whereas constant-spread traversing often is called electrical mapping or resistivity profiling.

Constant-Spread Traverse

Let's first examine a constant-spread traverse across a vertical contact. Figure 5-23(a) diagrams the essential elements of the traverse. The distance from the spread center to the contact is x, which is positive to the right of the contact and negative to the left. All electrodes are equally spaced a distance a apart. The entire spread is moved at equal increments along a line perpendicular to the strike of the contact. Table 5-5 contains results for a given selec-

(a)   Section

(b)   Map

(c)

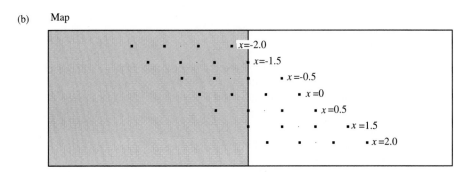

Figure 5-23   Apparent resistivity for a constant-spread traverse oriented perpendicular to a vertical contact. (a) Section view illustrating parameters used in Table 5-5. (b) Map view designating critical positions of electrode spreads. (c) Graph of values in Table 5-5.

tion of resistivities, electrode spacing, and traversing increment. A casual examination of the apparent resistivities in the table reveals that the apparent resistivities do not vary smoothly, as was the case for horizontal interfaces. A plot of Table 5-5 values (Fig. 5-23(c))

TABLE 5-5 Constant-Spread Survey Across a Vertical Contact

ρ_1	**100.0**
ρ_2	**10.0**
Electrode spacing a (m)	**1.0**
x-increment (m)	**0.25**

k-factor	-0.82

x (m) (spread center to vertical contact)	Apparent Resistivity ($\Omega \cdot$ m)
-2.5	94.959
-2.3	92.680
-2.0	88.669
-1.8	80.704
-1.5	63.237
-1.3	64.352
-1.0	65.989
-0.8	68.538
-0.5	72.557
-0.3	64.463
0.0	55.000
0.3	40.731
0.5	13.538
0.8	13.146
1.0	13.401
1.3	13.565
1.5	13.676
1.8	11.930
2.0	11.133
2.3	10.732
2.5	10.504

confirms this impression. Actually, there are four points on the curve where the value and direction of the slope suddenly change.

In trying to develop a sense of why these slope reversals occur, it helps to plot electrode positions relative to the contact on a map view. Figure 5-23(b) illustrates several important arrangements. Note that the traverse typically is along a single line. The positions are staggered in the diagram for purposes of clarity. As you read through the following discussion, refer frequently to Figure 5-23(b) and (c). As expected, when the electrode spread is to the far left ($x = -5$) or far right ($x = 5$), apparent resistivities approximate the actual resistivities. As the spread is moved closer to the contact ($x = -2$), apparent resistivities decrease relatively quickly, because the spread is approaching a lower-resistivity medium. Slope reversals occur at $x = -1.5$, -0.5 , 0.5, and 1.5. These values represent spread positions where an electrode crosses from one side of the contact to the other. Therefore, we should be able to locate the contact quite closely by noting the positions of spread electrodes when reversals take place. Remember that Figure 5-23 is idealized, and it is very unlikely that field values of a and spread-center positions will locate the contact as closely as in our model.

Why is the change in apparent resistivity values much more pronounced between $x = -3$ to $x = -2$ than from $x = 3$ to $x = 2$? *Hint:* Note that the negative x-values are in the higher resistivity medium. Use the current density distribution model or a water-flow analogy to arrive at a qualitative explanation.

We strongly suggest that, if possible, you work with Table 5-5 to produce curves of your own for various values of the quantities in boldface. For best results use x-increments that are smaller than the value you select for electrode spacing. Although Table 5-5 is a dynamic table, it functions a bit differently than those you previously encountered. Because one of five equations must be selected depending on the electrode positions relative to the contact, it is necessary to employ a separate routine to make this selection. Therefore, you must first open **Macro5** before opening **Table 5-5**. You then may proceed to type values into the table as before with one exception. Because of the macro execution, recalculation times are not instantaneous. Because of these recalculation times, the default of automatic recalculation is disabled. To see the results of your model, hold down the **Command** key and press =. Finally, the equations used are not defined at $|x| = a$ and $|x| = 3a/2$. If you select values that result in these equalities, the words "not defined" will appear instead of apparent resistivities. To avoid this, we suggest selecting an a-spacing such as 2.01 instead of 2.0.

Of course, when working in the field, it would be extremely fortuitous for a traverse to be oriented perpendicular to the trace of a vertical contact. Fortunately, the main difference in an apparent resistivity curve for a traverse at an angle to the contact is a reduction in the size of the slope reversals. These slope reversal points, or cusps, become less and less pronounced as the angle between the traverse and contact trace decreases. For representative curves consult Van Nostrand and Cook (1966, p. 120). If a traverse is parallel to the contact, apparent resistivity values remain constant. Their magnitude depends on the distance of the traverse line from the contact.

Expanding-Spread Traverse

In an expanding-spread traverse the electrode center is fixed, and electrode spacings are systematically increased. Similarly to the previous example, we link the values in a dynamic table (Table 5-6) to a specific diagram (Fig. 5-24). In this example the spread center is at $x = -2.0$, and the traverse is perpendicular to the contact. The results in Table 5-6 are plotted

TABLE 5-6 Expanding-Spread Survey Across a Vertical
Contact

ρ_1	100.0
ρ_2	10.0
Electrode spacing a increment in m	0.25
Spread center to vertical contact x in m	$-$ 2.01

k-factor	-0.82

Electrode spacing a (m)	Apparent Resistivity ($\Omega \cdot$ m)
1.00	89.30
1.25	73.36
1.50	63.84
1.75	64.87
2.00	65.87
2.25	66.84
2.50	67.79
2.75	68.69
3.00	69.56
3.25	70.39
3.50	71.19
3.75	71.94
4.00	72.67
4.25	71.81
4.50	70.93
4.75	70.16
5.00	69.48
5.25	68.86
5.50	68.31
5.75	67.79
6.00	67.33

in Figure 5-24(c) which reveals two sharp slope discontinuities. Various expanding electrode spreads are illustrated in Figure 5-24(b). Once again the slope discontinuities can be correlated with electrodes crossing the contact. Comparison of the $a = 1.0$ and $a = 2.0$ spread geometries suggests that the slope discontinuity for an a-spacing of 1.25 correlates with the a current electrode moving across the vertical contact. For an a-spacing of 4.0 m, a potential electrode crosses the contact which corresponds to the second slope discontinuity. Although perhaps not as dramatic as the resistivity curve obtained by constant-spread traversing, the expanding-spread curve certainly is sufficiently distinctive both to establish the presence of a rapid lateral change in resistivity and to establish reasonably accurately the location of the change.

A traverse oriented parallel to a vertical contact and located close to the contact produces a curve that is remarkably similar to a two-layer curve. How can we explain this behavior when a constant-spread traverse taken in the same orientation produces constant apparent resistivity values? This seeming contradiction is easily explained if we remember that equipotential surfaces in a homogeneous, isotropic material are hemispherical in shape. When the electrodes are close to the contact but are spaced very close together relative to the contact distance, the effect of the material on the opposite side of the contact will be minimal. The situation is analogous to that illustrated in Figure 5-14(a). However, as electrode spacing is increased, a greater percentage of current flow not only penetrates more deeply but spreads laterally as well. If a material with different resistivity is encountered in this lateral direction, current density will be affected. Thus, apparent resistivity values will change also.

If we use the configuration in Figure 5-24(a), the resistivity to the right of the contact is low. Assume we place our electrodes as shown, but rotate them 90° so they are parallel to the contact. An initial reading for a small a-spacing would be close to 100 $\Omega \cdot m$. As we increase electrode spacing, apparent resistivity will decrease as more current flows in the lower resistivity material and current density between the potential electrodes decreases. This behavior continues as electrode spacing increases. The extent to which apparent resistivity values approach the resistivity of the material across the contact depends on the distance of the traverse from the contact. The final curve for such a traverse looks much like that in Figure 5-15(b).

TWO VERTICAL CONTACTS, HEMISPHERICAL STRUCTURES, AND DIPPING INTERFACES

Unfortunately, in resistivity surveying it seems that all basic structural configurations require derivation and application of separate equations. Many of these are treated in detail by several of the basic resistivity texts cited at the end of this chapter. Our goal in this section is simply to point out some of these other subsurface geometries and to emphasize that the material you've just absorbed equips you to predict apparent resistivity curves for these geometries, if only in a generalized way.

Two Vertical Contacts

First, let's briefly consider the structure illustrated in Figure 5-25(a). This feature could just as easily be termed a vertical dike, but we prefer to emphasize that it is also adequately described as two vertical contacts. Use of the word *dike* conjures up igneous terrain, but this feature could represent a rapid lateral variation in sediment type or a saturated, high-porosity zone of brecciated rock. More important, however, is the emphasis on the idea of *two* vertical contacts. As we already are familiar with the effects of one vertical contact, we might ask if we can predict the effect of two vertical contacts on apparent resistivity curves. If the width of the feature delimited by the vertical contacts is large compared to electrode spacing, then as we traverse the contacts with a constant-spread electrode arrangement, we

(a)   Section

(b)   Map

(c)

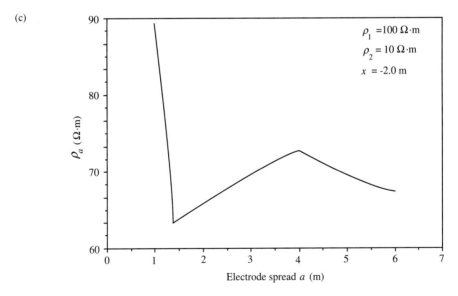

Figure 5-24 Apparent resistivity for an expanding-spread traverse oriented perpendicular to a vertical contact. (a) Section view illustrating parameters used in Table 5-6. (b) Map view illustrating electrode positions for various a-spacings. (c) Graph of values in Table 5-6.

(a)

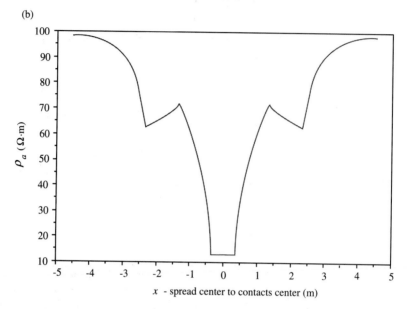

(b)

x - spread center to contacts center (m)

Figure 5-25   Constant-spread traverse across a vertical dike. (a) Diagram of relationships.
(b) Qualitatively derived sketch of likely curve shape.

should (1) see a curve with features similar to that in Figure 5-23(c) if we are proceeding from high-resistivity material to low-resistivity material and (2) observe a mirror image of the first portion of the curve as we now proceed from low resistivity material to high-resistivity material (Fig. 5-25(b)).

Of course, this is much too simplified an approach to yield any quantitative results if such a vertical feature actually is traversed. However, if at some time you do observe a curve with cusps that appear to be mirror images, you should at least recognize the possibilities and then seek a more theoretical treatment in the electrical resistivity literature. Our point here is that resistivity curves come in a seemingly endless variety of shapes, and you cannot begin to remember more than a few basic features. But if you grasp the essential features and behavior of a few curve types (single horizontal interface, multiple horizontal interfaces, vertical contact), you certainly will be more effective in the field when required to make decisions concerning electrode spacing, traverse type, and traverse direction.

Hemispherical Structures

A number of subsurface configurations are better approximated by a hemispherical form than vertical contacts (Fig. 5-26). An abandoned river channel filled with gravel incised into floodplain deposits of silts and clays or a sinkhole in limestone filled with dissimilar material are two common examples. Let's assume we are sufficiently fortunate to cross the center of such a form with a constant-spread traverse and an a-spacing that is substantially smaller than the diameter of the hemispherical form. Can you sketch a representative curve for such a traverse?

Actually, for rough approximation purposes, we can think of the hemispherical form as two vertical contacts that do not extend very far downward. As we cross each contact, we expect to see curves somewhat similar in form to vertical-contact curves. However, the effect of the ρ_2-material is tempered by the presence of ρ_1-material beneath it. Nevertheless, the curve shape is quite similar to that produced by two vertical contacts. If the traverse cuts the hemispherical form but crosses nearer to its edge, the bilaterally symmetric curve form still will be evident, but the cusps will be more rounded and less pronounced.

Dipping Interfaces

Because of the number of permutations involved with traverses across dipping interfaces, we will confine this brief discussion to one specific comparison. Our standard will be the vertical contact (or vertical interface) curve in Figure 5-23(c), which is reproduced in Figure 5-27(b). How will this curve change if we lower the dip to 15°, as diagramed in Figure 5-27(a)?

If the traverse proceeds from left to right, apparent resistivity values approaching ρ_1 (100 $\Omega \cdot$ m) will be encountered first. As electrodes cross the contact separating ρ_1-material from ρ_2-material, we should expect similar behavior as observed for the vertical contact. However, due to the 15° dip, only a thin wedge of ρ_2-material is present near the contact. This will reduce the effect of this lower-resistivity material, so that apparent resistivity values will be higher relative to those for the vertical contact. Also, as the electrode spread moves farther to the right away from the contact, the decrease in apparent resistivity values will not occur as quickly, because ρ_1-material still is not far from the surface. If we sketch a qualitative representation of this behavior, our curve should appear somewhat similar to that

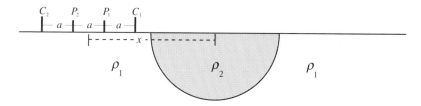

Figure 5-26 Hemispherical sink with diameter substantially greater than electrode spacing a.

(a)

(b)

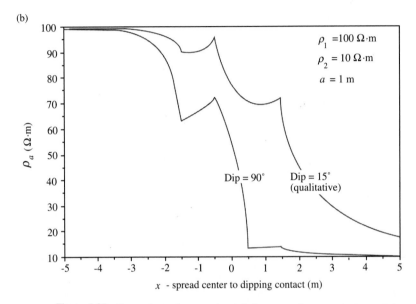

Figure 5-27 Comparison of apparent resistivity curves for a constant-spread traverse across an interface dipping 90° and one dipping 15°.

in Figure 5-27(b). Cusps still are present due to electrodes crossing a boundary between materials of different resistivities, but the total variation in apparent resistivity from cusp to cusp is less than for the vertical contact. This may result in greater difficulty in defining the cusp position, or its existence, and, therefore, in locating the position of the contact and the amount of its dip.

FIELD PROCEDURES

Up to this point we concentrated on developing a sufficient theoretical background for understanding the shape of qualitative and quantitative apparent resistivity curves connected to a variety of simple subsurface geometries. We have not as yet addressed the question of proceeding from actual field data to an interpretation of the subsurface. Before doing so, we need briefly to address resistivity equipment, possible electrode geometries to use in data collection, and some strategies for acquiring useful resistivity data.

Equipment

Equipment for a typical resistivity survey consists of an ammeter, voltmeter, power source, electrodes, and connecting wire (Fig. 5-28). Electrodes are metal stakes (copper, steel, or aluminum) with a cap to facilitate driving them into the ground. Wire to connect the electrodes to the power source and meters must be durable, light, of low resistance, and well insulated. The wire is placed on reels for ease in winding and unwinding as electrode spreads are increased; a typical length on an individual reel is rarely less than 100 m. Power is provided either from a motor-driven generator or by a series of batteries connected in series. A number of considerations dictate the exact source selected. Batteries are best for portability, and a generator often is used when depths of investigation, and therefore current demands, are large.

The exact design of the meters and allied circuitry varies and depends on the type of current used (direct current or alternating current) and also on portability requirements. Direct current causes potentials to develop at the electrodes due to electrochemical reactions between electrodes and soil. These potentials are included in potential measurements unless special electrodes (*porous pot*) are used. Commutators which frequently reverse the polarity of the direct current alleviate this problem, but a common trend in modern instruments is to utilize low-frequency alternating current.

The circuits to measure potential and current sometimes are in separate units which increases flexibility but reduces portability. A typical unit for standard shallow exploration exercises includes not only the power source but both ammeter and voltmeter. Obtaining a reading is straightforward with such instruments. Electrodes are placed in a selected pattern and connected to the instrument. A few preliminary adjustments are made as directed by the

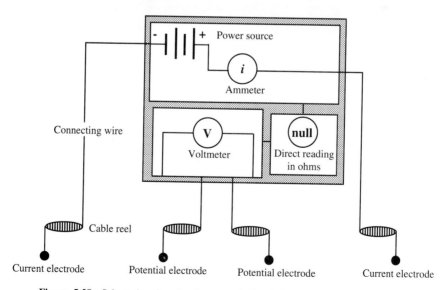

Figure 5-28 Schematic of main elements of electrical-resistivity surveying system including electrodes, power source, ammeter, voltmeter, and direct-reading device.

manufacturer, and current is then applied. Typically a meter needle is deflected from a zero or *null* position. Rotating a dial (that provides a direct readout in ohms) brings the needle back to the zero position. The ohmmeter reading is recorded, and electrodes are moved to the next position. Figure 5-29 pictures an instrument of this type. Readings can be taken quickly, but moving electrodes and wire for the 10 or more readings required for any individual traverse is time consuming.

Electrode Configurations

The electrode patterns used in resistivity surveying almost always are the Wenner, Schlumberger, or dipole-dipole. Many other arrangements have been tried, but these three are the survivors. The Wenner electrode geometry is illustrated in Figure 5-30(a). Spacing between all electrodes is equal and conventionally is denoted by the letter *a*. This configuration should appear familiar, as it is the pattern we have used in all previous examples in this chapter. In conducting an expanding-spread Wenner survey all electrodes are moved along a straight line after every reading such that the spacing between electrodes remains equal and takes on certain preselected values. Typically one endeavors to achieve values of *a* that are equally spaced on a logarithmic scale and ideally number six points per decade (e.g. 10.0, 14.7, 21.5, 31.6, 46.4, and 68.1). If you used ERModel, you may have noted that these are the intervals for which apparent resistivity values are calculated. ERSolve, which is discussed in the next major section of this chapter, assumes field readings are acquired at these same intervals. The main reason for such spacings is that shallow materials are always included in any resistivity reading. Larger and larger electrode spacings, therefore, are nec-

Figure 5-29   Photograph of a typical resistivity meter.

(a)

(b)

(c)

(d)

Figure 5-30 The most common electrode geometries used in resistivity surveying. (a) Wenner. (b) Wenner/Lee (Lee Partitioning). (c) Schlumberger. (d) Dipole-dipole.

essary in order to acquire a reasonable "electrical sample" of material at depth. Using the recommended electrode intervals produces equally spaced points on the horizontal axis of a log/log graph of the field data.

A symmetrical distribution of current and potential electrodes about a central point also is employed for the Schlumberger array (Fig. 5-30(c)), but the potential electrodes are spaced much more closely than the current electrodes. Spacings are selected to maintain the relationship $2L > 5MN$ and also follow the same numbering scheme as the Wenner array. Because of this particular geometry, meter sensitivity is exceeded after the current electrode spacing is increased several times. At this point the potential electrode spacing MN is increased to the next larger value in the numbering scheme (e.g., 14.7 m if the previous value was 10.0 m), and the current electrode spacing L is reduced by two intervals (e.g., to 100 m if the previous reading was 215 m). This recording procedure produces overlapping curve segments on a plot of apparent resistivity versus electrode spacing (Fig. 5-31).

The final common geometry is the dipole-dipole (Fig. 5-30(d)). In this arrangement the potential electrodes and current electrodes function independently. Both sets tend to be fairly closely spaced with a significant distance between the sets. The potential electrodes are placed relative to the current electrodes in one of several basic patterns (not all along the same line). Because cable lengths between the electrodes are short, it is much easier to place the potential electrodes at large distances from the current electrodes, thereby facilitating deep investigations. At the same time larger currents are necessary to reach these depths. This geometry is used much less in North America than the Wenner and Schlumberger spreads, although use in the Soviet Union, especially for petroleum exploration, has been significant. Because of this relative lack of use in North America, we will concentrate on the Wenner and Schlumberger arrays.

The Wenner electrode array has been used primarily in North American surveys because Frank Wenner employed that geometry early in the development of the resistivity method. Most European workers use the Schlumberger spread, as do many American investigators. Each geometry has advantages and disadvantages. One advantage of the Wenner array is that the larger potential electrode spacing places less demand on instrument sensitivity. A second "advantage" is the simplicity of the apparent resistivity equation (Eq. 5-20) when electrodes are equally spaced (Eq. 5-35). Prior to the present accessibility and capability of computing facilities, this simplicity reduced the complexity of many equations and, therefore, demanded much less effort in quantitative approaches to resistivity interpretation

Electrode spacing L (m) ——➤

Figure 5-31   Apparent resistivity values illustrating effect of changing MN spacing during Schlumberger expanding-spread traverse. MN spacing changed when L values equivalent to positions 1, 2, 3, and 4.

(which is the reason we used it in previous sections). Frequent use led to many examples and techniques in the literature, which furthered more use.

The Wenner geometry has several shortcomings. Because all electrodes must be moved for each reading, which is not the case with the Schlumberger method, the Wenner array requires more field time. Perhaps more serious is that it is more sensitive to local, near-surface lateral variations than the Schlumberger geometry. Of course, Wenner data can be analyzed directly, but Schlumberger data require some preinterpretation processing, as the overlapping curve segments must be smoothed and new values obtained from the smoother curve. On the other hand, the curve segments follow a specific pattern of over-laps, which is quite clearly illustrated in Figure 5-31. Increasing MN spacing when apparent resistivity values are rising results in offset of the new values downward (positions 1 and 2 in Fig. 5-31). When apparent resistivity values are decreasing, an increase in MN results in offset of the new values upward (positions 3 and 4 in Fig. 5-31). If the overlaps depart from this pattern, lateral inhomogeneity is indicated and, if not too severe, can be corrected for in the smoothing process.

The susceptibility of the Wenner electrode geometry to near-surface lateral variations increases the likelihood that these may be interpreted as variations at depth. One approach to reducing the possibility of such an erroneous conclusion is to place a third potential elec-trode at the center point of the electrode array. Readings are taken as usual, but in addition to the normal potential difference reading at P_1P_2, readings are also taken for P_1P_L and P_LP_2. This approach usually is referred to as the *Lee modification*. All apparent resistivities are calculated using the Wenner equation. If no lateral variations are present, the Lee apparent resistivity curves should exhibit the same form and should mimic the Wenner curve. The Lee values will plot at half-values of ρ_a, which makes visual interpretation easier. (*Note*: When using the Lee partition array independently, 2π should be changed to 4π in the Wenner equation (Eq. 5-35)). Departures from lateral homogeneity should be evident from departures in the similarity of the Lee values. Once such variations are identified, appropri-ate adjustments can be made to the exploration plan. Figure 5-32 illustrates a case where the Lee modification values are very similar and almost perfectly mimic the Wenner appar-ent resistivity values. Hence, lateral variations are not significant.

Because quantitative interpretation of Wenner and Schlumberger curves are very similar, we will generally continue to provide examples based on the Wenner array for rea-sons cited previously. Both ERModel and ERSolve provide options to use either array, so you may select either depending on the category of field data you have available. If you have an opportunity to gather data on your own, we suggest you take the time to use both arrays in order to better judge their strengths and weaknesses.

Surveying Strategies

As in the case of seismic exploration, resistivity survey goals should be specifically defined before field work begins. Equipment must be selected that is adequate for the task at hand. And, if at all possible, independent geologic controls should be available. It is very difficult to derive detailed geologic information from resistivity surveys unless such control exists.

Figure 5-32 Apparent resistivity values obtained with a Wenner array near Hatfield, Massachusetts. Note that the Lee modification values are very similar and mimic the Wenner values closely.

The next step is to locate sampling sites and to decide whether to use constant-spread or expanding-spread traverses.

If the goal of the effort is primarily to seek information about apparent resistivity variations with depth in order to define the location of a high resistivity-contrast interface, then expanding-spread traverses are the obvious choice. An example is a survey interested in mapping depth to a bedrock surface which is overlain by saturated silty sands. Of course, either Wenner or Schlumberger can be used. In either case, you always should run another traverse at right angles to the first in order to establish whether lateral variations are present. Remember that an expanding-spread traverse oriented parallel to a vertical contact can produce a curve almost identical to that for a two-layer case, whereas a traverse perpendicular to the contact reveals cusplike discontinuities that establish the presence of the contact without doubt.

If survey goals are to map lateral variations in resistivity, perhaps due to the presence of a gravel bar, then constant-spread traversing is required. Normal practice dictates that two or three electrode spacings be used at each station in order to develop three-dimensional information. Apparent resistivities for each electrode spacing are placed at each array center and contoured. If three spacings are utilized at each site, three separate contour maps are produced. Results of constant-spread surveys are difficult to interpret quantitatively, especially without some vertical control. For this reason, at least one, and ideally more, fully implemented expanding-spread surveys are included in order to develop a sense of resistivity variation with depth and to provide a basis for quantitative analysis. If you have a good grasp of the principles covered to this point and well-defined survey goals, traverse design should be reasonably straightforward.

A major question that often arises is the largest electrode spacing to use. In other words, What is the largest spacing that will assure adequate sampling of the subsurface at the depth of interest? In the past many explorationists incorrectly equated electrode spacing directly with depth. Even at this stage you should realize that such a relationship is not correct. This topic is important, so we return briefly to a number of diagrams discussed previously in order to review several relationships. First, let's reexamine Figure 5-11. This graph demonstrates, for the case of a single horizontal interface, that the amount of current penetrating beneath the interface (and therefore providing a sufficient "sample" of the lower material) depends not only on electrode spacing but also on the relative values of ρ_1 and ρ_2. In cases where the depth of the interface is equal to electrode spacing ($z = a$) and ignoring extreme resistivity contrasts, the percent of current penetrating beneath the interface can vary from 20 percent to 80 percent.

Similar lessons are available from Figures 5-19 and 5-20. Figure 5-19 demonstrates that increased electrode spacings are required to completely define curves for interfaces at increasing depths. Resistivity contrast is held constant for each curve. Graphs such as these suggest that three decades of readings (from 1 m to 1000 m) are sufficient to define curves for most shallow interfaces. Study Figure 5-20. Interface depth is constant for each curve. At an a-spacing of 100 m the curve for $\rho_2 = 20\ \Omega \cdot m$ is completely defined. At this same spacing, however, the curves for $\rho_2 = 200\ \Omega \cdot m$ and $500\ \Omega \cdot m$ are barely distinguishable and are still rising at such a rate that they offer no insight into the resistivity value of the lower layer except that it is high. However, many experienced explorationists find the rule that a-spacing is equal to depth to be useful in a variety of geologic settings. It is perfectly acceptable to use this rule to gain a feel for the meaning of curve form during field work as long as you remember this approximation is imperfect and eventually subject your field data to detailed computer analysis.

The best advice we can offer with regard to electrode-spacing decisions is to plot apparent resistivity values as readings are taken. Unless you are positive that the form of the curve is sufficient to meet the goals of your survey, keep acquiring data until you reach the limits of your instrument, cable length, or available room. Of course, if many traverses are required, then time and expense may constrain the number of readings you can accumulate. More often that not, we find that we run out of room or cable before obtaining as many readings as we would like.

Other Considerations

A number of factors can affect resistivity measurements in addition to the resistivities of the rocks and sediments beneath the surface, and you should be aware of the most significant of these in field-survey design. Some factors are obvious. Buried pipelines or cables may be good conductors and, if so, will affect current flow. Electrode contact must be good, and resistances should be low and uniformly maintained. If dry surficial materials are present, wetting the ground around the electrode often is all that is required.

Topography affects apparent resistivity measurements because topography affects current flow. Perhaps the most obvious example of this is the vertical contact. Check Figure 5-23 and replace the ρ_1-material with air, which we assume to have infinite resistivity. This provides the example of the vertical cliff, which admittedly is an extreme case, but which does illustrate that current flow would be affected and that apparent resistivities measured near the cliff would also be affected. Slight undulations and elevation changes are not a problem, but rugged topography should be avoided if at all possible. No topographic corrections are available, as there were for seismic work.

Finally, many rocks and some sediments often possess different resistivities depending upon the measurement direction. An example is a rock unit containing a single, very well oriented fracture set filled with water. Traverses parallel to the strike of the fractures will measure lower resistivities than traverses normal to the fractures. Although this electrical anisotropy is not as serious an interpretation problem as buried objects and rugged topography, you should be aware that it can affect apparent resistivities. You generally attempt to collect data along two perpendicular traverses. Variations between curves for the two traverses normally reflect lateral variations in materials, but keep in mind that anisotropy in a given unit may be responsible in some cases.

QUANTITATIVE INTERPRETATION OF APPARENT RESISTIVITY CURVES

An exclusively quantitative interpretation of apparent resistivity curves often is difficult because of the wide variations in resistivity possessed by geologic materials and the difficulty in developing theoretical expressions for apparent resistivities of all but the simplest geometries. Quantitative interpretation is best developed for cases in which layering is horizontal, so it is to these situations that we will confine our analysis. First, however, we turn to a consideration of the resistivities of sediments and rocks.

Electrical Resistivities of Geologic Materials

In the shallow subsurface, electricity is conducted almost entirely by the fluid present. Thus, the resistivities of sediments and rocks in this environment are controlled by the amount of water present and its salinity. Because this conduction is electrolytic and is related primarily to ions moving through the fluid, the cation exchange capacity of clay minerals increases conductivity. Although clay minerals are the most important in this regard, all fine-grained minerals possess an exchange capacity to some degree. This means that increasing silt or clay content in poorly sorted rocks or sediments will reduce resistivities. In saturated materials increasing porosity will reduce resistivities. Well-sorted materials have greater porosities than poorly sorted ones, but in well-sorted rocks or sediments with well-rounded grains, grain size alone does not influence porosity (Fetter, 1980, p. 60–62). In addition, irregularly shaped grains will pack less tightly than sphere-shaped grains and consequently will have higher porosities. In general, finer-grained sediments will have higher porosities (and lower resistivities) than coarse-grained sediments. Based on the preceding comments, however, it seems prudent to apply this relationship only in the most general way.

All the following *reduce* resistivities: increasing water content, increasing salinity of water, increasing clay content, and decreasing grain size. Assuming that water is available to fill voids, resistivity is lowered by increasing porosity, increasing number of fractures, and increasing weathering. Conversely, resistivities are raised by increasing compaction and lithification.

Because these factors vary so much in the natural environment, it is not surprising that resistivities vary greatly and that it will be difficult to correlate resistivities with source materials in the absence of other geologic information. Indeed, Zohdy et al. (1974, p. 9) note: "No other physical property of naturally occurring rocks or soils displays such a wide range of values." Although direct correlation will not be as straightforward as it was with seismic velocities, some very generalized relationships may often be established. Bedrock almost always has higher resistivities than the saturated sediments lying above. Generally, the unsaturated sediments above the water table will have higher resistivities than the saturated sediments below the water table. However, the only reliable way to correlate resistivities with local geology is by utilizing independent geologic information (such as an expanding-spread traverse located near a carefully-logged drillhole). In this way you can establish correlations and carry them forward into unknown terrain.

The following list of resistivities is based on personal experience and is adjusted to include values provided by numerous sources. These values are supplied primarily to provide a starting point for interpretation, but beware. For every observation that falls within the values on this list, there probably is an observation that provides an exception.

Material	Resistivity ($\Omega \cdot$m)
Wet to moist clayey soil and wet clay	1s to 10s
Wet to moist silty soil and silty clay	Low 10s
Wet to moist silty and sandy soils	10s to 100s
Sand and gravel with layers of silt	Low 1000s
Coarse dry sand and gravel deposits	High 1000s
Well-fractured to slightly fractured rock with moist-soil-filled cracks	100s
Slightly fractured rock with dry, soil-filled cracks	Low 1000s
Massively bedded rock	High 1000s

A thorough discussion of factors controlling resistivities and a good presentation of resistivity ranges for common geologic materials is presented in Ward (1990, p. 148–151).

Empirical Methods

There are at least two empirical methods cited frequently in texts dealing with interpretations of expanding-spread apparent resistivity curves, namely the *Moore Cumulative Resistivity Method* (Moore, 1945) and the *Barnes Layer Method* (Barnes, 1952). Normally, neither method should be used except for preliminary analysis or when the analytical methods described in the following sections cannot be utilized because the required materials are not available.

When using the Moore method, one collects apparent resistivity data at electrode spacings that are evenly spaced on a *linear* scale using a Wenner spread. For each *a*-spacing a cumulative resistivity value is calculated that is the sum of all apparent resistivities collected at all previous electrode spacings. These cumulative resistivities are plotted on the vertical axis; *a*-spacings are plotted on the horizontal axis. Straight lines are fit to the data points. The horizontal coordinates of the points of intersection of these lines are assumed to give the depths to horizontal interfaces.

The Moore method has succeeded in some locales and failed in others. It is possible to check this approach by using a program such as ERModel to generate apparent resistivity values for known models and then to interpret the data using the Moore method. If your curiosity leads you to such an investigation, note that ERModel plots data that are equally spaced on a *logarithmic* scale, whereas the Moore method requires data equally spaced on a *linear* scale. Zohdy et al. (1974, p. 46) report that the method appears to work well for two-layer cases where the resistivity contrast is not too great. If this condition does not hold, the depth to the interface can be in error by as much as 50 percent. These investigators regard the application of the Moore method to data from three or more layers as highly question-able. In addition, the method does not provide resistivities for the derived layering.

You realize of course, that each time electrode spacing is increased the volume of earth that affects an apparent resistivity determination is increased. A basic premise of the Barnes method is that each measured apparent resistivity represents the average resistivity in a layer that extends from the surface to a depth equal to the electrode spacing. This premise leads to the following: when the electrode spacing is increased in equal intervals, the depth representing the average resistivity also increases in equal intervals. This assump-tion is not accurate, as you should realize at this point. Once again, you could test this method by generating data using ERModel, plotting and interpreting it using the Barnes procedure, and comparing the results with your initial model. As was the case for the Moore method, the Barnes method works for certain subsurface configurations but not for others.

Both methods function best when data are collected at equal electrode-spacing inter-vals on a linear scale. The analytical methods to be discussed next assume that data are col-lected at electrode-spacing intervals that are equally spaced on a logarithmic scale. Because these analytical methods are preferred for quantitative interpretation, using either the Moore or the Barnes method will require additional field time to collect data or to interpolate exist-ing data.

Analytical Methods—Curve Matching

In discussing curve-matching techniques we restrict examples to the case of the single hori-zontal interface and an expanding-spread traverse. However, these techniques can be extended to any situation for which the appropriate equations exist (such as Eq. (5-36)). Curve-matching techniques as described here have essentially been supplanted by the com-puter techniques that are described in the following section. Nevertheless, we believe that understanding this approach and working with and observing many different types of curves provides useful exposure and practice. Some authors recommend curve matching as a preliminary step to approaching the computer.

Consider the following. If we have a field curve for which apparent resistivity is plot-ted against electrode spacing and can match this curve with a curve calculated by ERModel for a given model data set, we then have a solution to our field data. The actual subsurface geology must be very close to the model on which the theoretical curve is based. Therefore, given a field curve, all we have to do is find a similar theoretical curve and a solution is at hand. Unfortunately, since we are working with three variables (ρ_1, ρ_2, and depth to the interface z_1), there exist an infinite number of combinations, which create an infinite number of curves through which to search.

It turns out that there exists an elegant but simple solution to this problem. If we choose a model and use ERModel to calculate apparent resistivity data and then divide ρ_a-values by ρ_1, and electrode-spacing values by z_1, we create a curve such as that labeled as $k = -0.4$ in Figure 5-33(a). If we now choose a new model with different values of ρ_1, ρ_2, and z_1 but constrain the resistivity values so we maintain the condition $k = -0.4$, our new curve plots in the same position as our first curve! In fact, any two-layer curve for which $k = -0.4$ plots at this same position. Therefore, we are able to create a family of curves, termed *master curves*, for the two-layer case. A selection of four of many possible master curves is illustrated in Figure 5-33(a).

Once master curves are available, they can be used to analyze field data. Field data, such as those in Figure 5-33(b), are plotted on transparent paper at the same scale as the master curves. The transparency is then placed over the master curves and moved about, keeping the graph axes parallel, until the field data lie on one of the master curves (Fig. 5-33(c)). The lines labeled "resistivity index" and "depth index" are located on the master curve sheet, and their intersections with the field-data graph axes are noted. These intersections provide values for ρ_1 and z_1 (see Fig. 5-33(c)). We also note the k-value (see Eq. (5-

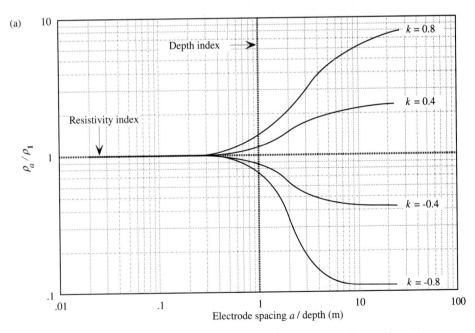

Figure 5-33 Basic procedures for curve matching. (a) Master curves for several possible k-values. (b) Field data plotted on the same scale as (a). (c) Answer obtained by superimposing field-data curve on master curves.

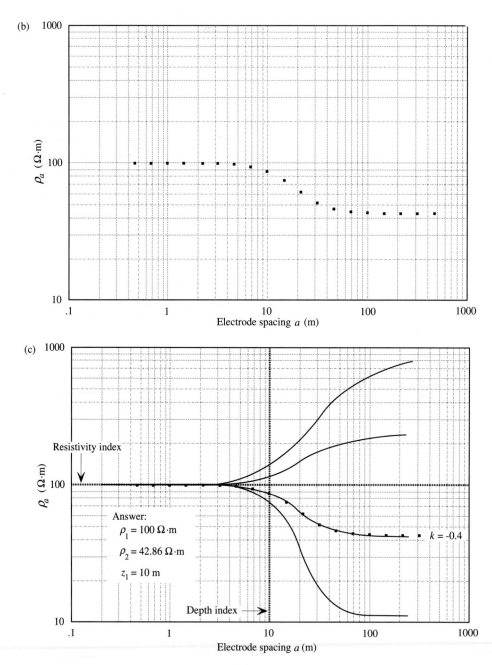

(b)

ρ_a ($\Omega \cdot$m)

Electrode spacing a (m)

(c)

ρ_a ($\Omega \cdot$m)

Resistivity index

Answer:

$\rho_1 = 100\ \Omega \cdot$m

$\rho_2 = 42.86\ \Omega \cdot$m

$z_1 = 10$ m

$k = -0.4$

Depth index

Electrode spacing a (m)

Figure 5-33 *(continued)*

298

25)) of the curve that best fits the field data. In cases where a field curve does not lie exactly on a master curve representing an integer k-value, the k-value is interpolated. Because we now know ρ_1 and because

$$\rho_2 = \left(\frac{1 + k}{1 - k} \right) \rho_1 \tag{5-37}$$

it is straightforward to determine ρ_2. Figure 5-33(c) provides values of $\rho_1 = 100 \ \Omega \cdot m$, $k = -0.4$, and $z_1 = 10$ m. Equation (5-37) then provides a value of 42.86 $\Omega \cdot m$ for ρ_2.

Generating master curves for three, four, or five layers basically follows the same procedure but is more complicated. Many sets of curves are produced and more searching is required to find proper fits between master curves and field data. Several sets of master curves have been published (Mooney and Wetzel, 1956; Orellana and Mooney, 1966; and Fläthe, 1963) and are available at many libraries for those interested in working with this approach.

Two specific problems occur when attempting to interpret field data based on comparisons with theoretical curves representing three or more layers. These problems are known as *equivalence* and *suppression*. Equivalence means that various subsurface models can generate curves that are very similar (equivalent). In other words, a given interpretation is not necessarily unique. You already observed one case of equivalence in Figure 5-22, in which a three-layer curve was very similar to a two-layer curve. Another case is documented in Figure 5-34(a), in which two three-layer models generate quite similar curves. Given the noise inherent in field data, one of these likely could not be distinguished from the other.

Suppression occurs when the presence of a thin layer in a multilayer sequence cannot be recognized on an apparent resistivity curve. Suppression is controlled by the relative thickness of a layer (thickness of the layer / depth of the layer) and the resistivity contrast between the thin layer and adjacent layers. Figure 5-34(b) illustrates curves produced by a three-layer and a five-layer model that are very similar. Your major observation should center on the form of the Model 1 curve. Even though the model consists of five layers, the curve essentially is identical in form to a common type of three-layer curve.

Note that the problems of equivalence and suppression are also present when fitting model curves to field data by computer. We now turn to this approach.

Analytical Methods—Automated Curve Matching

As noted at the beginning of our discussion of curve matching, computer analysis of field curves largely has supplanted the curve-matching technique. In performing a computer-based analysis the geophysicist enters field observations and an initial model that he or she believes is responsible for the observed values. The initial model should produce a curve of the same type and as close in form as possible to the observed curve. Hence, if one is inexperienced in interpreting resistivity curves, a preliminary step using curve matching often saves computer time and results in more realistic solutions.

After field data and an initial model are entered, the computer program calculates an apparent resistivity curve based on the model and compares this curve to the observed

(a)

(b)

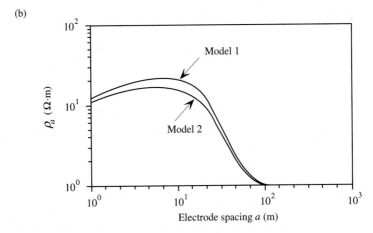

Figure 5-34 ERModel curves illustrating equivalence and suppression. (a) An example of equivalence. Model 1: layer 1 (3 m, 10 $\Omega \cdot$ m); layer 2 (10 m, 100 $\Omega \cdot$ m); layer 3 (1 $\Omega \cdot$ m). Model 2: layer 1(3 m, 10 $\Omega \cdot$ m); layer 2 (20 m, 50 $\Omega \cdot$ m); layer 3 (1 $\Omega \cdot$ m). (b) An example of suppression. Model 1: layer 1 (1 m, 10 $\Omega \cdot$ m); layer 2 (1 m, 40 $\Omega \cdot$ m); layer 3 (1 m, 10 $\Omega \cdot$ m); layer 4 (7 m, 40 $\Omega \cdot$ m); layer 5 (1 $\Omega \cdot$ m). Model 2: layer 1 (1 m, 10 $\Omega \cdot$ m); layer 2 (13 m, 20 $\Omega \cdot$ m); layer 3 (1 $\Omega \cdot$ m).

curve. Depending on the variation between the two, adjustments are made to the model according to the mathematical procedure used in the computer algorithm, and a new model curve is calculated. This process continues until both curves are as similar in shape as specified by the person guiding the analysis or until the computer program is unable to make the curves any more similar. Depending upon the results of the automated interpretation, the geophysicist may accept the final model or may initiate another analysis by specifying a new model.

The program ERSolve that is included with this text follows this general procedure. It performs analysis for either Wenner or Schlumberger data. It is possible to fix certain model parameters so they cannot be changed by the computer program during the analytical procedure. Full details about this program are in Appendix C.

A slightly different computer-based approach is suggested by Zohdy (1989). Zohdy notes that apparent resistivity curves tend to be out of phase with true resistivity-depth curves and that the values of observed apparent resistivities are always less than or equal to the actual resistivities. These and other properties of apparent-resistivity curves for horizontally stratified materials lead to the following iterative approach. The observed apparent resistivities and electrode spacings are used as the beginning model (i.e., the apparent resistivities are treated as true resistivities and the electrode spacings are used as depths. Of course, we realize this is not the case, but this assumption is made only to generate an initial curve. A theoretical curve is computed for this model. The theoretical curve is compared to the observed curve. If the agreement is not good (say within 2 percent), model depths are uniformly decreased by 10 percent, and a new theoretical curve is calculated. This continues until good agreement is reached. Resistivities used in the model then are adjusted by a similar interactive procedure until observed and computed curves agree to within set conditions. The final model represents a solution to the observed data.

Other computer-based procedures will doubtless appear. Because of equivalence and suppression and because of the many factors that affect curve details, a thorough knowledge of resistivity principles and as much experience as possible must go hand-in-hand with computer-derived solutions for field measurements.

APPLICATIONS OF ELECTRICAL-RESISTIVITY SURVEYING

In terms of the material discussed up until this point you may wonder if electrical resistivity surveying is a viable tool for exploration purposes. Although it is true that resistivity interpretation is not particularly straightforward, especially when compared to the techniques presented for seismic work, resistivity surveying can be extremely valuable. In some cases it is the method of choice simply because of cost. For example, in many areas it is possible to determine reasonably accurate depths to bedrock using resistivity. This may be a more cost-efficient procedure than investing in relatively expensive seismic equipment. However, resistivity applications are most generally used when other geophysical methods simply cannot supply the desired information. It is these cases that we emphasize in the following descriptions.

Applications Related to Aquifers

Many communities that depend on confined aquifers for their water supplies obviously want to prevent contamination of this supply or to have some idea of the extent of the aquifer for planning purposes. In many cases a buried aquifer (sand and gravel) is overlain by a clay layer that by and large prevents infiltration of solutions from above. Therefore any information about the lateral distribution of the clay-gravel contact, its depth, and the clay thickness is invaluable. Refraction seismic methods often cannot map the contact because of a velocity inversion, and reflection methods, if they can detect the contact, may be too expensive or elaborate for use.

In Chapter 2 we introduced the Whately study and noted that resistivity methods also were used. In this case resistivity was effective in mapping bedrock depths which tended to correlate with the location of a buried aquifer. Figure 5-35(a) illustrates a typical apparent resistivity curve from the Whately area. The steep downward segment of the curve is due to a thick clay layer (approximately 50 m), and the last upward segment reveals the presence of bedrock. Analysis of the curve yields a bedrock depth of 71 m, which is abnormally deep for the local area, but a well less than 1 km away along the trend of the buried aquifer gives

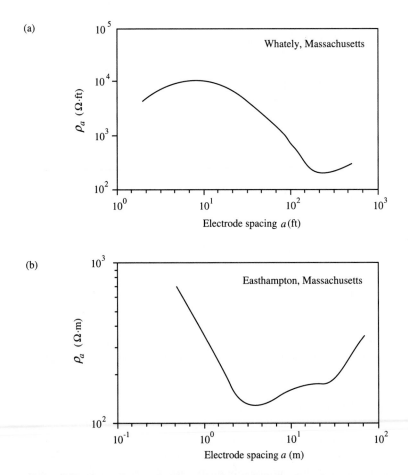

Figure 5-35 Expanding-spread, Wenner-configuration field curves. In both cases the presence of a thick, low-resistivity clay layer and the presence of high-resistivity bedrock are evident. (a) Whately, Massachusetts. (b) Easthampton, Massachusetts.

a bedrock depth of 75 m. Although the gravel aquifer cannot be detected on the curve, the bedrock depth suggests it should be present.

Easthampton, Massachusetts, also depends for most of its water supply on a confined aquifer. The aquifer is a glacial sand and gravel deposit that rests for the most part on Triassic sedimentary rocks. The sand and gravel deposit is overlain by glacial lake clays which vary in thickness, thinning to zero in the recharge area where the aquifer is exposed at the surface. In 1981 a group of Smith College students and faculty conducted a study, supported by funds from the Shell Oil Co. Foundation, to outline the thickness and extent of the aquifer. The study utilized both seismic refraction and resistivity methods and accomplished its objectives. Figure 5-35(b) is an expanding-spread profile from this study. This particular profile yielded a clay thickness of 31 m which, when compiled with many other such determinations, produced a contour map for clay thickness and distribution.

In some situations it might be of interest to map the depth of the water table. If the surface deposits are homogeneous, then the lower resistivity of the water-saturated material will impart a sufficiently great resistivity contrast for analysis. However, if the near-surface materials are not homogeneous, it will then be very difficult or impossible to map the water table with any precision. Moreover, the extent to which the capillary fringe is developed also will affect how closely the actual depth of the water table can be determined.

Resistivity surveys have been used successfully at a number of sites to map buried stream channels. Most of these surveys utilized both constant-spread and expanding-spread traversing to gather sufficient information to locate a buried stream channel. Shallow reflection work might be able to produce superior results but would entail at least equivalent field time as well as more expensive equipment. Figure 5-36(a) is an apparent resistivity map from a study in the Penitencia, California, area which is northeast of San Jose (Zhody et al., 1974, p. 47–50). This map was produced from many constant-spread traverses using an *a*-spacing of 6.1 m. Contours of equal apparent resistivity values define a zone of high resistivity oriented approximately east-west. A geologic section (Fig. 5-36(b)) based on four expanding-spread traverses, the apparent resistivity profile along the section line *BA*, and three boreholes reveal that the high-resistivity trend on the apparent resistivity map is a zone of gravel and boulders that define the location of a buried stream channel.

Applications Related to Contamination

Some of the most important successes of the resistivity method are based on contamination of normal groundwater. Usually, contaminated groundwater, whether affected by leachate from a landfill or by saltwater intrusion, has a greater conductivity than ordinary groundwater. Therefore, materials containing the contaminated water will possess lower resistivities than materials containing unaffected groundwater. If the water table is fairly shallow and the subsurface is relatively homogeneous, it is possible to map the extent of the contamination.

Ideally, traverses are implemented *before* contamination occurs. Such data document resistivity variations due to lateral and vertical variations in materials. Deviations from these values, based on traverses taken after contamination, provide a good basis for delineating the horizontal and vertical extent of the problem. Figure 5-37 summarizes the man-

(a)

(b)

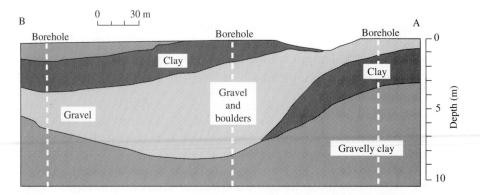

Figure 5-36 Mapping a buried stream channel using resistivity. (a) Apparent resistivity map, Penitencia, California, prepared using constant-spread traverses with electrode spacing = 6.1 m. (b) Geologic section along line *BA* based on resistivity and borehole information. (Modified from Zohdy, A. A. R., Eaton, G. P., and Mabey, D. R., 1974, Application of surface geophysics to ground water investigations: Techniques of Water-Resources Investigation of the United States Geological Survey, Book 2, Chapter D1, p. 49–50.)

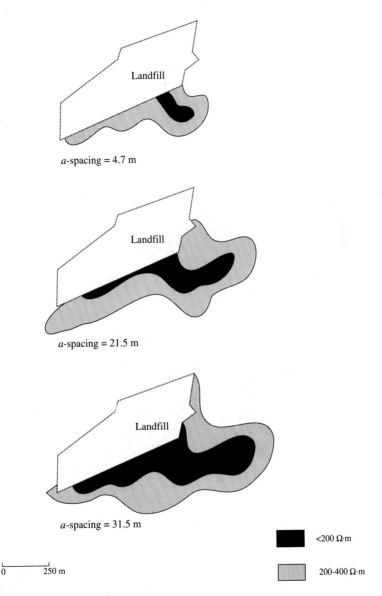

a-spacing = 4.7 m

a-spacing = 21.5 m

a-spacing = 31.5 m

| | <200 Ω·m |
| | 200-400 Ω·m |

0   250 m

Figure 5-37 Generalized diagram of leachate plumes associated with landfills. Apparent resistivity contours determined by constant-spread traversing. Note the expanding form of the leachate plume with depth.

ner in which resistivity can define the leachate plume associated with contaminated waters leaking from landfills. It is based on patterns observed in studies of several landfills located at various sites throughout the United States. The resistivity contours are determined from variations in apparent resistivities compared to normal values associated with the landfill site.

Applications in Mapping Karst and Geologic Structures

Hubbert (1932, 1944) published two classic studies demonstrating the application of constant-spread traversing in defining karst relationships and mapping fault locations. Figure 5-38 illustrates a constant-spread profile over karst topography in Hardin County, Illinois (Hubbert, 1944). Low apparent resistivity values are associated with clay at the surface, whereas somewhat higher apparent resistivity values are associated with masses of limestone at shallow depths which are overlain by clay. The highest apparent resistivity values correlate with limestone at the surface or large voids (caverns), which have infinite resistivity.

Hubbert (1932) also located faults between blocks of sandstone and limestone and a shear zone in sandstone by a constant-spread traverse (Fig. 5-39). Note the similarity of the apparent resistivity curves in the vicinity of faults to the theoretical curves for vertical contacts (Figs. 5-23 and 5-25). If the 30-m station interval was reduced, the increased station density likely would result in curves with even more similarities to theoretical curves. Both Hubbert's studies confirm that, given sufficient resistivity contrast and relatively straightforward geology, constant-spread traversing is capable of locating vertical geologic contacts and similar features of interest.

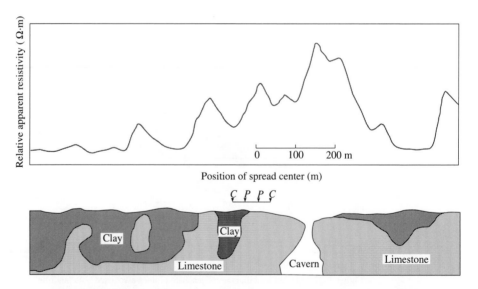

Figure 5-38   Apparent resistivity curve for a constant-spread traverse over karst topography, Hardin County, Illinois. Wenner array, *a*-spacing = 30.48 m, spread-center spacing = 30.48 m. Apparent resistivity highs correspond to limestone and voids; apparent resistivity lows correlate with clays. (Modified from Hubbert, M. K., 1944, An exploratory study of faults in the Cave in Rock and Rosiclare districts by the earth resistivity method *in* Geological and geophysical survey of fluorspar areas in Hardin County, Illinois: United States Geological Survey Bulletin 942, Part 2, p. 73–147.)

Figure 5-39 Apparent resistivity curve for a constant-spread traverse over faults in Illinois. Wenner array, *a*-spacing = 30.48 m, spread-center spacing = 30.48 m. Curve form is similar to that predicted over vertical contacts. (Based on data in Hubbert, M. K., 1932, Results of earth-resistivity survey on various geologic structures in Illinois: American Institute of Mining and Metallurgical Engineers Technical Publication 463, p. 16.)

Other Applications

Resistivity surveys have been used to outline new geothermal fields in areas of known geothermal potential. Geothermal waters are better conductors than normal waters due to their greater content of dissolved salts.

Because of the typically high resistivity contrast between rock and sediment, resistivity surveys are often able to delineate buried stone walls and foundations. Because surveys are also sensitive to differences in moisture content of materials, they are often capable of differentiating between excavated and unexcavated ground. For both these reasons resistivity has been a frequent tool in archaeological studies.

The complexities inherent in resistivity interpretations and the nonuniqueness of potential solutions to observed data have tended to obscure the value of resistivity surveys. Although the method does have these drawbacks, when combined with seismic surveys or the other electrical methods detailed next, it can be an effective exploration tool. However, when using resistivity to explore subsurface geologic relationships, independent control from well logs is mandatory. Its capability for use as an inexpensive monitoring tool has not been developed as fully as possible, but this likely will change as our society is faced with more and more contamination and the need to locate and remedy the problem.

OTHER ELECTRICAL METHODS

As noted at the beginning of this chapter, the following descriptions of additional electrical and electromagnetic methods are purposely brief. We include them mainly to provide a sense of applications, limitations, advantages, and how they may be similar to or differ from electrical resistivity methods.

Induced Potential (IP)

Explorationists initially involved with the development and application of electrical resistivity surveying noticed that the potential difference does not immediately decline to zero after current is turned off. Rather, a large percentage of the potential disappears immediately, but the remainder gradually decreases over an interval of time, usually several seconds. Subsequent work demonstrated that certain subsurface materials are capable of electrical polarization and function analogously to a rechargeable battery. Applied current induces a buildup of charges which, when the current is removed, gradually discharge, producing a flow of current and potential differences.

This decay of potential with time is referred to as *induced potential in the time domain* and is the phenomenon we discuss next. One can also investigate the effect of different frequencies of alternating current on apparent resistivity values. This approach is termed IP in the frequency domain. A common measure of the effect of IP in the time domain is *chargeability*, which is defined as the ratio of the area under the decay curve to the potential difference measured before the current was turned off. It is possible to demonstrate a theoretical relationship between apparent resistivity and chargeability. This is extremely valuable because it permits us to derive quantitative relationships between subsurface geometries and apparent chargeabilities if we know the appropriate equations for apparent resistivity, such as those that were presented for the horizontal interface.

Because of this specific relation between apparent resistivity and apparent chargeability, field procedures and data interpretation techniques are similar. Electrode arrays typically are the Schlumberger or dipole-dipole, and apparent chargeability is plotted against electrode spacing. Curve matching is commonly employed. Historically, IP has found much more application in the exploration for base metals rather than in typical electrical resistivity applications such as groundwater studies. One reason is that equipment tends to be bulkier and data acquisition slower. Field work is more expensive, and considering the relative lack of interpretive techniques, IP tends not to be cost effective unless the economic potential is substantial.

Equipment for conducting induced potential studies is similar to electrical resistivity equipment, with two major exceptions. A timing circuit is required to switch the current on and off, and a recording unit is necessary in order to preserve the decrease of potential with time. Standard electrical-resistivity equipment cannot, therefore, be used for IP studies, but equipment designed for IP work can be used to measure apparent resistivities.

Induced potential studies can be valuable assets to electrical-resistivity studies. Based on our discussion of interpretive procedures for resistivity, it is apparent that curves that appear due to a single horizontal interface may actually contain the signature of several layers but cannot be differentiated. It is quite possible that one or more of these "hidden" layers possess a different chargeability. If apparent resistivity and apparent chargeability curves are plotted on the same graph, such distinctions may permit more informed, and more correct, interpretations of the subsurface. A good, recent elaboration of this method is presented in Ward (1990).

Spontaneous Potential (SP)

Natural potential differences at the surface may arise from electrochemical reactions at depth which cause currents to flow. The most common example of this type of phenomenon is associated with ore bodies part of which are above the water table. Oxidation takes place above the water table while reduction takes place below it. These processes result in different charge concentrations, which result in current flow. Sulfide and graphite bodies are by far the most common source of significant potentials of this sort, so it is not surprising that most SP surveys are conducted to locate ore bodies, especially sulfide ones.

Equipment demands are fairly straightforward. Two electrodes, wire, and a millivolt-meter constitute the normal setup. In this case porous-pot electrodes must be used, for metal electrodes themselves produce a self-potential. Porous-pot electrodes consist of a ceramic container with holes. A copper tube is surrounded by a solution of copper sulfate which slowly leaks through the holes into the ground. This solution disseminates the charges that otherwise would build up if the ground was in direct contact with a copper electrode.

Several characteristics conspire to negate the possibility of wide application of the self-potential method. Although surveys are quick and efficient and equipment demands are minimal, quantitative interpretation is difficult, and depths of sources are difficult to model. Self-potentials are easily masked by other sources, such as telluric currents. Because the water table is an important component of most sources, its typically shallow depth limits surveys to the same shallow terrain. And finally, of course, sources are limited to the specific types mentioned. SP exploration, therefore, is not a common shallow method except where one desires a relatively fast, inexpensive tool in searching for shallow sulfide bodies. A detailed discussion of SP exploration can be found in Corwin (1990).

Telluric and Magnetotelluric Methods

Natural electric currents flow in the earth which are induced by the flow of charged particles in the ionosphere. At any given moment these currents follow uniform patterns over large areas. However, currents are flowing at depth, and based on information from this chapter, it seems reasonable to expect these currents to describe patterns based on the conductivity and geometry of the materials through which they flow. Quite naturally, we measure the potential differences produced by this current flow.

Because telluric currents vary with time, field operations include a base station at which two pairs of electrodes are placed at right angles. Voltage is continuously monitored. Two additional pairs of electrodes then are moved along a grid. Comparison between the base station readings and readings taken at grid locations yields a potential difference that is due to the subsurface effect on current density.

Magnetotelluric methods combine the measurement of magnetic field intensity variations with voltage measurements. Both sets of data are used to compute apparent resistivity functions which then are compared with theoretical curves to determine thicknesses and true resistivities. A major requirement of this approach is a magnetometer capable of measuring weak magnetic field variations that fluctuate frequently.

Most surveys of this type have been in Europe and the Soviet Union. Survey objectives typically are directed at more regional objectives than we have mentioned to date, such as determining the geometry of sedimentary rock sequences in major depositional basins.

Electromagnetic Surveying (EM)

In electromagnetic surveying alternating current is passed through a wire loop, which produces a local magnetic field termed the *primary field*. This in turn produces current flow in the earth, which produces a secondary magnetic field that varies in amplitude and phase from the primary. The secondary field is measured by a receiver coil in which current flow is induced by the secondary field. Because the primary field is known, it is possible to isolate the secondary.

Because current flow in the subsurface is governed by the distribution of conductivity, the characteristics of the secondary field should provide information about the distribution of conducting material. Unfortunately, quantitative interpretation is fairly complex, but signatures of good conductors are sufficiently clear that the method is well suited to reconnaissance surveys.

Because no electrodes are involved, field work is fairly rapid and even can be airborne. Equipment demands are not unreasonable, so this method is relatively inexpensive. However, depth penetration is poor and good conductors must be present. These limitations have confined electromagnetic surveying primarily to ore exploration, although some studies have been directed at groundwater. For a good, modern review of groundwater applications of EM consult McNeill (1990). EM applications have also been developed for archaeological studies and have enjoyed some successes (Frohlich and Lancaster, 1986).

PROBLEMS

5-1 Determine the potential difference between the two potential electrodes for cases (a) and (b). Assume a current of 0.6 ampere.

(b)

C_1 ▽ Source

40 m

Sink ▽ C_2

4 m

6 m

$\rho = 40\ \Omega \cdot m$

6 m

Depth

P_1 ▽

▽ P_2

5-2 Construct the current-flow lines beneath the interface in (a) and (b).

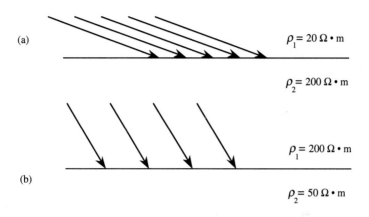

(a)

$\rho_1 = 20\ \Omega \cdot m$

$\rho_2 = 200\ \Omega \cdot m$

$\rho_1 = 200\ \Omega \cdot m$

(b)

$\rho_2 = 50\ \Omega \cdot m$

5-3 Calculate the potential at P_1 due to a current at C_1 of 0.6 ampere. The material in this section view extends to infinity in all directions. The bold line represents an interface between ρ_1- and ρ_2-material.

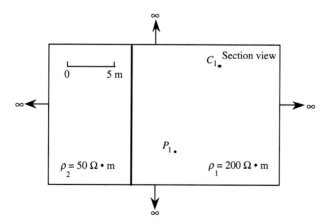

∞

C_1 • Section view

0   5 m

∞ ←   → ∞

P_1 •

$\rho_2 = 50\ \Omega \cdot m$ $\rho_1 = 200\ \Omega \cdot m$

∞

5-4 Let A and B represent two different geologic sections, and let H_1 and H_2 represent the thicknesses of the first and second layers in a three-layer sequence. It is well known that one type of equivalence occurs for a three-layer case when $\rho_1 < \rho_2 > \rho_3$, $\rho_{1A} = \rho_{1B}$, $\rho_{3A} = \rho_{3B}$, $H_{1A} = H_{1B}$, and $\rho_{2A}H_{2A} = \rho_{2B}H_{2B}$. Demonstrate that this is true.

5-5 Assume the same general situation as explained in Problem 5-4. Does equivalence exist if $\rho_1 > \rho_2 < \rho_3$? Document your position.

5-6 For each of the following subsurface models sketch an appropriate apparent resistivity curve on the designated graph. The general shape of the curve is what is important. Do not use ERModel to generate the curves; rather, base your curves on what you know about current penetration, current density, and measured resistivities.

(a)

Layer	Thickness (m)	Resistivity ($\Omega \cdot$ m)
1	1	100
2	5	10
3	5	100
4	Infinite	1

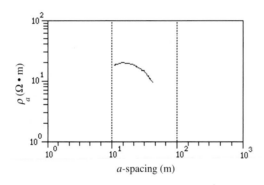

(b)

Layer	Thickness (m)	Resistivity ($\Omega \cdot$ m)
1	10	100
2	10	10
3	Infinite	100

(c)

Layer	Thickness (m)	Resistivity ($\Omega \cdot$ m)
1	5	1
2	10	100
3	5	1
4	Infinite	100

5-7 The following data were acquired using a constant-spread, Wenner traverse (a-spacing = 3 m). Interpret the data as completely as possible.

Horizontal Position (m)	ρ_a ($\Omega \cdot$ m)	Horizontal Position (m)	ρ_a ($\Omega \cdot$ m)	Horizontal Position (m)	ρ_a ($\Omega \cdot$ m)
5.00	20.05	19.00	22.55	33.00	858.94
6.00	20.06	20.00	25.32	34.00	872.33
7.00	20.07	21.00	28.45	35.00	880.40
8.00	20.08	22.00	27.97	36.00	885.58
9.00	20.10	23.00	27.11	37.00	889.07
10.00	20.12	24.00	242.39	38.00	891.51
11.00	20.15	25.00	460.00	39.00	893.26
12.00	20.19	26.00	572.39	40.00	894.57
13.00	20.24	27.00	580.25	41.00	895.55
14.00	20.32	28.00	541.30	42.00	896.31
15.00	20.44	29.00	519.88	43.00	896.90
16.00	20.61	30.00	660.54	44.00	897.38
17.00	20.91	31.00	785.22	45.00	897.76
18.00	21.45	32.00	834.74		

5-8 Refer to Figures 5-9 and 5-30(c) and Eq. 5-20. Derive an apparent resistivity equation for the Schlumberger array (see the derivation of Eq. 5-35 for the Wenner array).

5-9 Using ERModel or Table 5-4 create data for a k-value of -0.6. Using this data develop a master curve similar to those illustrated in Figure 5-33(a). Using at least three different models, demonstrate that all two-layer models with a k-value of -0.6 plot on the master curve.

5-10 Interpret the following data, which were obtained with a Wenner, expanding-spread traverse.

Electrode Spacing (m)	ρ_a ($\Omega \cdot$ m)	Electrode Spacing (m)	ρ_a ($\Omega \cdot$ m)	Electrode Spacing (m)	ρ_a ($\Omega \cdot$ m)
1.00	984	14.68	72	215.44	327
1.47	955	21.54	53	316.23	432
2.15	883	31.62	63	464.16	554
3.16	742	46.42	87	681.29	686
4.64	533	68.13	124	1000.00	818
6.81	311	100.00	174		
10.00	150	146.78	241		

5-11 Interpret the following data, which were obtained with a Schlumberger, expanding-spread traverse.

Electrode Spacing (m)	ρ_a ($\Omega \cdot$ m)	Electrode Spacing (m)	ρ_a ($\Omega \cdot$ m)	Electrode Spacing (m)	ρ_a ($\Omega \cdot$ m)
1.00	108	14.68	307	215.44	293
1.47	121	21.54	245	316.23	381
2.15	148	31.62	168	464.16	479
3.16	191	46.42	122	681.29	580
4.64	244	68.13	125	1000.00	675
6.81	295	100.00	162		
10.00	323	146.78	220		

5-12 The following data were gathered with a Wenner, expanding-spread traverse in an area with thick alluvial deposits at the surface. What is the likely depth to the water table?

Electrode Spacing (m)	ρ_a ($\Omega \cdot$ m)	Electrode Spacing (m)	ρ_a ($\Omega \cdot$ m)
0.47	198	6.81	84
0.69	160	10.00	82
1.00	140	14.68	92
1.47	112	21.54	101
2.15	95	31.62	100
3.16	84	46.42	102
4.64	79		

5-13 The following data were gathered with a Wenner, expanding-spread traverse in an area of thick deltaic sands. Bedrock depths are greater than 30 m. What is your best estimate of the depth to the water table in this area?

Electrode Spacing (m)	ρ_a ($\Omega \cdot$ m)	Electrode Spacing (m)	ρ_a ($\Omega \cdot$ m)
0.47	2590	6.81	8753
0.69	3288	10.00	7630
1.00	4421	14.68	4805
1.47	5198	21.54	2160
2.15	6055	31.62	995
3.16	6686	46.42	584
4.64	7782		

5-14   The following data were gathered with a Wenner, expanding-spread traverse in an area of dune sands underlain by lake clays which in turn are underlain by Triassic sedimentary rocks. Estimate a value for clay thickness.

Electrode Spacing (m)	$\rho_a\ (\Omega \cdot m)$	Electrode Spacing (m)	$\rho_a\ (\Omega \cdot m)$
0.69	1298.90	10.00	360.67
1.00	1398.06	14.68	240.11
1.47	1306.98	21.54	191.51
2.15	1153.02	31.62	153.81
3.16	925.27	46.42	116.38
4.64	762.40	68.13	98.03
6.81	554.13	100.00	86.08

REFERENCES CITED

Barnes, H. E., 1952, Soil investigations employing a new method of layer value determination for earth resistivity investigation: Highway Research Board, Bulletin 65, p. 26–36.

Corwin, Robert F., 1990, The self-potential method for environmental and engineering applications *in* Ward, Stanley H., ed., Geotechnical and environmental geophysics; Volume 1: Review and tutorial: Society of Exploration Geophysicists Investigations in Geophysics No. 5, p. 127–145.

Fetter, C. W., Jr., 1980, Applied hydrogeology: Columbus, Ohio, Merrill, 488 p.

Fläthe, H., 1963, Five-layer master curves for the hydrogeological interpretation of geoelectric resistivity measurements above a two-story aquifer: Geophysical Prospecting, v. 11, p. 471–508.

Frohlich, Bruno, and Lancaster, Warwick J., 1986, Electromagnetic surveying in current Middle Eastern archaeology: Application and evaluation: Geophysics, v. 51, p.1414–1425.

Hubbert, M. K., 1932, Results of earth-resistivity survey on various geologic structures in Illinois: American Institute of Mining and Metallurgical Engineers Technical Publication 463, p. 9–39.

Hubbert, M. King, 1940, The theory of groundwater motion: Journal of Geology, v. 48, p. 785–944.

Hubbert, M. K., 1944, An exploratory study of faults in the Cave in Rock and Rosiclare districts by the earth resistivity method *in* Geological and geophysical survey of fluorspar areas in Hardin County, Illinois: United States Geological Survey Bulletin 942, Part 2, p. 73–147.

Keller, George V., and Frischknecht, Frank C., 1966, Electrical methods in geophysical prospecting: Oxford, Pergamon Press, 517 p.

McNeill, J. D., 1990, Use of electromagnetic methods for groundwater studies *in* Ward, Stanley H., ed., Geotechnical and environmental geophysics; Volume 1: Review and tutorial: Society of Exploration Geophysicists Investigations in Geophysics No. 5, p. 191–218.

Mooney, Harold M., 1958 , A qualitative approach to electrical resistivity interpretation: Pure and Applied Geophysics, v. 40, p. 164–171.

Mooney, H. M, and Wetzel, W. W., 1956, The potential about a point electrode and apparent resistivity curves for a two-, three-, and four-layer earth: Minneapolis, Minnesota, University of Minnesota Press, 145 p., 243 plates.

Moore, W., 1945, An empirical method of interpretation of earth resistivity measurements: Transactions of the American Institute of Mining and Metallurgical Engineers, v. 164, p. 197–223.

Orellana, Ernesto, and Mooney, H. M., 1966, Master tables and curves for vertical electrical sounding over layered structures: Madrid Interciecia, 150 p., 66 tables.

Telford, W. M., Geldart, L. P., and Sheriff, R. E., 1990, Applied geophysics, 2nd ed.: Cambridge, Cambridge University Press, 770 p.

Van Nostrand, Robert G., and Cook, Kenneth L., 1966, Interpretation of resistivity data: United States Geological Survey Professional Paper 499, 310 p.

Ward, Stanley, H., 1990, Resistivity and induced polarization methods *in* Ward, Stanley H., ed., Geotechnical and environmental geophysics, Volume 1: Review and tutorial: Society of Exploration Geophysicists Investigations in Geophysics No. 5, p. 147–189.

Zohdy, A. A. R., 1989, A new method for the automatic interpretation of Schlumberger and Wenner sounding curves: Geophysics, v. 54, p. 245–253.

Zohdy, A. A. R., Eaton, G. P., and Mabey, D. R., 1974, Application of surface geophysics to ground-water investigations: Techniques of Water-Resources Investigation of the United States Geological Survey, Book 2, Chapter D1, 116 p.

SUGGESTED READING

Mooney, Harold M., 1958 , A qualitative approach to electrical resistivity interpretation: Pure and Applied Geophysics, v. 40, p. 164–171.

Zohdy, A. A. R., Eaton, G. P., and Mabey, D. R., 1974, Application of surface geophysics to ground-water investigations: Techniques of Water-Resources Investigation of the United States Geological Survey, Book 2, Chapter D1, 116 p.

6

Exploration
Using Gravity

Because an object on the earth's surface is attracted by the mass of the earth, the gravity exploration method detects variations in the densities of subsurface materials by measuring gravity at the surface and analyzing the differences in the recorded values. Using gravity to explore the subsurface appears to be very straightforward compared with some of the difficulties we encountered when deriving relationships for electrical resistivity surveying. However, if gravity exploration were a panacea, the gravity method would be at the beginning of this text, and the text itself would be much thinner.

One difficulty becomes apparent if we consider the magnitude of the variations we must measure. An average value for gravity on the earth's surface is 980 cm/s^2, but variations as small as 0.00001 percent of average Earth gravity need to be detected in routine surveys. This demands instruments with great sensitivity. In addition, factors other than subsurface density changes cause gravity to vary. All these causes must be identified and corrections for their effects applied to our data. For example, changes in elevation of the land surface lead to variations in gravity, as observation points will be closer to or farther from the center of the earth and more or less material will be between the observation point and the earth's center. Elevations of observation points must be known with reasonable accuracy—often to at least 0.25 m. This requirement adds time and cost to any survey but cannot be ignored if the data are to be useful.

Another factor affecting exploration using gravity is the complexity of the subsurface. All the material beneath an observation site affects a gravity value. Therefore, we are constantly are attempting to remove the effects due to bodies in which we have no interest in order to isolate the effects of items of interest. This is a task for which we typically achieve only degrees of success, never complete success. Computing the effect of subsurface masses of varying shapes and densities is not too difficult, and surely is easier than determining apparent resistivities for complex shapes, but we face a similar problem in that a given curve of observed gravity values can be produced by numerous subsurface configurations.

Although the approach of this chapter is similar to that of those that precede it, and the same general topics are considered, you should concentrate on those tasks and requirements that strive to provide the most accurate values of gravity possible and to then process these values so what remains is related to bodies of interest in the subsurface.

FUNDAMENTAL RELATIONSHIPS

Gravitational Acceleration

Newton's law of universal gravitation states that there is a force of attraction F between two particles with masses m_1 and m_2 separated by a distance r which is represented by the relationship

$$F = G \frac{m_1 m_2}{r^2}$$ (6-1)

where G is the universal gravitation constant. The value of G was first determined in 1798 by Lord Cavendish; the present value, which was determined in 1942, is equal to

$$6.6732 \times 10^{-11} \, \text{N} \cdot \text{m}^2/\text{kg}^2 \quad \text{in SI units}$$

or

$$6.6732 \times 10^{-8} \, \text{dyn} \cdot \text{cm}^2/\text{g}^2 \quad \text{in cgs units}$$

If we assume that the earth is spherical, then the force exerted by the earth on a spherical body with mass m resting on the earth's surface is

$$F = G \frac{mM}{R^2}$$ (6-2)

where M is the earth's mass and R is its radius. This also assumes that density varies only with distance from the center of the earth (that the earth is spherically symmetrical) and that R is large compared to the size of the object. Values that we will use in future calculations include

Earth radius (at equator): 6.367×10^8 cm
Earth radius (at 45° latitude): 6.378×10^8 cm
Earth mass: 5.976×10^{27} g

Force is also given by Newton's second law of motion, which states that

$$F = ma$$ (6-3)

where a is acceleration. If we define g as the gravitational acceleration, then when the acceleration is caused only by the gravitational attraction of the earth, we can write

$$F = mg = G\frac{mM}{R^2}$$

and

$$g = \frac{GM}{R^2} \tag{6-4}$$

The value g, is expressed as m/s^2 (SI) or cm/s^2 (cgs). In geophysics the normal unit of gravitational acceleration is the Gal (in honor of Galileo), which is 1 cm/s^2. Because the variations in gravitational acceleration g (or gravity for short) in which we are interested are so small, we often use the milliGal (1 mGal = 0.001 Gal) for exploration purposes. The gravity unit gu often is used instead of the mGal. It is equal to 0.000001 m/s^2, which is 0.1 mGal.

MEASURING GRAVITY

Ultimately we are interested in determining the acceleration due to gravity at various sites around the world. Instruments designed to measure gravity directly perform *absolute* measurements. Such devices are elaborate, difficult to transport, and require considerable time to set up and use. Somewhat easier to design and use are instruments that determine the difference in gravity between two observing stations. These obtain *relative* differences in gravity and are the instruments used in exploration surveys. Because this type of instrument is the one you are most likely to encounter, we discuss relative measurements first.

Relative Measurements Using a Pendulum

The classic instrument for measuring both absolute and relative gravity is the pendulum. An ideal, simple pendulum suspends a material point that is dimensionless from a massless string that does not stretch and is perfectly flexible. For such a device the period T is related to the length of the string l and gravitational acceleration by the relationship

$$T = 2\pi\sqrt{\frac{l}{g}} \tag{6-5}$$

Such an ideal pendulum cannot be constructed, as should be obvious from its description. For a real pendulum Eq. 6-5 becomes

$$T = 2\pi\sqrt{\frac{K}{g}} \tag{6-6}$$

where K is a constant that represents the characteristics of a particular pendulum system. If we are able to determine T and K accurately, then we have a measured or observed value for gravity which we refer to as g_{obs} Unfortunately, K cannot be determined accurately, and absolute values measured in this fashion have an accuracy of no better than 1 mGal, and often are not that good. In routine surveys we need to know g_{obs} to at least 0.1 mGal.

However, all is not lost. If we use the same pendulum at two observation sites, say x and y, then we have

$$g_{obs_x} = \frac{4\,\pi^2 K}{T_x^{\,2}} \quad \text{and} \quad g_{obs_y} = \frac{4\,\pi^2 K}{T_y^{\,2}} \tag{6-7}$$

Both values will be inaccurate due to the imprecision of K, but since the same value of K enters into both measurements, the gravity values will vary from the true value of gravity by the same factor. Therefore, they can still be used to determine the relative difference in gravity (Δg) by accurately measuring T at each site as

$$\Delta g = g_{obs_x} - g_{obs_y} \tag{6-8}$$

Precision is on the order of 0.1 mGal, but pendulum-based instruments not only are bulky but require substantial amounts of time in order to acquire a measurement. They have been superseded by the gravimeter, which is very portable. Reading times are short, and precision is on the order of 0.01 mGal.

Relative Measurements Using a Gravimeter

As in the case of the pendulum, the principle upon which gravimeter design is based is quite simple. If a mass is placed on a spring and this assembly is moved from one position to another, the spring will lengthen or shorten a small amount due to variations in gravity. However, the small changes in gravity from site to site result in only very small displacements of the spring. If we employ a very weak spring, it will extend only 1 μ (micron) for a gravity increase of 1 mGal. In order to accurately measure changes for such small gravity variations, a spring several tens of meters in length is required. The length of such a spring is clearly impractical, but it helps to illustrate another factor that enters into gravimeter design. A mass suspended from such a long spring will oscillate up and down for quite a long time after being disturbed. Such behavior does not facilitate rapid readings, so a high degree of stability is crucial if field work is to progress at an acceptable rate.

A great deal of ingenuity, careful design, and much trial and error produced various combinations of special springs attached to a mass at the end of a beam. These assemblies typically are held together by various hinges and levers. Not only do the springs have to be short and sensitive, but the design must ensure oscillation periods of only a few seconds. A critical component of most modern gravimeters is the *zero-length* spring. Tension is placed on such a spring during manufacture so that the spring is much more sensitive than normal and amplifies the displacements caused by small variations in gravity. A schematic example of the interior of one gravimeter (the Worden) utilizing an arrangement of this type is illustrated in Figure 6-1. Figure 6-2 is a photograph of the actual gravimeter.

The instrument is nulled by turning a calibrated dial until a beam produced by a pointer (Fig. 6-1) is brought to the center of the magnifying eyepiece. The setting of the dial is noted. The instrument is then moved to a new location and leveled. Because the change in gravity produces a movement of the mass attached to the zero-length spring and the

Figure 6-1   A highly diagrammatic representation of the interior of a Worden gravimeter.

pointer, the beam no longer is in the center of the eyepiece. The instrument is then nulled again by turning the calibrated dial. The amount of restoring force necessary to move the beam back to the center of the eyepiece is proportional to the change in gravity from the one observation point to the other. After the second nulling operation is performed, there is a change in the dial reading. Before shipping, the instrument was calibrated by following this same procedure at two or more locations for which accurate values of gravity are available. This procedure results in a dial constant (e.g., 0.0869 mGal/dial division) which, when multiplied by the change in dial reading, gives the gravity variation in milliGals. The Worden, as well as other similar gravimeters, is capable of measuring changes in gravity of 1 unit in 100,000,000. By following standard procedures carefully, a precision of 0.01 mGal can be obtained.

What is truly amazing is that the spring-and-lever assembly of the Worden is constructed from quartz and is small enough to fit into an average coffee mug. Problems, of course, do exist when using such a delicate instrument. The assembly is placed in a sealed flask at the factory to minimize temperature changes, which affect mechanism constants, and to maintain a constant air pressure. The instrument must be handled with great care, kept vertical, and must not be subjected to sudden accelerations. The major problem during routine use is what is termed meter *drift*. Over time the physical properties of the springs change very slightly. Because the springs are so sensitive, these changes are large enough to

Figure 6-2 Photographs of the Worden gravimeter.

affect readings, so procedures must be devised to check for and correct for these variations. Corrections for instrument drift are described in a following section that discusses field procedures.

Absolute Measurements

As noted previously, pendulums can be used to measure gravity directly but not with a precision as high as is desirable. Today most absolute measurements use what is called the *falling-body* method. The first accurate determinations using this approach were made in 1952, and by the 1960s the availability and sophistication of lasers and electronic measuring and timing devices enabled a precision of 0.01 mGal to be obtained.

All methods essentially measure the amount of time t it takes an object to fall a distance z and utilize the relationship

$$z = \frac{gt^2}{2} \tag{6-9}$$

One such approach throws a body vertically upward. The object passes two established levels on its way up and again on its way down. If the positions of the levels are z_1 and z_2, and if t_1 and t_2 are the times the object passes these levels going up, and t_3 and t_4 are the times on the way down, then it is straightforward to demonstrate that

$$g = \frac{8(z_2 - z_1)}{(t_4 - t_1)^2 - (t_3 - t_2)^2} \tag{6-10}$$

We need only to measure accurately the distance $z_2 - z_1$ and to measure accurately the four times. Such instruments clearly are not appropriate as field instruments, but because gravimeters are able to achieve the same level of precision in determining relative variations in gravity, we really require only a few sites where gravity is known accurately.

Excellent detailed descriptions of instruments for determining absolute and relative gravity are presented in Robinson and Coruh (1988) and Tsuboi (1979). If you are interested in more detail about design and operating principles, you should consult these sources.

International Gravity Standardization Net 1971 (IGSN71)

If a gravity survey is searching for shallow anomalous masses in a restricted geographic area, it does not need access to an accurate gravity value. Exploration methods utilize variation from a norm, so such a survey could assume a value for g, make accurate relative determinations, calculate absolute values based on the assumed g, and then map the departures from this value. However, once we decide to compare gravity values from various areas, especially on a worldwide basis, we require a network of observation sites at which gravity is known accurately.

Work to establish such a worldwide network began in the 1950s under the leadership of Professor George P. Woollard of the University of Wisconsin. This survey, based on very painstaking measurements using pendulums and gravimeters, was completed in 1963. Further refinements, primarily due to more accurate values from instruments utilizing the free-fall approach, necessitated adjustments to the 1963 determinations. Values from more than 1800 sites around the world were published in 1971 as a unified, internally consistent network which is referred to as the *International Gravity Standardization Net 1971 (IGSN71)*. Every gravity survey should be tied to one of these sites.

ADJUSTING OBSERVED GRAVITY

If the earth had perfect spherical symmetry and did not rotate, then gravity due to earth mass would be the same everywhere on the surface. Of course, we know this is not so; therefore, it is essential to identify the reasons that gravity varies, so that we may correct for them in our quest to use gravity to explore the subsurface. One factor clearly is position on the earth's surface, as the absolute value of gravity at the equator is 978.0 Gal, and at the poles gravity is 983.2 Gal.

Variation in *g* as a Function of Latitude

The rotation of the earth produces an outward-directed centrifugal force that acts in a direction opposite to gravity and, therefore, diminishes the measured value of *g*. The effect of this centrifugal force is greatest at the equator and diminishes to zero at the poles of the earth's rotational axis. As a direct consequence of this force, *g* is greater at the poles than at the equator by 3.4 Gal.

The long-term behavior of the earth is that of a fluid, so the earth's rotation produces centrifugal effects that cause its shape to be an *ellipsoid of revolution*. The fact that the earth is not spherical but is an ellipsoid of revolution means that it is flattened at the poles. In other words the length of the earth's radius is greater at the equator than at the poles. This distance factor causes *g* to increase from equator to pole by 6.6 Gal, because the surface is closer to the center of mass (the center of the earth) at the poles. However, since radius length is greater at the equator, more mass is positioned between the surface and the earth's center at the equator than at the poles. This mass factor causes *g* to decrease by 4.8 Gal from equator to pole.

The effect of rotation, the distance factor, and the mass factor result in a net increase in gravity by 5.2 Gal as one travels from the equator to the poles. This explains the variation of gravity from 978.0 Gal at the equator to 983.2 Gal at the poles. Recall that our ultimate goal in gravity exploration is to obtain values of gravity for which variations are entirely due to subsurface density distributions. Therefore, we must derive an equation that removes position on the earth's surface as a cause of gravity difference. Differences that remain in our measured gravity values then will be due to circumstances other than position.

Correcting for the Latitude Effect

Consider an imaginary, regular surface that corresponds to the true, gross shape of the earth. This is the surface we referred to previously as an ellipsoid of revolution, and it coincides closely with sea level over the open oceans. This surface also often is referred to as the *reference ellipsoid*, which is the term we adopt for this discussion. In detail, however, the reference ellipsoid does not coincide exactly with the sea-level surface. If we could crisscross the continents with canals open to the oceans, the water level in these canals would also be at sea level. The surface passing through the sea-level surface around the world is known as the *geoid* and contains broad undulations that are minor compared to its overall dimensions. Although the reference ellipsoid does not fit the geoid exactly, it approximates it to a degree that is acceptable for most gravity exploration.

We now need an equation that corrects for position on the reference ellipsoid. Equation 6-11 specifies such a relationship.

$$g_n = g_e \left(1 + A\sin^2 \phi - B\sin^2 2\phi \right) \text{ cm/s}^2 \qquad (6\text{-}11)$$

Because the derivation of this equation is quite lengthy and is not relevant to what follows, we do not include it here. A good development is presented in Tsuboi (1979, p. 68–78). *A* and *B* are constants that take into account Earth's angular velocity of rotation, its size, and its ellipticity. Values for *A* and *B* are determined from careful analysis of satellite orbits and from thousands of measurements of g at various latitudes. Note that longitude does not enter into the computations. Latitude is represented by ϕ, g_e stands for gravity on the equator at sea level, and g_n represents normal or standard gravity. The value for g_e is determined from actual measurements at coastal sites on the equator and from carefully adjusted values from other sites at various latitudes. Current values for A, B, and g_e were adopted by the International Association of Geodesy in 1967. When we place these values into Eq. 6-11, the result is the *geodetic reference system formula of 1967* or *GRS67*, which is presented as Eq. 6-12.

$$g_n = 978.03185\left(1 + 0.005278895\sin^2\phi - 0.000023462\sin^4\phi\right) \; \text{cm/s}^2 \qquad (6\text{-}12)$$

Normal gravity g_n represents the value of gravitational acceleration that we should observe when measuring gravity at various positions on the earth's surface, as Eq. 6-12 corrects for position. Typically, the value of gravity that is measured g_{obs} (observed gravity) does not match the value predicted by the geodetic reference system formula. The difference, $g_{obs} - g_n$, is thus anomalous and is called the *gravity anomaly*.

How accurately do we need to know our position on the earth's surface when making a gravity measurement? Table 6-1 provides an answer. At a latitude of 50° the variation in normal gravity is 88.91 mGal for a degree of latitude. If we take the length of a degree of latitude to be 111 km (it actually varies from about 110.5 km at the equator to 111.5 km at the poles), this variation translates to 1 mGal in 1248.5 m. If we want to maintain an accuracy of 0.1 mGal in our work, we must be able to locate our position to within 125 m. On a map with a scale of 1 : 25,000 this translates to a location within 0.5 cm.

Are positioning requirements more severe at low latitudes than at middle latitudes? At high latitudes than at middle latitudes?

TABLE 6-1 Latitude, Normal Gravity, and Position Requirements

Latitude	**50**	Normal gravity (mGal) 981068.6407
Latitude	51	Normal gravity (mGal) 981157.5477

Length of 1° of latitude (km)	111
Variation in g per degree (mGal)	88.91
Meters per milliGal	1248.50
MilliGal per meter	0.000800964
Position (m) for 0.1 mGal accuracy	125
Position (m) for 0.05 mGal accuracy	62

Of course, most of our measurements will not be obtained at sea level, and, as we noted in the introduction to this chapter, variations in elevation substantially affect observed gravity. Once again corrections must be applied to observed gravity before such data are useful.

Elevation Correction 1: The Free-Air Correction

Let's again return to the ideal earth, which is spherical and nonrotating. We've already established that the gravitational acceleration imparted by the earth is

$$g = \frac{GM}{R^2} \tag{6-4}$$

If we are interested in how gravity varies with elevation, we can determine the vertical gradient by taking the first derivative of Eq. 6-4. Thus,

$$\frac{dg}{dz} = \frac{dg}{dR} = -2\frac{GM}{R^3} = -g\frac{2}{R} \tag{6-13}$$

If we substitute representative values at 45° latitude for g at sea level and R, the result is -0.3086 mGal/m. This value tells us what we already know—that gravity decreases as we increase our distance from the earth's center, but now we have an idea of the magnitude of the decrease. Equation 6-13 gives us a good approximation of the value we seek, but remember the simplicity of our initial assumption. As in the case of the latitude correction, we must use a more adequate representation of the true character of the earth and take a derivative of an equation that includes factors for rotation and ellipsoidal shape. Because the equation and its derivative are complex they are not presented here. The final relationship (Grant and West, 1965, p. 237–238) is

$$\frac{dg}{dR} = -0.3086 - 0.00023\cos 2\phi + 0.00000002\,z \tag{6-14}$$

where ϕ is latitude, z is elevation, and the units are mGal/m. In any given survey of modest extent the latitude adjustment (second term) in Eq. 6-14 is very small. The correction for the vertical gradient due to increasing distance from the ellipsoid surface (third term) becomes significant only in mountainous regions where there is extreme variation in elevation. In practice, both terms are usually ignored, and the value of 0.3086 mGal/m is the only value used. Note that this correction considers only elevation differences relative to a datum and does not take into account the mass between the observation point and the datum. For this reason the correction is termed the *free-air correction*.

Normally, the datum used for gravity surveys is sea level. Because gravity decreases 0.3086 mGal for every meter above sea level, the free-air correction is added to observed gravity. If the elevation of the observation point is below sea level, the free-air correction is subtracted. If the datum chosen is not sea level, the same procedure holds relative to the selected datum. Once again it is important to note that if we want to maintain an accuracy of 0.1 mGal in our final gravity values, we must know the elevation of an observation point to at least 33 cm.

The value remaining after adding the free-air correction FA_{corr} and subtracting normal gravity from observed gravity is termed the *free-air anomaly*. We can express the free-air anomaly Δg_{FA} as

$$\Delta g_{FA} = g_{obs} - g_n + FA_{corr} \qquad (6\text{-}15)$$

Let's turn now to an analysis of how to handle the mass lying between an observation station and the selected datum.

Elevation Correction 2: The Bouguer Correction

Consult Figure 6-3. If a gravity observation point is at A, which is at sea level, no free-air correction is necessary. If a station is at B, then a correction is applied for the difference z between sea level and the elevation of station B. At this point we have not corrected for the mass lying between B and sea level, assuming sea level is our datum. Even if there are no subsurface density variations between stations A and B, the gravity at B will be greater due to the presence of the material between B and sea level. Recall once again that our purpose is to apply a correction for all factors that cause gravity to vary from one station to another, so that in the end the only reason for gravity differences must be subsurface density variations.

The usual approach to correcting for the excess material beneath B is to compute the attraction of an infinite slab of density ρ and thickness z. This correction is called the *Bouguer correction* in honor of Pierre Bouguer (1698–1758), who was the first to attempt to measure the horizontal gravitational attraction of mountains. A cursory examination of Figure 6-3 clearly indicates that this is an approximation, but an additional correction that we present in the next section addresses this concern.

We now proceed to develop the Bouguer correction. Although this is a somewhat lengthy derivation, we include it here because we will use one of the equations again when computing gravity anomalies over buried objects of tabular shape. Figure 6-4 illustrates the critical relationships. Our goal is to derive a relationship for g due to a very thin rod. We do this by summing the individual contributions of very small volume elements. We then use the relationship derived for the rod to compute g for a thin sheet. This is achieved by summing the individual contributions of an infinite number of thin rods. Because an infinite slab is simply an assembly of a number of infinite sheets, this provides the desired result. In order to use this approach, we employ integral calculus. Even if you are not familiar with this approach, work through the derivation, as you will be able to follow most steps except for the evaluation of the integral.

Figure 6-3   The infinite slab used to correct for excess mass beneath point B.

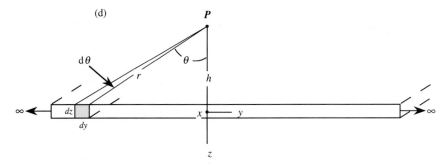

Figure 6-4   Relationships and notation used to derive the Bouguer correction.

We first want to determine the small value of gravity Δg at point P due to a very small volume element of volume $dx\ dy\ dz$ and density ρ. We can use the general form of Eq. 6-4 to state the relationship for g along the direction r, which gives us Δg_r.

$$\Delta g_r = \frac{G(\rho dx dy dz)}{r^2} \tag{6-16}$$

Because we are always interested in the vertical downward attraction of gravity Δg_z, we multiply Eq. 6-16 by $\cos \theta$ to produce

$$\Delta g_z = \frac{G(\rho dx dy dz)}{r^2} \cos \theta \qquad (6\text{-}17)$$

The attraction of a thin rod is the sum of all these components, so we write

$$g = \int_{x=-\infty}^{x=\infty} \frac{G(\rho dx dy dz)}{r^2} \cos \theta \qquad (6\text{-}18)$$

Before evaluating the integral, we note that $r = m / \cos \theta$. Also (see Fig. 6-4(b)), because the length of arc for a small angle $d\theta$ equals the radius r times the change in angle $d\theta$, it is straightforward to demonstrate $dx = r\, d\theta / \cos \theta$. Using these relationships we show that

$$\frac{G(\rho dx dy dz)}{r^2} \cos \theta = \frac{G(\rho dy dz) r d\theta \cos \theta (\cos \theta)^2}{m^2 \cos \theta}$$

$$= \frac{G(\rho dy dz) m d\theta \cos \theta (\cos \theta)^2}{m^2 \cos \theta \cos \theta}$$

$$= \frac{G(\rho dy dz)}{m} \cos \theta d\theta$$

Using this result, we adjust Eq. 6-18 so that

$$g = \int_{-\frac{\pi}{2}}^{\frac{\pi}{2}} \frac{G(\rho dy dz)}{m} \cos \theta \, d\theta \qquad (6\text{-}19)$$

which evaluates to

$$g = \frac{2 G \rho dy dz}{m} \qquad (6\text{-}20)$$

Equation 6-20 gives us g at P over a thin rod, as illustrated in Figure 6-4(a).

The next step in our derivation is to sweep out a thin sheet using a thin rod (i.e., we use Eq. 6-20 to calculate g over a thin sheet). Figure 6-4(c) is an end view of our thin rod. If lines 1, 2, 3, 4, 5, 6, 7, and 8 are of equal length, then g is the same at the end of every line. Line 1 is equivalent to m in our original diagram, and we will refer to line 2 as r. Clearly $m = r$, and g is the same at positions 1 and 2. This gives us the same view of the thin rod as we see in Figure 6-4(d), where line 2 now is labeled as r. Because we know g due to the thin rod (Eq. 6-20), we use this to write

$$\Delta g_r = \frac{2 G(\rho dy dz)}{r} \qquad (6\text{-}21)$$

The next several steps mimic exactly those that we just listed, so following the same procedure, we can say that

$$\Delta g_z = \frac{2 G(\rho dy dz)}{r} \cos \theta \qquad (6\text{-}22)$$

$$g = \int_{\frac{\pi}{2}}^{\frac{\pi}{2}} 2G(\rho dz) \, d\theta \tag{6-23}$$

and

$$g = 2\pi G(\rho dz) \tag{6-24}$$

Thus, the Bouguer effect, or g due to a slab of infinite extent and thickness z is $2\pi G(\rho z)$, and the value of the correction is $0.04193\rho z$ mGal/m.

Because the Bouguer correction is supposed to remove the effect of the additional mass above the datum which adds to observed gravity, the correction is subtracted from g_{obs} If an observation station is at a point below the datum, such as C in Figure 6-3, the Bouguer correction must be added to observed gravity. Following the same line of reasoning as we used in the case of the free-air correction, residual gravity remaining after the Bouguer correction B_{corr} is referred to as the *Bouguer anomaly* (Δg_B) and is equal to

$$\Delta g_B = g_{obs} - g_n + FA_{corr} - B_{corr} \tag{6-25}$$

Since it can be confusing to remember whether the Bouguer and free-air corrections should be added or subtracted, it is best to record elevation differences from a datum as positive when above the datum and negative when below the datum. If this procedure is followed, the differences can be substituted directly into Eq. 6-26, which is an expanded version of Eq. 6-25, and the correction will always be in the proper sense.

$$\Delta g_B = (g_{obs} - g_n + 0.3086 z - 0.04193\rho z) \quad \text{mGal} \tag{6-26}$$

Note that the units in Eq. 6-26 are in milliGals, so g_{obs} and g_n also must be in milliGals.

Because the majority of topographic maps in the United States still present elevations in feet, many explorationists find it more convenient to retain feet as a unit when correcting gravity readings. In this case the proper form of the Bouguer anomaly equation becomes

$$\Delta g_B = (g_{obs} - g_n + 0.09406 z - 0.01278\rho z) \quad \text{mGal} \tag{6-27}$$

Because Bouguer anomaly values are the ones most often used for interpretation in gravity surveys, Eq. 6-26 is presented in its complete expanded form as Eq. 6-28. This incorporates the GRS67 formula and the free-air and Bouguer corrections.

$$\Delta g_B = \left\{ \begin{array}{l} g_{obs} - \left[978031.85\left(1 + 0.005278895 \; \sin^2\phi + 0.000023462 \; \sin^4\phi \right)\right] + \\ \left[(0.3086 - 0.04193\rho)z \right] \end{array} \right\} \quad \text{mGal} \tag{6-28}$$

Table 6-2 provides an opportunity to study the process of adjusting observed gravity values (gravity reduction). By selectively changing values for observed gravity, latitude, elevation, and Bouguer density, you can study the effect of each of these on the final anomaly values. It is also instructive to change the amount of error for elevation or latitude (or both) in order to see the extent to which the Bouguer anomaly would change if such errors were present in your data. Remember that we are striving to keep the accuracy of our final values to within 0.1 mGal and would like to do better if possible.

The most vexing problem in the gravity reduction process is selecting an appropriate value for density in the Bouguer correction. Many surveys simply select the value of 2.67

TABLE 6-2 Example of Gravity Reduction

Observed gravity	980658.67	Observed gravity	**980,658.67**
Normal gravity	980674.39	Latitude (°)	**45.62**
Free-air correction	30.93	Elevation (m)	**100.24**
Bouguer correction	11.22	Bouguer density (g/cm3)	**2.67**
Free-air anomaly	15.22		
Bouguer anomaly	4.00	Elevation error (m)	**0.33**
		Bouguer anomaly error	0.06
		Latitude error (°)	**0.01**
		Bouguer anomaly error	0.90

(All gravity values are in mGal.)

g/cm³, which is the average density for crustal rocks. In many areas such a value is appropriate and is useful if we want to compare gravity values from numerous surveys over an extended area. If a survey is of limited extent and has a specific target, 2.67 g/cm³ may not be appropriate. Selecting the best value is possible only by extensive sampling and determining densities for a large number of specimens or by employing special field procedures during the gravity survey. An example of this latter method is presented in the section addressing field procedures.

Elevation Correction 3: The Terrain Correction

As you may already have sensed, the Bouguer correction is somewhat crude in its assumption of an infinite slab. Figure 6-5 illustrates the Bouguer slab for a gravity observation station at *B*. In order to reduce observed gravity at *B* to sea level we subtract the effect of the mass in this slab. Note that at *x* no rock or sediment is present, as the shaded area is above the land surface. This means we have overcorrected with respect to region *x*, because a mass effect was subtracted where no mass existed in the first place. Therefore, we must *add* a small amount to observed gravity to adjust for this overcorrection.

If we consider the shaded region at *y* in Figure 6-5, we see that this presents a differ-

Figure 6-5   An illustration of terrain corrections required at regions *x* and *y* due to the oversimplified nature of the Bouguer correction applied at point *B*.

ent problem. The mass at y exerts an attraction on the gravimeter mass at B. The vertical component of this attraction reduces observed gravity at B. Since observed gravity is reduced, we must *add* a small amount to observed gravity to negate the effect of y. In both cases the correction factor is added to observed gravity if the observation station is above sea level (assuming sea level is the datum being used). These corrections that account for the undulations of topography above and below the elevation level of an observation point are referred to as the *terrain correction*. If TC represents the terrain correction, the Bouguer anomaly formula becomes

$$\Delta g_B = g_{obs} - g_n + FA_{corr} - B_{corr} + TC \qquad (6\text{-}29)$$

The question now becomes one of deciding whether a terrain correction is necessary and how to go about the process in an area of irregular topography. Certainly the most widely used method is one proposed by Hammer (1939) that many geophysicists still employ. Hammer's approach considers the gravity effect of a ring (Fig. 6-6(a)). If the ring has a thickness z, an inner radius R_i, and an outer radius R_o, the equation expressing the gravitational attraction of the ring at a point at its center and on the same level as the top (or bottom) is

$$g_{ring} = 2\pi G\rho \left[R_o - R_i + \left(R_i^2 + z^2 \right)^{1/2} - \left(R_o^2 + z^2 \right)^{1/2} \right] \qquad (6\text{-}30)$$

(a)

(b)

Observation point

Figure 6-6 The basic approach to representing masses for the purpose of terrain corrections. (a) A cylindrical ring with inner radius R_i, outer radius R_o, and thickness z. A section of the ring is referred to as a *sector*. (b) A sector representing a topographic feature above the level of a gravity observation point.

By dividing the ring into an equal number n of compartments or sectors, the attraction of each sector becomes g_{ring}/n. Because we are interested in the effect of a portion of land above (or below) the level of our observation point (Fig. 6-6(b)), we treat that segment of land as a ring sector, and z becomes the difference in elevation between the level of the observation point and the top (or bottom) of the ring sector.

Hammer calculated the sizes of ring radii and sectors that gave the most accurate results at increasing distances from the gravity observation point in question. In practice a template consisting of concentric circles is drawn on a clear overlay. Each pair of adjacent circles outlines a ring and is referred to as a *zone*. Each zone is divided into varying numbers of sectors. The rings are drawn to the scale of the map using Hammer's original ring radii. The overlay is placed on a topographic map so that the center of the circles are located on the observation point for which observed gravity is being reduced. Figure 6-7 illustrates a simplified template placed on a topographic map, with only a few contours shown for purposes of clarity. The average elevation in each sector then is noted. For example, in sector H9 the average elevation is about 500 m. The absolute value of the difference between the average elevation and the elevation of the gravity station is computed. Note that sign is not important, as the terrain correction is always added to observed gravity. The next step typi-

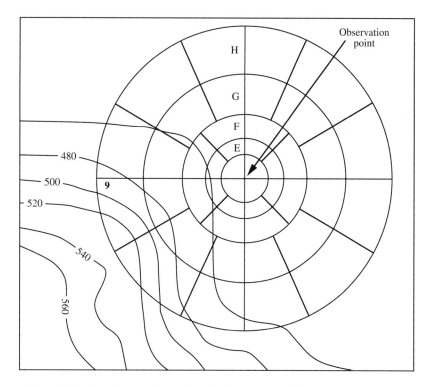

Figure 6-7   A terrain-correction template placed on a topographic map with the center at a gravity-observation point. Most contours are not shown for purposes of clarity. Only zones E, F, G, and H are illustrated.

cally is to consult a table to determine the gravitational effect of the sector. This process is repeated for each sector in a ring. All sector effects are then added together to determine the total terrain correction for the gravity station.

It's obvious that this is a very labor-intensive (and boring) chore. In fact, it is the most time-consuming aspect of obtaining and reducing gravity values. There are various ways to partially automate the process. One example that is useful for our purposes is to employ a spreadsheet instead of a look-up table. Table 6-3 is intended to provide an example of how this might be done but is oversimplified and would need to be expanded to serve such a purpose. It is useful, however, to work with Table 6-3 in order to get a sense of the magnitude of terrain corrections for various degrees of topographic relief. Seven of the 12 zones defined by Hammer are included in the table. The term *equation constant* refers to $2\pi\rho G$ in Eq. 6-30 and is adjusted so that the results of calculations are in milliGals. In Table 6-3 the number of sectors and the ring radii are designated for each zone. The elevation of the gravity station is entered in the appropriate cell. Next, the average elevation is entered into the designated cell for a sector in the zone in question. The value of the sector correction in milliGals can then be observed.

Usually, corrections for zones B and C are determined in the field. The remainder are determined from topographic maps. For purposes of illustration, the identical value for the average elevation in each sector was placed in the spreadsheet in order to illustrate how a 5-m difference between station elevation and sector elevation relates to the size of the sector correction. As you should realize, the significance of such an elevation difference diminishes rapidly with increasing distance from the gravity station. In Table 6-3 each sector correction is multiplied by the number of sectors in the zone in order to compute a total terrain correction. This is done just to give you a sense of the size of the terrain correction for various topographic-relief factors. Of course, in an actual case the average elevation in each sector of a zone would vary, sometimes by quite a lot.

Approximately what average elevation must be present in each sector of zone H in order to result in total zone correction of 0.25 mGal?

If digitized versions of topographic maps are available for a region in which a gravity survey is taking place, the terrain-correction process can be almost wholly automated. It's still somewhat rare to have such a resource available at a suitable scale for the precision required. In some cases terrain corrections are ignored because of the time involved. Is this justified? It depends. The best approach is to undertake a few corrections in an area in order to gain an appreciation of their magnitudes. Once these are known, they should be compared with the objectives of the survey in order to determine the effect of neglecting to make the corrections.

TABLE 6-3 Terrain Corrections

Elevation of gravity station	120.00	Elevations in meters; distances in meters; gravity in milliGals
Equation constant	0.11	

	Zone B	Zone C	Zone D	Zone E	Zone F	Zone G	Zone H	Zone I
Sectors	4.00	6.00	6.00	8.00	8.00	12.00	12.00	12.00
Ri	2.00	16.64	53.34	170.08	390.14	894.89	1529.49	2614.57
Ro	16.64	53.34	170.08	390.14	894.89	1529.49	2614.57	4453.74
Average elevation	115.00	115.00	115.00	115.00	115.00	115.00	115.00	115.00
Sector correction	0.074	0.009	0.003	0.001	0.000	0.000	0.000	0.000
Total for zone	0.30	0.06	0.02	0.00	0.00	0.00	0.00	0.00

Total Terrain Correction = 0.38

The Isostatic Anomaly

The general flotational equilibrium of lithospheric segments is known as *isostasy*. Most geologists and geophysicists are aware that high mountains have extensive "roots" of lower density material that extend into the higher-density mantle. The mountains appear to be floating in a manner that is very similar to icebergs. This flotational equilibrium is revealed in Bouguer anomalies that are negative over high areas and positive over the oceans. Near sea level the average is close to zero. The negative Bouguer anomalies arise due to the excess of low-density material at depth in mountainous regions when compared to rocks at depth in areas of low relief near sea level.

It is possible to correct for the isostatic effect by assuming a certain compensating mechanism and computing a correction. Residual gravity left after applying an isostatic correction I_{corr} is referred to as the isostatic anomaly Δg_I and is expressed as

$$\Delta g_I = g_{obs} - g_n + FA_{corr} - B_{corr} + TC \pm I_{corr} \qquad (6\text{-}31)$$

For exploration purposes the Bouguer anomaly generally is the value used, so the isostatic correction is not applied. If variations in Bouguer anomalies are due to isostatic effects in an exploration area, they will be present as gradual, regional trends due to their deep-seated causes. As such, they may be removed during the process of anomaly separation to be discussed in a following section. If so, the isostatic effect is deleted directly as part of a regional trend and not by considering isostasy as a cause and computing a compensating mechanism.

BASIC FIELD PROCEDURES

Most of the topics discussed in this section were already mentioned during the preceding discussion. The purpose of the following material is to provide additional detail and to give some insights into how certain field requirements and procedures are typically accomplished. This is not intended to be an exhaustive survey, and more detailed discussions should be consulted before commencing a full-scale effort involving many gravity readings. What is presented here certainly is sufficient for beginning efforts involving a small area and a limited number of measurements.

Drift and Tidal Effects

If we placed a gravimeter in one position and took readings every hour or so, the values we obtained would vary. This variation is due to two causes. One is instrument *drift*, which is caused by small changes in the physical constants of gravimeter components. The other is due to tidal effects, which are governed by the positions of the sun and moon relative to Earth.

Due to the extreme sensitivity of gravimeters, instrument drift affects readings to such an extent that it cannot be ignored. Neither can the changes be predicted, so a meter must be returned to a reference point every so often to establish if drift occurred and, if so, its magnitude. The simplest technique is to obtain a reading at a selected site, visit several other sites, and then return to the original site. This reading sequence is referred to as *looping*. Figure 6-8 illustrates the results of such a loop.

The sequence in the loop is: base - 1 - 2 - 3 - 4 - base. If we return to base every 1 or 2 hours, we can assume linear drift. The drift rate is then simply established by dividing the difference in the initial and final base values by the elapsed time between base station readings. The remaining stations in the loop are then adjusted using this drift rate. For example, in Figure 6-8 all stations are referred to the original base reading. The drift rate is illustrated by the solid line connecting the two base readings. Dotted lines parallel to the solid line are drawn through stations 1, 2, 3, and 4 in order to illustrate how these values should be adjusted to correct for the observed drift. Note that at this stage in the process we have not yet changed gravimeter dial readings into gravity values. Since we are dealing only with relative changes in *g* among stations, we can continue to work with dial divisions if we wish.

Tidal variations produce an effect on gravimeter mass that normally varies within a range of ± 0.15 mGal from a mean value and can have a rate of change as high as 0.05 mGal/h. Because these are substantial values relative to the 0.01 precision of most gravimeters, a correction is clearly called for. Since tidal effects can be predicted accurately, it is relatively straightforward to devise a computer program to produce values for any location at any time. Figure 6-9 is a representative curve for tidal variations produced by such a program (Rudman, Ziegler, and Blakely, 1977).

Observe the scale lengths for 1- and 2-h time durations in Figure 6-9. The tidal variation curve is essentially linear within 1-h time segments. Thus, if we construct a drift curve such as in Figure 6-8 and return to base every hour, such a curve will remove both the tidal effect and instrument drift with good precision. However, returning to a base station every hour is not only inconvenient but clearly reduces significantly the number of gravity stations that can be visited in an interval of time. It simply is not very efficient. We can increase the return interval to 2 h, but, with the help of the 2-h scale in Figure 6-9, it

Figure 6-8   A typical drift curve for gravity observations at base and stations 1, 2, 3, and 4.

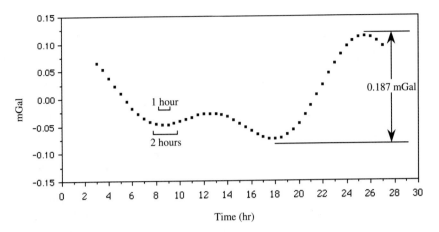

Figure 6-9   Variation of gravity values due to tidal effects. This is a representative curve for middle latitudes.

becomes evident that there are some times during the tidal cycle when our assumption of linear change does not hold. However, due to the relatively small gravity change due to tidal effects in a 2-h period, we could employ the standard drift-curve technique to correct for both tidal effects and instrument drift with only a slight loss in precision if we remain within the 2-h maximum. Many gravity surveys have used this approach.

   If it is impossible or extremely inconvenient to return to a base location every 2 h, then it is imperative to employ tidal-correction programs. Actually, due to the ready availability of such programs and the present computing power of personal computers, it seems only reasonable to utilize such programs for all gravity surveys. Figure 6-10 illustrates how gravimeter readings would vary at one position due to both the tidal effect and instrument drift. Given a standard loop consisting of base stations and gravity readings, one first removes the tidal variation effect by the appropriate program. What remains is therefore instrument drift, which is removed by the aforementioned procedure.

   The question now becomes the frequency with which we must return to base to correct for instrument drift, since we can predict the tidal effect as far into the future as we wish. The answer depends on instrument characteristics. Some instruments have very stable drift rates, whereas others are less predictable. The best course is to establish the drift behavior of the instrument used and then to plan a looping procedure accordingly. We tend to favor applying tidal corrections and returning to base every 2 to 3 h. Although this slows down field work somewhat, we are confident that we are maintaining a high-precision survey. But this approach demands a number of base stations throughout a survey area. We now turn to the question of how these base stations are established.

Establishing Base Stations

It is always preferable to tie gravity measurements to a location where absolute gravity is known accurately. Therefore, one needs to find the closest IGSN71 locality and to relate

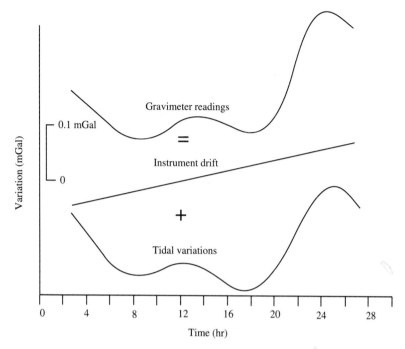

Figure 6-10   An illustration of how gravity values measured at a stationary position would vary due to a combination of tidal effects and linear instrument drift.

measurements to this IGSN71 value. Although it is not absolutely necessary to do this for limited, local surveys, as such surveys seek only variations in gravity, you should always make such a link simply in order to add to the research base by making your measurements part of the system that everyone else uses.

Figures 6-11 and 6-12 help to illustrate the standard procedure. For simplicity Figure 6-12 assumes linear drift rates and time intervals short enough so that tidal variations are included in the drift curves as linear changes. The looping procedure begins at the IGSN71 station. After a reading is obtained there, the Base-1 position is occupied and is followed by a return to the IGSN71 station. This sequence provides the relative gravity difference between the two stations after correcting for drift. Because we know the absolute gravity value for the IGSN71 station, we now have an absolute gravity value for Base 1. We now return to Base 1 and use a similar loop to determine absolute gravity at Base 2. Following this loop, we determine absolute gravity at Base 3. The final operation uses Base 3 to acquire a value for the IGSN71 station. Rarely will the value be the same as the established value, and the difference provides an indication of the amount of error in our measurements that provided values for Bases 1 to 3. It is hoped that the accumulated error is small, so that each base value can be adjusted slightly. One adjustment technique is simply to allocate the error equally among all the bases. For surveys with many such base loops, the best procedure is to tie bases together through many interconnected loops and to allocate error through more sophisticated statistical techniques (Nettleton, 1976, p. 84–87).

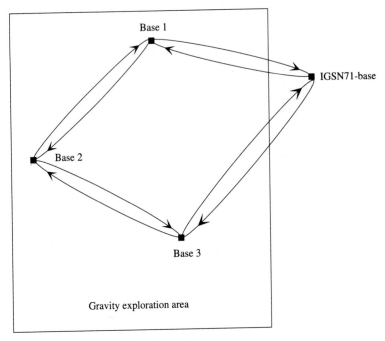

Figure 6-11 Establishing absolute gravity stations in an exploration area by referring them to an IGSN71 station and following a looping procedure.

Determining Elevations

We've already seen that elevations must be known to within 25–30 cm in order to maintain Bouguer anomaly values that are accurate to better than 0.1 mGal. Normally, such control can be achieved only by surveying procedures. These normally include using a rod and

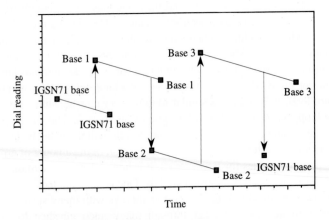

Figure 6-12 Drift curves resulting from the process illustrated in Figure 6-11.

transit and consume considerable time. Because this increases expense, elevation requirements may preclude many gravity surveys that otherwise make good sense in terms of acquiring useful information.

Topographic maps show the position of *benchmarks*, which are points where elevations have been measured to a high order of precision. The benchmarks are brass cylinders encased in concrete and are highly reliable as starting points for transit surveys. Other points (referred to as *spot elevations*) are indicated on most topographic maps and are points where elevations have been determined to the nearest foot (assuming the typical nonmetric map). Even though these points are located at easily identified locations such as road intersections, the exact point is not marked in the field, so considerable error in elevation is possible. For some low-precision surveys, it is possible to interpolate between contour lines, but the accuracy of elevations will be no better than half the contour interval. In rugged areas or for regions where topographic control is not available, some surveys use aneroid barometers. Because the barometers are affected by pressure variations, a looping procedure similar to that used when measuring relative gravity differences could be used. Since such a procedure would be very time consuming, a better solution would be to employ a stationary recording altimeter to record pressure variations that occur during a day's work. However, elevations determined in this manner normally are subject to errors of a meter or more.

If you are involved in a gravity survey taking place in a relatively urban environment, be sure to consult with city and town engineering departments. Such departments often have large-scale topographic maps used for road constructions and the like that contain many surveyed elevations possessing the required accuracy. Although tracking these maps down may take several days, the time savings can be enormous.

The best hope as an alternative to the time-consuming rod-and-transit surveys is the relatively new Global Positioning System (GPS). GPS surveying uses satellites to achieve horizontal and vertical controls on the order of 1 cm. Although the equipment is still too expensive for many to afford, the speed and ease make this an attractive alternative, and GPS likely will become the dominant approach in the not too distant future.

Determining Horizontal Position

Horizontal positioning can usually be determined sufficiently accurately by employing topographic maps with a scale of 1:25000 or larger and by careful location of gravity stations. In the absence of suitable computer digitizing capabilities, latitudes are best measured by using a clear plastic overlay on which is inscribed a grid with cells at 0.1-min intervals. This makes quick work of reading latitudes to 0.01 min, which results in a Bouguer accuracy on the order of 0.02 mGal. If more accuracy is needed, either larger-scale maps must be used, positions must be surveyed from benchmarks, or GPS must be employed.

Selecting a Reduction Density

Many gravity surveys select a density of 2.67 g/cm^3 for the Bouguer reduction. This enables values to be combined with other surveys for production of regional gravity maps but may not be suitable for some purposes. Of course, the best reduction values are determined by

sampling rocks and sediments from the survey area and determining densities in the laboratory. In most cases, however, surveys cover so large an area so that such sampling is not
practical, or exposures are not sufficiently numerous to permit reliable and representative
samples to be gathered.

In such cases an approach detailed by Nettleton (1942) can be attempted. This procedure requires detailed gravity profiles over topographic features such as hills. Field measurements are reduced to Bouguer anomaly values using different reduction densities,
which usually range from 1.8 to 2.8 g/cm³ in 0.1 intervals. The reduced values are plotted
against position on a graph that also includes a topographic profile. The idea is to minimize
the correlation of gravity with topography because the search is for gravity anomalies aris-
ing from subsurface density variations. Therefore, the density chosen for the final Bouguer
reduction is the one that produces the gravity profile that illustrates the least correlation
with topography.

The main problem with this approach occurs when there exists a correlation of topography with subsurface structure. In such cases we end up removing the very anomaly that
the survey is intended to locate. In this time of inexpensive computer access and good contouring software and graphics output, the best approach probably is to use several Bouguer
reduction densities including the standard of 2.67. Several contour maps can be prepared,
and if we are confident that there is no subsurface control of topography, then the map illustrating the least correlation of topography with gravity values is the one chosen for interpretation purposes.

Survey Procedure

Table 6-4 contains gravity data obtained by an undergraduate geophysics class at Smith
College. The purpose of the traverse was an attempt to map bedrock topography under the
floodplain of the Connecticut River near Northampton, Massachusetts. The first step in the
project was to determine elevations by rod and transit at preselected locations on a 1:24000
topographic map in order to obtain relatively equidistant gravity stations. Next, a base station
position was selected and linked to an IGSN71 location. Gravity stations were then occupied.

At each station the date, time, and dial reading of the gravimeter was recorded. This
was repeated until three consecutive readings were within 0.1 dial division in order to make
sure the meter was stable. The base station was occupied every 60 to 90 min in order to
allow for drift corrections without separately calculating tidal corrections.

After all stations had been occupied, the data were reduced. Sea level was selected as
a datum both because elevations in the study area averaged only about 40 m above sea level
and because the gravity stations would be added to a regional database. Because of the short
time loops, tidal variations were assumed to be linear. Therefore, standard drift curves were
used to correct for both instrument drift and tidal effects. Terrain corrections were calculated by overlaying templates on topographic maps of the region. The Bouguer reduction density was 2.67 g/cm³. Because the Bouguer anomaly values in Table 6-4 show some substantial variations, we will attempt to interpret them in a later section. Stay tuned.

TABLE 6-4 Data for a Gravity Traverse on the Connecticut River Floodplain

Station	Latitude (deg. min.)	Elevation (ft)	Dial Reading	Terrain Correction (mGal)	Normal Gravity (mGal)	Observed Gravity (mGal)	Free-air Anomaly (mGal)	Bouguer Anomaly (mGal)
SC base	42. 19.05	173.65	1,603.10	0.23	980,387.63	980,370.56	− 0.73	−9.20
SC1	42. 24.77	177.00	1,750.32	0.30	980,396.25	980,383.34	3.74	−4.80
SC2	42. 22.85	310.00	1,666.90	0.53	980,393.31	980,376.10	11.95	−0.94
SC3	42. 24.16	468.00	1,545.17	0.37	980,395.31	980,365.53	14.24	−4.26
SC4	42. 24.70	169.89	1,752.25	0.22	980,396.13	980,382.63	2.48	−5.90
SC5	42. 24.54	131.00	1,767.15	0.17	980,395.88	980,383.92	0.37	−6.72
SC6	42. 24.62	130.00	1,767.78	0.16	980,396.00	980,383.97	0.20	−6.86
SC7	42. 24.62	130.00	1,774.10	0.29	980,396.00	980,384.52	0.75	−6.18
SC8	42. 24.50	130.00	1,779.99	0.19	980,395.81	980,384.85	1.27	−5.76
SC9	42. 24.41	129.00	1,779.25	0.17	980,395.69	980,384.79	1.24	−5.78
SC10	42. 24.65	167.00	1,759.14	0.23	980,396.00	980,383.05	2.75	−5.51
SC11	42. 24.45	154.00	1,724.35	0.18	980,395.81	980,380.94	− 0.38	−8.25
SC12	42. 24.59·	158.00	1,706.48	0.17	980,395.88	980,379.39	− 1.62	−9.64
SC13	42. 24.59	158.00	1,702.38	0.24	980,395.88	980,379.04	− 1.98	−9.92
SC14	42. 24.58	161.00	1,711.49	0.19	980,395.88	980,379.83	− 0.90	−9.00
SC15	42. 24.61	314.00	1,679.86	0.25	980,396.00	980,376.37	9.91	−3.41
SC16	42. 24.57	310.00	1,683.79	0.32	980,395.88	980,376.72	10.00	−3.11

GRAVITY EFFECTS OF SIMPLE GEOMETRIC SHAPES

We have now arrived at the point where we can measure gravity and apply corrections to remove many factors that cause gravity to vary. Ideally, we are left with values (Bouguer anomalies) that are due to density variations within the subsurface. Our ultimate goal is to explain these anomalous values by invoking valid models of the subsurface. The first step toward realizing this goal is to document the gravity effects of simple shapes and to thoroughly understand these effects before proceeding to a more advanced analysis.

Rock Densities

Before addressing these gravity effects, it makes sense to spend a moment considering rock densities, as these will figure prominently in the analysis to come. The total variation in rock densities is quite small relative to the other physical properties we have discussed, and, as in the case of electrical resistivities, considerable overlap exists. The values in Table 6-5 consist of a compilation of density ranges from several sources. These should be utilized only in a very general way, such as answering some of the questions we ask when considering the gravity effects of various models. Clearly, you will need to establish much more precise values for any area in which you work.

The bulk density of rocks and sediments is controlled by the densities of the minerals present, the amount of open space in the rock or sediment, and the degree to which fluids fill these spaces. If you examine the densities for some of the common minerals in Table 6-5, the densities of the rocks listed make good sense. Use Table 6-5 in order to experiment with various combinations of density and porosity. For example, if you select a clay mineral with a density of 2.7 g/cm^3 and gradually bury it to greater and greater depths, so that the porosity is reduced from 80 percent to 40 percent to 0 percent, you in effect observe a den-

TABLE 6-5 Densities of Common Geologic Materials

Material	Density (g/cm3)
Quartz	2.65
Feldspars	2.5–2.7
Micas	2.8
Calcite	2.72
Clay minerals	2.5–2.8
Unconsolidated sediments	1.7–2.3
Sandstones	2.0–2.6
Shales	2.0–2.7
Limestones	2.5–2.8
Dolomites	2.3–2.9
Granitic rocks	2.5–2.8
Basaltic rocks	2.7–3.1
Metamorphic rocks	2.6–3.0

Density of mineral	**2.7**
Percent porosity	**50**
Density of saturated material	1.85

sity change from a mud to shale to metamorphic rock. The equation utilized in Table 6-5 to calculate the density of saturated material is

$$\rho_{sm} = \rho_{min} \left(\frac{1 - p_\%}{100} \right) + \frac{p_\%}{100} \tag{6-32}$$

where ρ_{sm} is the density of saturated material, ρ_{min} is the density of the mineral of which the material is composed, and $p\%$ is the percent porosity.

Gravity Effect of a Sphere

Due to its simplicity we begin our analysis with the sphere. Although we admit that few subsurface forms are spherical in shape, the effect of a roughly equal dimensional ore body is quite similar to that of a sphere, so our initial model is not entirely inappropriate. Because we seek to explain Bouguer anomalies in terms of density variations, we speak in terms of the *density contrast* ρ_c when considering gravity effects of models. The density contrast is the density of the model minus the density of the remaining material, which we assume to be homogeneous. Thus, a sphere with density of 2.0 g/cm³ in material with a density of 2.6 g/cm³ has a density contrast of -0.6 g/cm³.

Derivation of an equation for the gravity effect of a sphere is relatively straightforward because the effect is the same as if all the mass were concentrated at the sphere's center. Figure 6-13 illustrates the pertinent notation. Depth to the center of the sphere is z, sphere radius is R, distance from the observation point to the sphere's center is r, and g_z represents the vertical component of gravity due to the attraction of the sphere. The point $x = 0$ on the surface is directly over the center of the sphere. Given a density contrast ρ_c, the

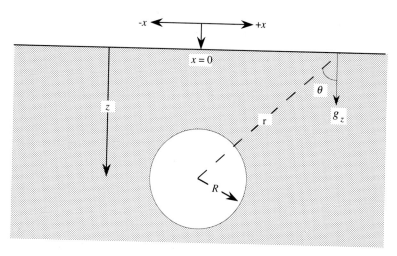

Figure 6-13 Notation used in derivation of the gravity effect of a buried sphere. The same notation is used for a traverse at right angle to the strike of a horizontal, infinitely long cylinder.

excess mass due to the sphere is equal to $4\pi R^3 \rho_c/3$. Using Eq. 6-4 we see that the gravitational attraction of the sphere in the direction of r is

$$g_{sphere(r)} = \frac{Gm}{r^2} = \frac{G\,4\,\pi R^3 \rho_c}{3\,r^2} = \frac{G\,4\,\pi R^3 \rho_c}{3\left(x^2 + z^2\right)} \tag{6-33}$$

We seek the vertical component of $g_{sphere(r)}$ because gravimeters respond only to gravity in the vertical direction. In general, unless noted otherwise, the component in the vertical direction g_z is implied. Therefore,

$$g_{sphere} = \frac{G\,4\,\pi R^3 \rho_c}{3\left(x^2 + z^2\right)}\cos\theta = \frac{G\,4\,\pi R^3 \rho_c}{3}\,\frac{z}{\left(x^2 + z^2\right)^{3/2}} \tag{6-34}$$

Now that we have the required equation, let's study some attributes of gravity over a buried sphere. Table 6-6 uses Eq. 6-34 and allows us to directly compare gravity over two spheres, so we can vary parameters in each and observe the differences. In case you are interested in making similar computations for other equations we present, Eq. 6-35 illustrates units conversion, so g is in mGal when density is in g/cm^3, distances are in meters, and G is in dyne $\bullet$ cm^2/g^2.

$$g(\text{mGal}) = 6.67 \times 10^{-8}\,\frac{\text{dyne·cm}^2}{g^2} \cdot \frac{g}{cm^3} \cdot \frac{m}{1} \cdot \left(\frac{g\cdot cm/s^2}{\text{dyne}} \cdot \frac{100\ cm}{m} \cdot \frac{\text{gal}}{cm/s^2} \cdot \frac{1000\,\text{mGal}}{\text{Gal}}\right)$$

$$\tag{6-35}$$

As you probably would predict after studying Figure 6-13 and Eq. 6-34, maximum gravity is directly above the sphere's center, and the Bouguer anomaly curve is symmetrical about this position. Especially note the form of the two curves in Figure 6-14(a) which represent the values in Table 6-6. The curve for the shallower sphere not only possesses higher gravity values, as one would expect, but is much sharper in form. Gravity values decrease at a much greater rate along a traverse extending away from a sphere's center for the shallower sphere than for the deeper sphere. Later, we will use this characteristic to estimate depths to sources of certain anomalies.

Also note that we could change the radius in one sphere in Table 6-6, and if we changed the density contrast in the other sphere an appropriate amount, gravity values would be identical. This is obvious from the equation for the sphere, as the gravity effect is due to the excess mass of the sphere. Even if we know the depth of the sphere, we cannot tell its size unless we also know the density contrast very accurately (usually we don't). For the examples in Table 6-6 we use a positive density contrast. If we substitute negative values and keep all other parameters constant, the Bouguer anomaly curves in Figure 6-14(a) will have exactly the same shape but will be inverted.

If you are able to utilize Table 6-6 as a dynamic table, it is very instructive to vary parameters, predict the results, and then graph the table values to study curve shapes and the magnitude of Bouguer values.

TABLE 6-6 Gravity Over a Sphere

	Sphere A		Sphere B
Sphere radius (m)	**200**	Sphere radius (m)	**200**
Depth to sphere center (m)	**500**	Depth to sphere center (m)	**1000**
Density contrast (g/cm3)	**0.4**	Density contrast (g/cm3)	**0.4**
Horizontal increment	**100**	Horizontal increment	**100**

Horizontal Position (m)	Gravity (mGal)	Horizontal Position (m)	Gravity (mGal)
−1200	0.0203	−1200	0.0235
−1100	0.0253	−1100	0.0272
−1000	0.0320	−1000	0.0316
−900	0.0410	−900	0.0367
−800	0.0532	−800	0.0426
−700	0.0702	−700	0.0492
−600	0.0938	−600	0.0564
−500	0.1264	−500	0.0640
−400	0.1703	−400	0.0716
−300	0.2255	−300	0.0786
−200	0.2862	−200	0.0843
−100	0.3372	−100	0.0881
0	0.3576	0	0.0894
100	0.3372	100	0.0881
200	0.2862	200	0.0843
300	0.2255	300	0.0786
400	0.1703	400	0.0716
500	0.1264	500	0.0640
600	0.0938	600	0.0564
700	0.0702	700	0.0492
800	0.0532	800	0.0426
900	0.0410	900	0.0367
1000	0.0320	1000	0.0316
1100	0.0253	1100	0.0272
1200	0.0203	1200	0.0235

Gravity Effect of a Horizontal Cylinder

It is not too difficult to envision geologic structures that are relatively equidimensional in cross section and are many times longer than wide. In such cases we can model the gravity effect by employing a horizontal cylinder with infinite length. Even though geologic forms do not possess infinite length, the gravity effect diminishes so quickly with distance from an observation point that the infinite length of our model still provides a quite sufficient representation for many actual subsurface configurations.

Fortunately, we have already done most of the work in developing an equation for the

(a)

(b)

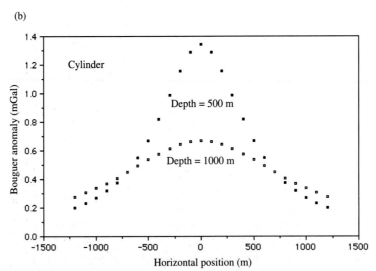

Figure 6-14 Bouguer gravity anomalies over a sphere and a horizontal cylinder. (a) Anomaly curves for spheres with centers at depths of 500 m and 1000 m. (b) Anomaly curves for horizontal cylinders with centers at depths of 500 m and 1000 m.

horizontal cylinder. Equation 6-20 is the equation for a thin rod that was an intermediate result during our derivation of an equation for the Bouguer correction, and we adopt that form for our not-so-thin cylinder. We can use the same notation as for the sphere, so we need pretend only that Figure 6-13 is a cross section through a horizontal cylinder. The section is oriented perpendicular to the axis of the cylinder. Because the area of the circle in

Figure 6-13 equals πR^2, substituting into Eq. 6-20 gives us, for gravity in the direction of r,

$$g_{cylinder(r)} \;=\; \frac{G\,2\,\pi R^2 \rho_c}{r} \tag{6-36}$$

Thus, the equation for the vertical component becomes

$$g_{cylinder} \;=\; \frac{G\,2\,\pi R^2 \rho_c}{r}\cos\theta \;=\; \frac{G\,2\,\pi R^2 \rho_c z}{r^2}$$

and

$$g_{cylinder} \;=\; G\,2\,\pi R^2 \rho_c \,\frac{z}{\left(x^2 + z^2\right)} \tag{6-37}$$

Table 6-7 uses Eq. 6-37 to calculate the gravity effect of a horizontal cylinder. Because the equations for the sphere and cylinder are similar in many respects, we might predict that the anomaly curves would be similar also. A quick glance at Figure 6-14(b), which is a graph of the values in Table 6-7, would prove us correct as far as curve form is concerned. The major difference is the greater magnitude of the gravity values for the cylinder, as much more mass is present in this case than if we select the same size radius for the sphere and the cylinder.

Most other relationships remain the same, such as broader and more attenuated curves as the cylinder is buried to greater depths. This might be a good time to explore another practical use of these equations and tables. Let's assume we decide to explore for a buried ore deposit and that the lower half of a horizontal cylinder is an appropriate prelimi- nary model. Bedrock density is 2.3 g/cm³, and the ore has a density of 3.0 g/cm³. The preci- sion of the Bouguer anomaly values measured by the survey will be 0.1 mGal but no better. Seismic information suggests that ore depths are on the order of 100 m. How large will the deposit have to be before we can reliably detect it? Figure 6-15 is a plot of results for cylin- ders of various radii. Gravity values for each curve represent one-half the values for the respective cylinder. Real values would be lower still, as our crude assumptions essentially are representing the deposit with the lower half of a cylinder. Therefore, a deposit would have to possess a "radius" of at least 40 m in order to be detected. Although admittedly somewhat crude, the real point of this exercise is to demonstrate how even simple models can be used to gain some idea of the size of mass excesses we are likely to detect under stat- ed conditions.

> Bouguer anomaly curves over the sphere (Fig. 6-14(a)) and the horizontal cylinder (Fig. 6-14(b)) both have similar shapes. Given good field values, do you believe it is possible to distin- guish between these two shapes as possible sources? Why?

TABLE 6-7 Gravity Over a Horizontal Cylinder

	Cylinder A		Cylinder B
Cylinder radius (m)	**200**	Cylinder radius (m)	**200**
Depth to cylinder center (m)	**500**	Depth to cylinder center (m)	**1000**
Density contrast (g/cm3)	**0.4**	Density contrast (g/cm3)	**0.4**
Horizontal increment	**100**	Horizontal increment	**100**

Horizontal Position (m)	Gravity (mGal)	Horizontal Position (m)	Gravity (mGal)
− 1200	0.1984	− 1200	0.2748
− 1100	0.2296	− 1100	0.3034
− 1000	0.2682	− 1000	0.3353
− 900	0.3163	− 900	0.3705
− 800	0.3767	− 800	0.4089
− 700	0.4531	− 700	0.4500
− 600	0.5496	− 600	0.4930
− 500	0.6705	− 500	0.5364
− 400	0.8177	− 400	0.5781
− 300	0.9861	− 300	0.6152
− 200	1.1561	− 200	0.6448
− 100	1.2895	− 100	0.6639
0	1.3411	0	0.6705
100	1.2895	100	0.6639
200	1.1561	200	0.6448
300	0.9861	300	0.6152
400	0.8177	400	0.5781
500	0.6705	500	0.5364
600	0.5496	600	0.4930
700	0.4531	700	0.4500
800	0.3767	800	0.4089
900	0.3163	900	0.3705
1000	0.2682	1000	0.3353
1100	0.2296	1100	0.3034
1200	0.1984	1200	0.2748

Gravity Effect of a Vertical Cylinder

Another particularly useful model is the vertical cylinder. Let's assume we want to determine gravity directly above the axis of the cylinder (at point P in Figure 6-16). We can treat this derivation in a fashion similar to our Bouguer derivation. We select a small volume element on the upper margin of the cylinder with volume equal to $dr\,dz\,r\,d\phi$. The gravitational attraction of this volume element at point P (first in the direction s and then the vertical component) is

Figure 6-15 Bouguer anomaly curves for horizontal cylinders with various radii.

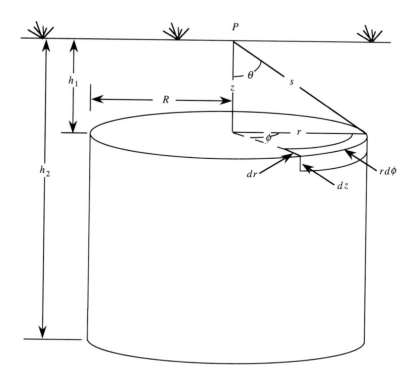

Figure 6-16 Notation used to derive the gravity effect of a vertical cylinder at a point P over its axis.

$$\Delta g_{vertical\ cylinder(s)} = \frac{G\rho_c\, dz\, (r d\phi)\, dr}{s^2} \tag{6-38}$$

$$\Delta g_{vertical\ cylinder} = \frac{G\rho_c\, dz\, (r d\phi)\, dr}{s^2}\cos\theta = \frac{G\rho_c\, dz\, (r d\phi)\, dr}{s^3}\, z$$

and

$$\Delta g_{vertical\ cylinder} = \frac{G\rho_c\, (zdz)(rdr)\, d\phi}{\left(r^2 + z^2\right)^{3/2}} \tag{6-39}$$

Next we determine the gravitational attraction of the cylinder by using the volume element to sweep out a thin ring, then a thin disc, and finally the cylinder. Equation 6-40 gives the formula for this integration.

$$g_{vertical\ cylinder} = \int_{h_1}^{h_2} dz \int_0^R dr \int_0^{2\pi} \frac{G\rho_c\, (z)(r)}{\left(r^2 + z^2\right)^{3/2}}\, d\phi \tag{6-40}$$

The result of this integration is Eq. 6-41.

$$g_{vertical\ cylinder} = G\,2\,\pi\rho_c\left(h_2 - h_1 + \sqrt{R^2 + h_1^2} - \sqrt{R^2 + h_2^2}\right) \tag{6-41}$$

Note that this equation gives gravity only at a point above the cylinder axis and is not valid for determining gravity at points away from the axis.

There are some other interesting results that arise from this derivation. When evaluating the integral from 0 to R, if we assume R goes to infinity, the result is an equation for an infinite slab (the Bouguer correction). Hence, we could have used this approach to determine the Bouguer correction. In addition, if we have a large cylinder with radius R_o and a smaller cylinder (with radius R_i) with an axis coincident with the larger cylinder and with coincident top and bottom, we can determine the effect of a ring by subtracting the effect of the smaller cylinder from the larger. Let P be on the top of the cylinder, so that $h_1 = 0$, and let the bottom of the cylinder be a distance z from the top, then

$$g_{ring} = G2\pi\rho_c\left[\left(z + \sqrt{R_o^2} - \sqrt{R_o^2 + z^2}\right) - \left(z + \sqrt{R_i^2} - \sqrt{R_i^2 + z^2}\right)\right]$$

and

$$g_{ring} = G2\pi\rho_c\left(R_o - R_i - \sqrt{R_o^2 + z^2} + \sqrt{R_i^2 + z^2}\right). \tag{6-42}$$

Equation 6-42 is the terrain correction.

Although the vertical cylinder equation is useful in finding the maximum gravity associated with such a form (which is over its axis) and for demonstrating how the terrain-correction equation can be developed, we often will want to know the gravity curves for traverses across vertical cylinders, as this geometry approximates a number of important geologic forms (e.g., salt domes). Unfortunately, it is much more difficult to derive an equation to determine gravity away from the axis of a vertical cylinder than it is for any of the forms considered thus far. In many cases, however, a thin rod may be substituted for the cylinder, so we now turn to the inclined rod and its special case, the vertical rod.

Gravity Effect of an Inclined Rod

Relationships and notation used to derive an equation for an inclined rod are illustrated in Figure 6-17. Rod cross section is assumed to be small relative to depth to the rod's top. The derivation follows a procedure very similar to that used for the Bouguer correction and the vertical cylinder, but because the resulting equation is complex we do not present it here (see Telford et al. 1976). However, the equation is used in Table 6-8 to calculate gravity over an inclined rod , and the result is presented in Figure 6-18. There are two important aspects to this curve. First, the maximum gravity effect is not at the $x = 0$ point, which is directly over the shallowest position of the rod, but is offset to the right of this point where more mass is present close to the rod. Second, the curve is not symmetrical, as in the case of the sphere, the vertical cylinder, and a vertical rod. Gravity increases only slowly as the inclined rod is approached from the $-x$-direction (in Fig. 6-17), but as distance to the rod is narrowed, curve slope increases markedly. After the curve maximum is passed, curve slope is less than that immediately before the maximum.

   In the case of a vertical rod (inclination = 90°), the equation becomes less complex and is

$$g_{vertical\ rod} = G\rho_c \pi R^2 \left\{ \frac{1}{\left(z^2 + x^2\right)^{1/2}} - \frac{1}{\left[(z + L)^2 + x^2\right]^{1/2}} \right\} \quad (6\text{-}43)$$

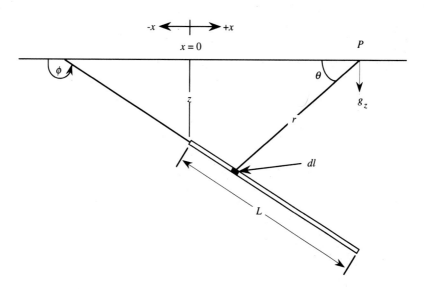

Figure 6-17   Notation used to derive the gravity effect of an inclined rod.

TABLE 6-8 Gravity Over Inclined and Vertical Rods

	Inclined Rod		Vertical Rod
Rod radius (m)	1	Rod radius (m)	1
Depth to rod top (m)	10	Depth to rod top (m)	10
Rod length (m)	100	Rod length (m)	100
Rod inclination (degrees)	135	Rod inclination (degrees)	**********
Density contrast (g/cm3)	0.4	Density contrast (g/cm3)	0.4
Horizontal increment (m)	2	Horizontal increment (m)	2

Horizontal Position (m)	Gravity (mGal)	Horizontal Position (m)	Gravity (mGal)
−24	0.009	−24	0.030
−22	0.010	−22	0.033
−20	0.011	−20	0.036
−18	0.013	−18	0.040
−16	0.014	−16	0.044
−14	0.016	−14	0.050
−12	0.018	−12	0.055
−10	0.020	−10	0.062
− 8	0.023	− 8	0.070
− 6	0.027	− 6	0.077
− 4	0.031	− 4	0.084
− 2	0.037	− 2	0.090
0	0.044	0	0.092
2	0.052	2	0.090
4	0.062	4	0.084
6	0.073	6	0.077
8	0.084	8	0.070
10	0.093	10	0.062
12	0.100	12	0.055
14	0.102	14	0.050
16	0.101	16	0.044
18	0.097	18	0.040
20	0.092	20	0.036
22	0.087	22	0.033
24	0.082	24	0.030

where πR^2 is the cross-sectional area of the rod. Example values for the same parameters as an inclined rod (except for inclination!) are presented in Table 6-8. Telford et al. (1976, p. 59) note that the vertical rod gives off-axis values within 6 percent of a vertical cylinder when the diameter is less than the depth to the top.

So far we have a few basic shapes we can use as models to provide information about the gravitational effects of simple geologic structures. However, one shape that is very common in the subsurface and for which we do not yet have a model is a sheetlike form, so it is to this that we now turn.

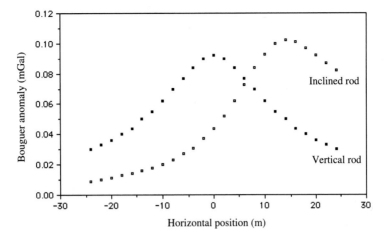

Figure 6-18   Bouguer anomalies over a thin vertical rod and a thin rod inclined at 135° (dipping at 45°). See Table 6-8 for the rods' parameters.

Gravity Effect of a Horizontal Sheet

When deriving the Bouguer effect, we created a thin slab by moving a thin rod from +∞ to -∞ (integrating from $\pi/2$ to $-\pi/2$; see Eq. 6-23). If instead we integrate from θ_1 to θ_2 (see Fig. 6-19(a)), we have

$$g = \int_{x=\theta_1}^{x=\theta_2} 2G(\rho\,dz)\,d\theta \qquad (6\text{-}44)$$

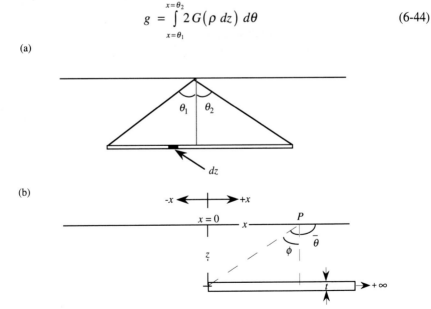

Figure 6-19   Relationships and notation for derivation of semi-infinite sheet relationships. (a) Diagram for a thin truncated slab. (b) Diagram for a semi-infinite sheet.

and

$$g = 2G\rho_c\left(dz\,\overline{\theta}\right) \tag{6-45}$$

where $\overline{\theta}$ represents the included angle in radians. If we keep the slab thin with respect to its depth, we can write

$$g_{truncated\ slab} = 2G\rho_c t\left(\overline{\theta}\right) \tag{6-46}$$

where t is the thickness of the truncated slab (Fig. 6-19(a)). Our goal in this section is to derive a relationship for a semi-infinite slab (Fig. 6-19(b)). For this case the included angle $\overline{\theta}$ is equal to $\pi/2 + \tan^{-1}(x/z)$ and we therefore have

$$g_{semi-\infty\ sheet} = 2G\rho_c t\left(\frac{\pi}{2} + \tan^{-1}\frac{x}{z}\right) \tag{6-47}$$

This approximation is within 2 percent if $z \geq t$ (Telford et al. 1990, p. 40). Table 6-9 uses Eq. 6-47 to calculate Bouguer anomaly values for semi-infinite sheets. In the case of this table the input parameters are for extremely shallow and thin sheets. Note that gravity values for such a model are quite small and that these sheets could not be detected with the normal accuracy of gravity surveys. Nevertheless, curve shapes are still valid, and it is primarily curve shapes and relationships that we are interested in at this point.

Study the thin semi-infinite sheet pictured in Figure 6-19(b) and try to predict the shape of a Bouguer anomaly curve for such a body. As you probably realize, such a curve has a maximum value above the sheet far removed from the edge location (+∞) and a minimum value of zero far removed from the edge in the opposite direction (-∞) (see curve for sheet A, Figure 6-20(b)). The value of gravity directly above the edge of the sheet is one-half the maximum value. These relations become quite clear by allowing x to assume very large positive values and very large negative values in the quantity ($\pi/2 + \tan^{-1} x/z$) from Eq. 6-47. The semi-infinite-sheet model also can be used to simulate a horizontal bed truncated by a high-angle fault where the displacement effectively removed the other half of the bed from consideration. Other models, such as a sill termination, are equally plausible.

By a minor bit of manipulation we can also use Eq. 6-47 to compute the effect of a vertical fault cutting a horizontal sheet where both displaced sheet segments are present. We simply compute the effect of each sheet as illustrated in Figure 6-20(a) using Eq. 6-47. Then, since we have to orient the sheets in opposite directions to simulate a faulted bed, we reverse the sense of the x-coordinate axes for one sheet and add the effects of both sheets. This is the scheme used in Table 6-9. Computations for each sheet are independent, but the signs of the "Horizontal Position" values are reversed. Therefore, we can add the values in the "Gravity" columns directly if we want to use the table to simulate a faulted sheet. The results in Table 6-9 are graphed in Figure 6-20(b). Here we see the curve for sheet A (assuming sheet B is not present), sheet B (assuming sheet A is not present), and sheets A and B (the sum of curves A and B for the geometry in Figure 6-20(a)). The purpose of this exercise is to demonstrate yet once again how we can devise models using the simple shapes already at our disposal.

Although interesting, the vertical fault model is somewhat limited. It is possible to derive an equation for a geometry similar to the semi-infinite sheet but with the added flexi-

TABLE 6-9 Gravity Over Semi-Infinite Sheets

Sheet A		Sheet B	
Depth to sheet (m)	4	Depth to sheet (m)	8
Sheet thickness (m)	1	Sheet thickness (m)	1
Density contrast (g/cm3)	0.4	Density contrast (g/cm3)	0.4
Horizontal increment (m)	2	Horizontal increment (m)	2

Horizontal Position (m)	Gravity (mGal)	Horizontal Position (m)	Gravity (mGal)
24	0.0009	− 24	0.0150
22	0.0010	− 22	0.0149
20	0.0011	− 20	0.0147
18	0.0012	− 18	0.0145
16	0.0013	− 16	0.0143
14	0.0015	− 14	0.0140
12	0.0017	− 12	0.0136
10	0.0020	− 10	0.0132
8	0.0025	− 8	0.0126
6	0.0031	− 6	0.0118
4	0.0042	− 4	0.0109
2	0.0059	− 2	0.0097
0	0.0084	0	0.0084
− 2	0.0109	2	0.0071
− 4	0.0126	4	0.0059
− 6	0.0136	6	0.0049
− 8	0.0143	8	0.0042
− 10	0.0147	10	0.0036
− 12	0.0150	12	0.0031
− 14	0.0153	14	0.0028
− 16	0.0155	16	0.0025
− 18	0.0156	18	0.0022
− 20	0.0157	20	0.0020
− 22	0.0158	22	0.0019
− 24	0.0159	24	0.0017

bility of varying fault angle. We do not derive the equation here but will use it to investigate some important relationships. A derivation can be found in Telford et al. (1990, p. 40–43). The equation for the faulted horizontal sheet is

$$g_{faulted\ sheet} = 2\,G\rho_c t \left[\pi + \tan^{-1}\left(\frac{x}{z_1} + \cot \phi \right) - \tan^{-1}\left(\frac{x}{z_2} + \cot \phi \right) \right] \quad (6\text{-}48)$$

where x is the horizontal position of the measurement location, t is sheet thickness, ϕ is fault inclination, z_1 is depth to upthrown block, and z_2 is depth to downthrown block. Note that the relative positions of the faulted sheet are constant (i.e., the upthrown block is always

(a)

(b)

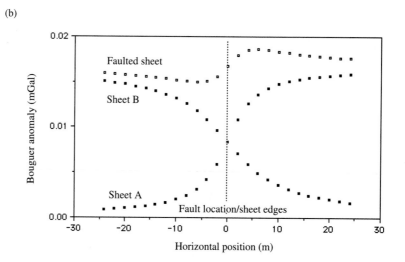

Figure 6-20 Using semi-infinite sheets to simulate a faulted horizontal bed. (a) Notation
used in Table 6-9 to derive gravity values. (b) Graph illustrating effects of sheet A, sheet B,
and sum of A and B (the faulted relationship shown in (a)).

toward the $+x$-coordinate axis). Table 6-10 is based on Eq. 6-48. Figure 6-21(b) illustrates
Bouguer gravity curves for fault inclinations of 60°, 90°, and 120° as defined in Figure 6-
21(a) and calculated by Table 6-10.

The input parameters in Table 6-10 are the same as in Table 6-9 (fault inclination of
90°). Note that the sums of the "Gravity" columns in Table 6-9 equal the values in the
"Gravity" column in Table 6-10, as should be the case, since both are computing the same
model. Let's use the gravity curve for the vertical fault as a basis of comparison for fault
inclinations of 60° and 120°. A ϕ-value of 120° models a high-angle reverse fault (Fig. 6-
21(a)). This geometry produces an overlapping of the sheets, so that double the excess mass
is present in the overlap zone. This doubling accounts for the pronounced maximum on the
120°curve (Fig. 6-21(b)) and the greater gravity values for this curve. In contrast, a ϕ-value
of 60° (Fig. 6-21(c)) models a high-angle normal fault which creates a gap where no sheet

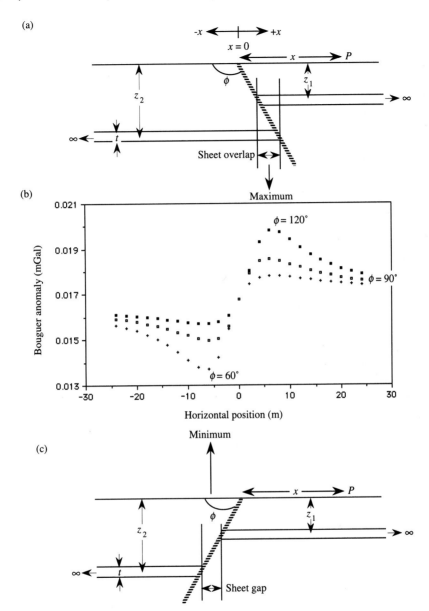

Figure 6-21   Gravity effects of a faulted horizontal sheet. (a) Fault inclination equals 120°
(60° dip to right). (b) Bouguer anomalies for fault inclinations of 60°, 90°, and 120°.
(c) Fault inclination equals 60° (60° dip to left).

is present. This mass deficiency leads to the distinct minimum on the 60° curve and
accounts for the lower gravity values of this curve (as compared to the 90° and 120°curves).

This concludes our analysis of the gravity effect of simple shapes. We could continue

TABLE 6-10   Gravity Over a Faulted Horizontal Sheet

Depth to downthrown sheet (m)	**8**
Depth to upthrown sheet (m)	**4**
Sheet thickness (m)	**1**
Fault inclination (degrees)	**90**
Density contrast (g/cm3)	**0.4**
Horizontal increment (m)	**2**

Horizontal Position (m)	Gravity (mGal)
− 24	0.0159
− 22	0.0159
− 20	0.0158
− 18	0.0157
− 16	0.0156
− 14	0.0155
− 12	0.0153
− 10	0.0152
− 8	0.0150
− 6	0.0150
− 4	0.0150
− 2	0.0156
0	0.0168
2	0.0179
4	0.0185
6	0.0186
8	0.0185
10	0.0183
12	0.0182
14	0.0180
16	0.0179
18	0.0178
20	0.0177
22	0.0177
24	0.0176

with numerous other shapes, but this does not seem warranted. Our goal in this section is to acquaint you with the general effect for a number of different shapes, to demonstrate that more complex shapes can be constructed from combining simple shapes, and to start you thinking about issues such as detectability of some models. Before considering some quite different issues in gravity exploration, we introduce a computer program that permits you to compute the gravity effect of some reasonably complicated models.

Using GravModel

GravModel calculates the gravity effect for solids of infinite strike length and polygonal cross section. Because each polygon can have up to 20 vertices, and several polygons can comprise a model, subsurface models of substantial complexity are possible. We do not derive here the equations used in the program, but the fundamental equation is derived following a procedure much like we employed to determine the Bouguer correction, the semi-infinite sheet, and the vertical cylinder. If we have a small cell of infinite length in the y-dimension with side lengths dz and $d\theta$, the gravity effect of such a cell is easily determined by integration. Thus, we have

$$g_{cell} = 2G\rho_c \int_{\theta_1}^{\theta_2} d\theta \int_{z_1}^{z_2} dz \qquad (6\text{-}49)$$

or

$$g_{cell} = 2G\rho_c(\theta_2 - \theta_1)(z_2 - z_1) \qquad (6\text{-}50)$$

The notation is defined in Figure 6-22(a). All we need now do is to represent a polygon such as that illustrated by Figure 6-22(b) as an accumulation of such cells. If a polygon has n such cells, then the gravity effect of the polygon is

$$g_{polygon} = 2G\rho_c n \, \Delta\theta \, \Delta z \qquad (6\text{-}51)$$

The main challenge is that of carrying out this summation given the coordinates of a polygon. A solution to this problem was published by Hubbert (1948) and was adapted for computer use by Talwani et al. (1959). Since the mathematics is fairly complex, the solution and computational steps are not given here.

Instructions for using GravModel are in Appendix D. If you can work with GravModel, attempt a number of polygons in various configurations. Always try to predict the rough form of the anomaly curve before computation begins. Be sure to understand why the curve has the shape it does. Remember that you are dealing with polygonal cross sections of bodies with infinite strike length. You can, therefore, simulate geological forms that have roughly constant cross-sectional shape along their length and with lengths that are several times cross-sectional width.

Other solutions are available to model shapes that do not fit within the constraints of the Hubbert-Talwani approach. Three-dimensional modeling programs exist that permit essentially any three-dimensional shape to be modeled and do not impose constraints such as infinite strike lengths on the model. The computational load of complex three-dimensional modeling algorithms is easily handled by large computers and is not beyond personal computers. The theoretical approach is not all that different from the one used by GravModel, but, of course, substantially more input data are required to define forms being modeled. If your mathematical background is good and you are interested in learning more about three-dimensional modeling, consult Plouff (1976).

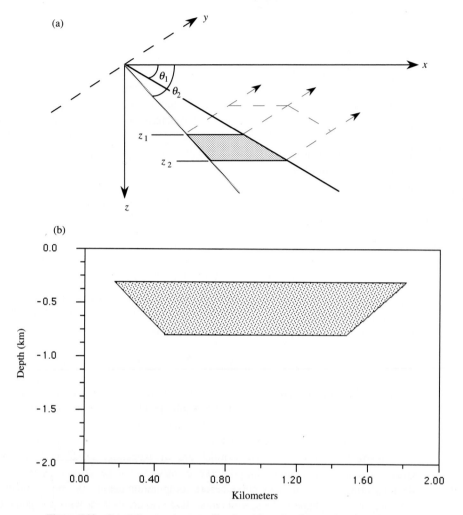

Figure 6-22   Calculating gravity anomalies due to irregular shapes using GravModel. (a)
Notation used to develop fundamental equation. (b) An example of a simple polygonal
shape drawn in GravModel.

ANALYZING ANOMALIES

Now that we are prepared to begin analysis of Bouguer anomaly curves, it is well to pause
for a moment to emphasize that we are faced with a somewhat different situation than in the
case of seismic and resistivity data. When looking at seismic data, we knew that the horizontal distance traversed by recorded wave energy was restricted to the length of the geophone spread and that the depth sampled was controlled by the shot-receiver distances, the
magnitude of the energy source, and the subsurface geometry. Similar controls existed in
resistivity surveying.

On the other hand, when measuring gravity, we know that the value recorded contains contributions from sources of interest at shallow depths as well as contributions from much deeper sources. Even mass excesses at considerable horizontal distances from an observation point affect measured values. Thus, deeper, larger sources may mask or obscure sources of interest. The theme of this section is directed at strategies for dealing with this problem.

Regionals and Residuals

Larger features produce Bouguer anomalies that are smooth over considerable distances. These smooth trends usually are referred to as *regional trends* or simply as *regionals*. Smaller, more local sources account for sharper anomaly shapes of more restricted areal extent. If regionals are removed from gravity data, the remaining or residual values are related to these local disturbances. These are referred to as *residual anomalies* or *residuals*. Of course, what is a regional and what is a residual depends on the exploration target. Once again we find that the goals of a survey must be firmly set before data are acquired. Station density must be sufficient to completely define residual anomalies and must be extensive enough to correctly describe regionals.

How can we determine regionals? The oldest method essentially involves construction of a series of profiles at regularly spaced intervals across a contoured Bouguer anomaly map. Each profile is a plot of Bouguer anomaly values against distance. A smoothly varying regional is then defined using as much geologic information as possible. Once defined, the regional trends along each profile are subtracted from the Bouguer values along the same profile. This produces a collection of residual values that can once again be contoured to define the form of the residual anomalies. The procedure is necessarily quite interpretative, as regional definition depends on the skill and knowledge of the geophysicist. However, when adequate geologic information and interpreter skill are available, the method can be quite useful.

As an example of this procedure, although a simple and somewhat contrived one, let's confine our analysis to a single profile (dangerous in the real world). For purposes of illustration we also will work with subsurface geology that we create. Figure 6-23 repre-

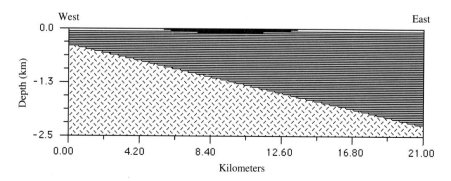

Figure 6-23 Model used for regional-residual discussion. A buried river channel is incised into a sedimentary rock sequence which overlies crystalline basement.

sents a model defined using GravModel. Resulting Bouguer anomaly values are plotted as small crosses (curve 1) in Figure 6-24(a). The model consists of a buried river channel filled with sands and gravels, a prism of sedimentary rock that thickens to the east, and crystalline basement that is present beneath the sedimentary sequence and abuts against the sedimentary rocks on both east and west margins. Our target is the buried river channel (density contrast = -0.4), so the regional in this case will be due to the sedimentary rocks (density contrast = -0.2). We assume we know enough about the geology of the area represented by the model to be aware of the general thickening of the sedimentary rocks to the east. Therefore, the smooth decline of Bouguer values to the east reflects this thickening and is subtracted to obtain residual values.

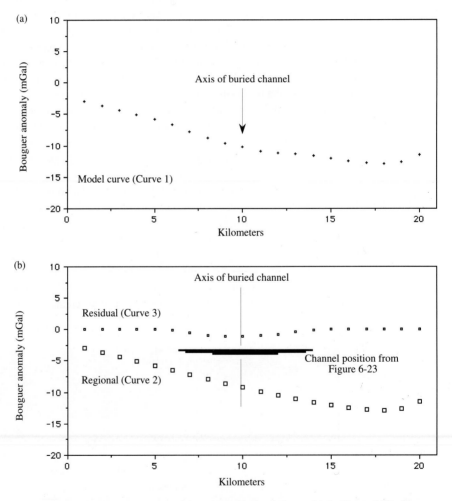

Figure 6-24 (a) Bouguer anomaly values calculated for model in Figure 6-23. (b) Regional and residual curves based on geology in Figure 6-23.

Because we are using GravModel to produce these curves, we simply subtract the effect of the sedimentary rocks, which is defined by the open squares (curve 2) in Figure 6-24(b). The residual (curve 3) clearly illustrates the small decline in gravity values due to the effect of the buried channel. Of course, the geology in this example is quite simple (and we know what it is!). The general regional trend (the west-east decline in Bouguer values) is quite obvious, but the increase in values at 19 km and 20 km (Fig. 6-24(a)) might be difficult to recognize as part of the regional. This increase is due to the termination of the sedimentary sequence at 21 km in the model. Of course, outcrops of crystalline basement at 21 km would explain the increase in values to a knowledgeable geophysicist, and these increases could then be included in the regional. This simple example illustrates the importance of both geologic information and interpreter skill in appropriately identifying regionals.

Let's add another complication to our model by assuming that a domelike mass (density contrast = 0.4) is present at depth. This added mass excess will complicate our previous curve by warping it as illustrated by curve 4 in Figure 6-25. Identifying an adequate regional is now more difficult, and the presence of the buried channel is more difficult to extract from the data. If we remove the effect of the sedimentary rocks as before, we still have a fairly complicated curve left. Since the general aspect of this curve is a smooth, convex-upward form, we might also elect to remove it as part of the regional field. Remember that we are dealing here with completely known quantities—try to imagine the difficulties that would be present when dealing with data from survey measurements. Finally, note how the scale of the target determines the regional-residual relationship. Instead of being interested in buried river channels, our goal might be to define the geometry and thickness of the sedimentary rocks in Figure 6-23. If so, we would need many more readings both to the east and west in order to search for regional patterns. The small decrease in gravity due to the channel likely would be treated as noise.

Figure 6-25 Effect of adding mass excess at depth to model in Figure 6-23.

Because material near the surface is often very inhomogeneous, it may be difficult or impossible to identify a regional with confidence. In these and other situations it makes sense to seek methods that define surfaces mathematically.

Trend Surfaces

Recall that our goal is to determine regional trends in our data. One approach is to generate *trend surfaces*. These are mathematical surfaces that represent a surface of best fit to data points distributed in *xyz*-space. Such surfaces are defined by polynomial functions, and the surface is fit to the data points using the method of least squares. Usually surfaces of various orders (a first-order surface is a plane, a second-order surface is a paraboloid, etc.) are computed in order to select one (or more) to represent the regional trend. Once a trend surface is chosen to represent the regional, it is subtracted from the Bouguer anomaly map to produce a residual map. The major deficiency of utilizing trend surfaces to define regionals is that the former is completely mathematically defined and, therefore, takes no geologic knowledge into account. This also could be looked on as a plus, since no bias enters into the determination. For good discussions of trend surfaces consult Robinson (1982, p. 72–81) and Maslyn (1984, p. 12–19).

The relation of trend surface to gravity data is initially somewhat difficult to visualize in three dimensions, so let's continue our string of two-dimensional examples. Figure 6-26 illustrates the Bouguer anomaly curve (curve 1) that we examined previously. Three curves are fit to these data: a first-order polynomial (straight line) and third- and fifth-order polynomials. The first-order polynomial curve provides a good generalized representation of the regional field. If these values were subtracted from curve 1, the anomaly effect of the buried river channel would be well delineated. Substantial residuals would also be present near the beginning and end of curve 1. A third-order polynomial gives a much better fit to the Bouguer data but would not define the residual anomaly due to the channel as well as the first-order curve. This residual anomaly still would be preserved, however, and the other residuals would be reduced in intensity. A fifth-order curve fits the data so well that the buried channel anomaly would all but disappear as a residual.

Typically, surfaces of order greater than four are not used to define regionals. In our two-dimensional example a little knowledge of local geology probably would allow us to define the true regional more precisely than the trend-surface approach. However, Bouguer maps are considerably more complex, and regionals are much more difficult to define, so trend surfaces may be the best approach if the Bouguer map is reasonably irregular. Figure 6-27 illustrates a Bouguer map of moderate complexity for 7½′ quadrangles in west-central Massachusetts. Figure 6-28 is a third-order trend-surface map computed from the data points used to produce Figure 6-27. A strong north-to-south decline in Bouguer values is quite well defined by this surface and provides a good indication of the regional for this area.

Upward and Downward Continuation

In the next chapter, which addresses magnetics surveying, we note that both gravity and magnetic fields are potential fields and explore some consequences of this property.

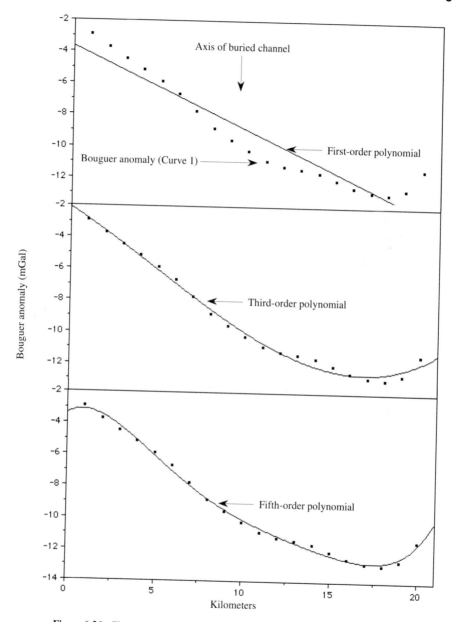

Figure 6-26   First-, third-, and fifth-order polynomials fit to Bouguer anomaly data.

Because potential fields are continuous, it is possible to determine mathematically how a given potential field would appear if it was measured from a different level. Therefore, we can project Bouguer anomaly values to various levels to enhance both regional and local anomalies.

Figure 6-27 Bouguer anomaly map for the Easthampton and Mt. Holyoke Quadrangles, west-central Massachusetts. Contour interval = 0.5 mGal.

Upward continuation projects a Bouguer anomaly map to a higher level, which results in a map illustrating how the gravity field would appear if we measured it from a higher level. If we measured the field from a higher level (greater elevation), local shallow sources would be suppressed and the regional field enhanced. Such a procedure provides a good idea of the regional anomaly for an area. We can simulate the effect of upward continuation by selecting the model in Figure 6-23 and moving it to greater depths. If we move the whole model 1 km deeper, the effect is the same as moving our instruments to a plane 1 km higher. In both cases we increase the distance from observational position to model coordinates.

Figure 6-29 portrays the results of increasing this distance. Once again the Bouguer anomaly curve for the model in Figure 6-23 is illustrated. Study the curve labeled "Up 2 km". Note that it is virtually identical in shape to the "true" regional as defined by curve 2 in Figure 6-24 and that no trace of the anomaly due to the buried channel is present. If we used the "Up 2 km" curve to define the shape of the regional in curve 1, a strong residual anomaly for the buried channel would result. Figure 6-30 is an upward-continuation map for the Bouguer anomaly map (Fig. 6-27) previously presented. Once again, considerable smoothing reveals a pattern very similar to that produced by a third-order trend surface (Fig. 6-28), which serves to reinforce strongly the shape of this area's regional. In this case the regional is due to a pronounced southward, as well as eastward, thickening of Triassic

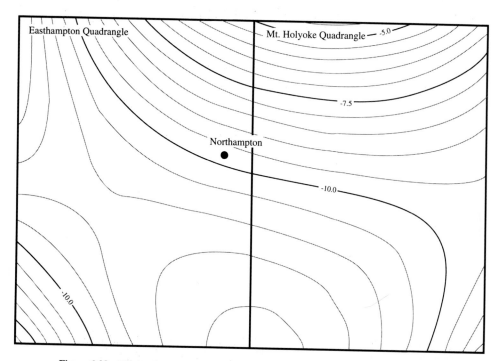

Figure 6-28 Third-order trend-surface map derived from Figure 6-27. Contour interval = 0.5 mGal.

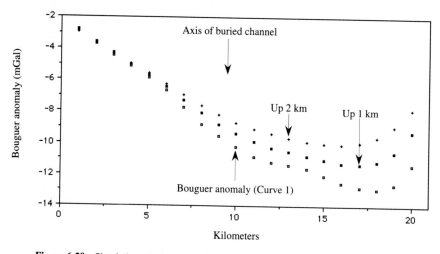

Figure 6-29 Simulation of upward continuation. Bouguer anomaly curve from model in Figure 6-23.

Figure 6-30   Upward-continuation map (1.5 km) derived from Figure 6-27. Contour interval = 0.5 mGal. Compare pattern with that in Figure 6-28.

and Jurassic sedimentary rocks in the Hartford Basin. The gross appearance of this rift basin is not too different from the sediment thickness pattern in our original model (Fig. 6-23).

Downward continuation is employed less frequently. It effectively brings the observation position closer to shallow sources, thereby enhancing them and subduing the anomalies related to deeper sources. This procedure is sometimes effective in outlining the sources of shallow anomalies but can lead to anomalous and wildly fluctuating values if the observation level is placed deeper than the top of the source.

Second Derivatives

The first derivative is a measure of slope. The second derivative provides the rate of change of slope, which therefore measures the curvature of the gravity field. Areas where the curvature of the field is greatest will have the greatest second-derivative values. This characteristic is useful for identifying the positions of small, shallow anomalies and suppressing regionals. When using second-derivative maps, it is important to remember once again that this is a strictly mathematical procedure and that relatively small changes in the gravity field can produce significant second-derivative values. Although second-derivative maps must therefore be used with care, they can bring to attention small anomalies that normally are hidden by strong regionals.

In order to provide an example of second-derivative use, we have calculated second-derivative values for the Bouguer anomaly curve (curve 1) that has functioned as our standard in the last few sections. Figure 6-31(a) illustrates this curve once again as well as contains an overlay of second-derivative values. Note that these values are constant for areas of constant slope and show the most variation when slope changes quickly. Figure 6-31(b) is a more detailed presentation of these second-derivative values. The position of the buried channel is superimposed on this graph in its correct spatial position in order to demonstrate

(a)

(b)

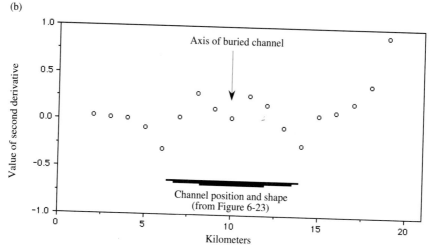

Figure 6-31 Example of using second derivative to remove regional effects and emphasize local anomalies. (a) Bouguer anomaly curve from model in Figure 6-23 with second-derivative curve superimposed. (b) Second-derivative values plotted on an expanded scale to emphasize value differences.

how effectively the second-derivative approach "extracts" this small anomaly from the strong regional. It's reasonably easy to visualize how well the position of this channel would be outlined on a second-derivative map.

Filtering

A gravity anomaly can be represented by summing an assemblage of cyclic curves that have different wavelengths, amplitudes, and starting positions. Once such an assemblage is generated for a given anomaly, wavelengths of interest can be enhanced and others suppressed by subtracting from the assemblage those we wish to suppress. This process is referred to as *wavelength filtering*. Clearly, filtering should be effective in enhancing anomalies due to local, shallow sources (short wavelengths) by removing regional effects (long wavelengths) or vice versa. Filtering is applicable to gravity maps as well as gravity profiles.

GRAVITY INTERPRETATION

Let's assume that we have conducted a gravity survey and are ready to interpret the data. How do we proceed? If we defined our objectives carefully, our data have good areal distribution and sufficient observational density to satisfy survey goals. The data are properly reduced and have the required accuracy. Regionals, if present, have been removed. Sufficient information on rock densities is available, and we have all relevant geologic information at our disposal. Normally, we then utilize a modeling program such as GravModel or one that accepts three-dimensional models. A model is constructed that fits the known geology as closely as possible. The Bouguer anomaly values calculated by the model are compared with measured values in order to evaluate the model. If agreement is not good, alterations are made to the model, again following constraints imposed by geologic information or other geophysical information, in order to improve the agreement between model and observed values.

When an acceptable fit is achieved between the model curve and the Bouguer anomaly curve, one possible geologic configuration is in hand. Note that, just as in the case of electrical resistivity, many different models can produce similar gravity values. There are no unique solutions. However, if sufficient geologic and geophysical controls are available, a good model fit likely provides a fairly close approximation to the actual subsurface geology. As geologic controls diminish, the number of possible geometries that will produce acceptable curves increases rapidly. A number of techniques are available that help to narrow the range of possible choices for model parameters. In this section we describe two approaches in order to give you a sense of what is possible.

Half-Maximum Technique

In our previous discussion of gravity over a buried sphere, we discovered that maximum gravity occurs directly over the center of the sphere and that there seems to be a relationship between the depth of the sphere's center and the relationship between maximum gravity and anomaly width. Let's see if we can develop a quantitative expression that directly utilizes these relationships.

Because gravity is a maximum over a sphere's center, we can write (using Eq. 6-34)

$$g_{\max_{sphere}} = \frac{G\,4\,\pi R^3 \rho_c}{3} \frac{z}{\left(z^2\right)^{3/2}} \tag{6-52}$$

because $x = 0$ over the sphere's center according to our coordinate system convention. At some horizontal position gravity declines to one-half its maximum value. If we use the notation $x_{(1/2)max}$ to refer to this special horizontal position, then the value of one-half the maximum gravity of the sphere is given by Eq. 6-34 at the special position $x_{(1/2)max}$. One-half the maximum is also given by one-half the value of Eq. 6-52. Therefore, we can equate these two statements, such that

$$g_{\frac{1}{2}\max_{sphere}} = \frac{1}{2}\left(\frac{G\,4\,\pi R^3 \rho_c}{3} \frac{z}{\left(z^2\right)^{3/2}}\right) = \frac{G\,4\,\pi R^3 \rho_c}{3} \frac{z}{\left(z^2 + x^2_{(\frac{1}{2})max}\right)^{3/2}} \tag{6-53}$$

Next, we extract from Eq. 6-53 the desired relationship between depth z and $x_{(1/2)max}$.

$$\frac{1}{2}\left[\frac{1}{\left(z^2\right)^{3/2}}\right] = \frac{1}{\left(z^2 + x^2_{(\frac{1}{2})max}\right)^{3/2}}$$

and

$$z = 1.305\, x_{(\frac{1}{2})max} \tag{6-54}$$

Let's check our results on the data in Table 6-6 (Sphere A), which is diagrammed in Figure 6-32. Maximum gravity is 0.3576 mGal, so half that value is 0.1788 mGal. Locating this value as best we can given the unit divisions in Figure 6-32, we find that this value occurs at a horizontal position of 380 m ($x_{(1/2)max}$) from the sphere's center at $x = 0$. Substituting 380 m into Eq. 6-54 results in a value of 496 m for the depth to the sphere's center. This compares very well with the value of 500 m which was used in Table 6-6, considering that horizontal and vertical scale coordinates in Figure 6-32 cannot be read very accurately.

If we have a gravity anomaly with a form suggesting the source is roughly equidimensional, then Eq. 6-54 provides a reasonably good value with which to begin our modeling exercise. If an anomaly suggests a source that can be approximated by a horizontal cylinder, then we can use

$$z = x_{(\frac{1}{2})max} \tag{6-55}$$

Similar expressions can be developed for other simple shapes (Telford et al., 1990). Although most other half-maximum expressions are not as simple as Eqs. 6-54 and 6-55, they are valuable in providing estimates of source depths.

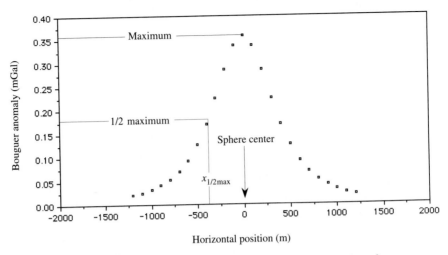

Figure 6-32 Diagram of gravity anomaly over a sphere illustrating locations of parameters used to determine depth to sphere's center.

Second-Derivative Techniques

Based on our application of the second derivative when discussing the regional-residual problem, it should not be surprising that this technique can be used to derive information about the anomaly source itself. Although a number of second-derivative techniques exist, we concentrate here on an approach developed by Stanley (1977) in order to present a specific example of this particular method. A truncated plate can be used to represent a geologic contact if the depth to the top of the plate t is small relative to the depth to the bottom of the plate T (Fig. 6-33). In such a case the equation for the second derivative is

$$\frac{d^2 g_{truncated\ plate}}{dx^2} = \frac{2Gp_c \sin\alpha \left(t \cos\alpha - x \sin\alpha\right)}{r_U^2} \tag{6-56}$$

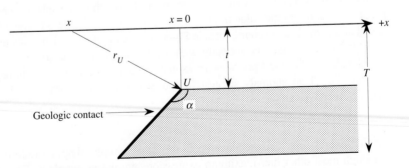

Figure 6-33 Notation used for Stanley's (1977) second-derivative technique over a geologic contact.

Stanley demonstrated that second-derivative curves over such forms have several characteristic points which can be used to solve for contact parameters. These points are the maximum value g''_{max}, which is at the horizontal position x_{max}; the minimum value g''_{min}, which is
at x_{min}; and $x = 0$, which is directly above the point U in Figure 6-33. This latter point is
defined by the intersection of the second-derivative curve and a line joining g''_{max} and g''_{min}.
Note that x_{max} and x_{min} are defined relative to the point $x = 0$. The dip of the contact a and
depth t (Fig. 6-33) are related to these parameters by the following equations (Stanley, p.
1233).

$$\frac{\left(g''_{max} + g''_{min} \right)}{\left(g''_{max} - g''_{min} \right)} = \cos \alpha \qquad (6\text{-}57)$$

and

$$x_{max} - x_{min} = \frac{-2t}{\sin \alpha} \qquad (6\text{-}58)$$

An example of Stanley's technique is presented in Figure 6-34. The gravity data from
which this graph is constructed were collected by Chandler (1978, p. 147) during a traverse
across a fault on the eastern margin of the Deerfield Basin (a Mesozoic rift valley) in west-
central Massachusetts. The graph establishes that $g''_{max} = 1.50 \times 10^{-6}$, $g''_{min} = -2.25 \times 10^{-6}$, x_{max}
$= -800$, and $x_{min} = 0$. Note that $x_{max} = -800$ ft is based on the location of $x = 0$ on the horizontal axis (see Fig. 6-34). Placing these values in Eqs. 6-57 and 6-58 yields $\alpha = 101.5°$ (dip =
78°) and depth $t = 408$ ft (124 m). These results provide a good starting basis for using a
program similar to GravModel in an attempt to develop a model of subsurface geology for
this region.

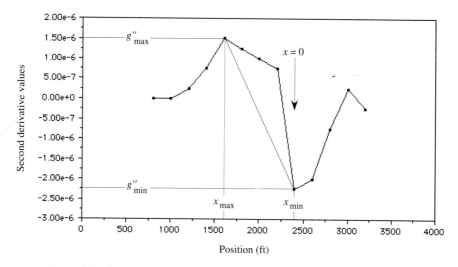

Figure 6-34 Second-derivative curve for a gravity traverse across a fault block near
Amherst, Massachusetts.

Revisiting Some Bouguer Anomaly Values

Earlier in this chapter we presented a sample of gravity data (Table 6-4) when discussing gravity survey procedures. These data are graphed in Figure 6-35(a). The purpose of the traverse was to gather information that would enable bedrock depths beneath the Connecticut River floodplain to be determined. Seismic data are available at two points along the western portion of the traverse and limit bedrock depths to no greater than 100 m. Mesozoic sedimentary rocks constitute basement along the extent of the traverse but thin rapidly to the east and west. Paleozoic metamorphic and igneous rocks are exposed at the surface approximately 1 km from both ends of the traverse. Density contrasts are -0.5 g/cm^3 for the sediments above bedrock and -0.2 g/cm^3 for the Mesozoic sedimentary rocks.

Figure 6-35(b) illustrates one possible model that provides gravity values that are quite close to the observed Bouguer anomaly values. The model follows all known geologic

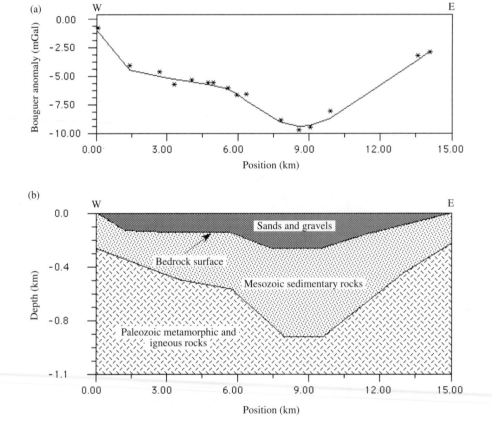

Figure 6-35 Gravity values (*) obtained across a buried river channel on the floodplain of the Connecticut River near North Hatfield, Massachusetts. (a) Measured values (*) compared with values (continuous curve) calculated for the model in (b).

and geophysical constraints and provides a reasonable solution. However, it is important to realize that it is not a unique solution. For instance, errors in the selected density contrasts would affect the size of the model polygons significantly. If the Mesozoic rocks beneath the central portion of the traverse are thicker than modeled, and assuming the density contrasts are correct, then the bedrock surface must be much shallower. Conversely, thinner Mesozoic rocks would necessitate thicker sands and gravels. For simplicity we assume that the Paleozoic rocks have an average density of 2.67 g/cm^3 (the Bouguer reduction density). If this assumption is in error, this will clearly have an effect on the present model. If anything, the Paleozoics are likely to have a positive density contrast, which would require increased thicknesses of sediments and/or sedimentary rocks in order to keep the same general model gravity values.

The purpose of this final presentation simply is to emphasize once again the nonuniqueness of a model that provides gravity values that match observed values and to point out how crucial additional independent evidence is in order to create solutions that most likely are similar to actual subsurface geology.

APPLICATIONS OF THE GRAVITY METHOD

One of the most common applications of gravity surveys in shallow exploration is mapping bedrock depth. Knowledge of bedrock depth is required in many engineering endeavors as well as in efforts to define aquifer geometries for purposes of planning groundwater use or identifying new groundwater sources. Other applications involving detailed gravity work using modern gravimeters include detection of sinkholes and other subsurface voids and establishing landfill boundaries. Recent efforts have been successful in demonstrating gravity variations due to removal of groundwater during pumping tests. The examples we present here were chosen because they represent several of these typical applications.

Bedrock Depths

Kick (1985) describes a survey concerned with bedrock depths in an account of exploration work in the urban setting of Reading, Massachusetts. The objective of the study was to determine depths to bedrock along two proposed routes for a tunnel needed to alleviate sewage transport problems. Engineering design studies required detailed bedrock depths, as the tunnel needed to be sited in bedrock rather than in the overlying soil, which would result in much higher costs.

Drilling was not feasible, due both to the urban environment and the highly irregular bedrock surface. Normally, seismic refraction (or electrical resistivity) would be the method of choice, but the local environment imposed widely paved areas, high noise levels, frequent obstructions such as buildings and fences, and underground pipe complexes. Therefore, gravity was selected because it would cause minimum disturbance to the surroundings and would be relatively unaffected by the urban factors. Readings could be obtained even inside buildings!

Detailed control was established by siting gravity stations at 40-ft intervals. In addition, stations were occupied well beyond the survey boundaries in order to adequately define regional variations. Careful surveying assured vertical control to ±0.1 ft and horizontal control to ±4 ft. Corrections for instrument drift and terrain effects were calculated carefully. Such attention to survey methods assured Bouguer values accurate to ±0.02 mGal. After Bouguer gravity was in hand, regionals were removed by manual smoothing. Also, tests indicated that bedrock density varied very little within the study area, so residuals were believed to be solely a function of soil thickness. The results of this survey are illustrated in Figure 6-36. Although most traverses don't have the degree of borehole control as the one illustrated, this traverse was chosen to demonstrate the remarkably good fit (with one exception) between bedrock depths calculated from Bouguer gravity and those confirmed by borehole information.

Subsurface Voids

A study by Butler (1984) at Medford Cave, Marion County, Florida, demonstrated how a carefully designed gravity survey can detect subsurface voids. This work utilized a

Figure 6-36 Bouguer gravity and depth to bedrock along a street in Reading, Massachusetts. (a) Bouguer anomaly curve. (b) Bedrock surface derived from Bouguer gravity. Depth at which wells encountered bedrock are shown by horizontal marks at well bottoms. Note that depths from Bouguer gravity are very close to those in the wells except at one location. (Modified from Kick, John F., 1985, Depth to bedrock using gravimetry: Geophysics: The leading edge of exploration: Society of Exploration Geophysicists, v. 4, p. 40.)

LaCoste-Romberg gravimeter to obtain 420 readings spread over an approximately 80 m by 80 m area. Standard tidal and drift corrections were applied to the data, as were Bouguer and terrain corrections. A planar regional field was determined by inspection and removed. Resulting residual gravity anomalies correlated well with features of the known cave system. Figure 6-37(a) illustrates a profile plotted from the residual map compared to a geologic section (Fig. 6-37(b)) along a traverse containing numerous boreholes. Gravity highs correlate well with limestone pinnacles, and lows correlate with depressions in limestone occupied by clay deposits. The most significant gravity low on the profile is due to a known void at depth. This study illustrates how careful field work can utilize gravity to detect limestone topography (sinkholes and caves) in an area of karst. Figure 6-37 also emphasizes the importance of borehole information in order to establish correlations between gravity highs and lows and subsurface features. By comparing Figure 6-37 with Figure 5-38, it is possible to visualize how a resistivity survey might be combined with a gravity survey to provide a powerful multitechnique approach to karst regions. Recall that in the area represented by Figure 5-38, high apparent resistivities correlated well with limestone pinnacles and voids, and apparent resistivity lows correlated with bedrock depressions containing clay. These studies emphasize that independent geologic information is critical (especially in gravity and resistivity work) for reliable interpretation of geophysical data and that applying multiple geophysical methods in an exploration exercise is often extremely rewarding in terms of constraining interpretations.

Landfill Geometry

Roberts et al. (1990) undertook a gravity survey of a multicomponent waste landfill in order to investigate the effectiveness of the method in establishing the lateral boundaries and the vertical extent of the landfill. The landfill is located on the Thomas Farm site in northwestern Indiana (Section 23; T23N, R5W). An evaluation of the effectiveness of the gravity study was possible due to the existence of prelandfill and postlandfill topographic maps. These maps supported construction of an isopach map (Fig. 6-38) that illustrated landfill thickness variations. Boreholes surrounding the landfill site provided thicknesses of glacial deposits and bedrock depths.

The survey consisted of approximately 200 gravity stations spaced at 5-m to 10-m intervals. Measurements were made with a LaCoste-Romberg gravimeter, and standard field corrections were applied to the data. Bouguer and terrain corrections were made using a density determined from a gravity profile along a traverse of marked topographic relief. Finally, a regional trend was defined and removed by fitting a third-degree polynomial surface to gravity stations located outside the landfill boundaries. The residual gravity anomaly map is presented in Figure 6-39. A comparison of Figures 6-38 and 6-39 reveals the close correlation between the landfill boundary and the fill thickness and the residual gravity. Roberts et al. demonstrated, in addition, that under suitable conditions gravity can be used to evaluate the gross nature of landfill material. We have mentioned previously, and emphasized, that independent geologic information is very important when evaluating gravity data. In order to focus further attention on this requirement, we conclude the discussion in this chapter with a quotation from Roberts et al. (p. 259): "However, the gravity method

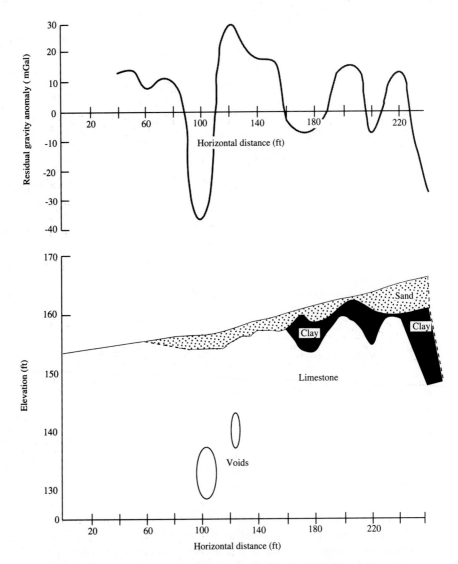

Figure 6-37 Gravity study at Medford Caves, Florida. (a) Profile from residual gravity map. (b) Geologic section determined from borehole information. Gravity highs correlate with limestone pinnacles and voids, and lows correlate with depressions in limestone occupied by clay deposits. (Modified from Butler, Dwain K., 1984, Microgravimetric and gravity gradient techniques for detection of subsurface cavities: Geophysics, v. 49, p. 1084–1096.)

depends on precise gravity and surface elevation measurements, careful computations, and constraining information obtained from collateral geologic and geophysical studies for the interpretations to be valid."

Figure 6-38 Isopach map of the Thomas Farm landfill, northwestern Indiana. (Modified from Roberts, R. L., Hinze, W. J., and Leap, D. I. 1990, Application of the gravity method to investigation of a landfill in the glaciated midcontinent, U.S.A. *in* Ward, Stanley H., ed., Geotechnical and environmental geophysics; Volume 2: Environmental and groundwater: Society of Exploration Geophysicists Investigations in Geophysics No. 5, p. 253–266.)

Figure 6-39 Residual gravity anomaly map of the Thomas Farm landfill, northwestern Indiana. (Modified from Roberts, R. L., Hinze, W. J., and Leap, D. I., 1990, Application of the gravity method to investigation of a landfill in the glaciated midcontinent, U.S.A. *in* Ward, Stanley H., ed., Geotechnical and environmental geophysics; Volume 2: Environmental and groundwater: Society of Exploration Geophysicists Investigations in Geophysics No. 5, p. 253–266.)

PROBLEMS

6-1. If a gravity determination is made at an elevation of 152.7 m, what is the value of the free-air correction (assuming sea level as the datum)? the Bouguer correction (assuming a 2.50 g/cm^3 reduction density)?

6-2. A gravity station at a location of 0 m is located in the center of an erosional basin. The floor of the basin has virtually no relief. An escarpment of a plateau is located at a distance of 450 m from the gravity station. The surface of the plateau has a relatively constant elevation of 400 m. Will terrain corrections be necessary?

6-3. Prepare a drift curve for the following data and make drift corrections. Convert your corrected data to milliGals. The data were collected by a gravimeter with a dial constant of 0.0869 mGal/dial division.

Station	Time	Reading in Dial Divisions
Base	11:20	762.71
GN1	11:42	774.16
GN2	12:14	759.72
GN3	12:37	768.95
GN4	12:59	771.02
Base	13:10	761.18

6-4. You are measuring values of g in a drill hole. Calculate the free-air and Bouguer corrections in measured g at positions 1, 2, 3, and 4. Sea level is the reduction datum. The mass loss due to the drill hole is negligible.

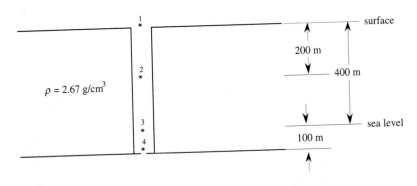

6-5. What is the radius of the smallest equidimensional void (such as a cave) that can be detected by a gravity survey for which Bouguer gravity values have an accuracy of 0.05 mGal? Assume the voids are in limestone (density = 2.7 g/cm^3) and that void centers are never closer to the surface than 100 m.

6-6. In a region of gentle, nonplunging folds ore deposits are localized in anticlinal crests of lime-stone beds. The center of the limestone bed in the crests is 1.5 km beneath the surface. The form of the deposit can be approximated by a horizontal cylinder with a radius of 0.25 km. Because of time constraints, a gravity survey can achieve an accuracy of only 0.1 mGal for Bouguer anomaly values. What is the smallest density contrast for which the ore deposits can be reliably located?

6-7. Determine the maximum value of the gravity anomaly due to the faulted basalt sill in the following diagram.

6-8. The curve in the following diagram represents a traverse across the center of a roughly equidimensional ore body. The anomaly due to the ore body is obscured by a strong regional anomaly. Remove the regional anomaly and then evaluate the anomaly due to the ore body, which has a density contrast of 0.75 g/cm^3. (Graph values are supplied in the table following the graph.)

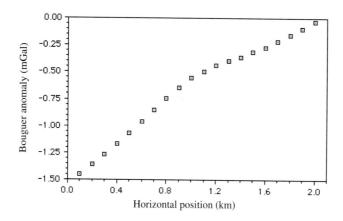

Horizontal Position (km)	Bouguer Anomaly (mGal)
0.1	-1.45
0.2	-1.36
0.3	-1.27

Horizontal Position (km) *(cont.)*	Bouguer Anomaly (mGal) *(cont.)*
0.4	-1.17
0.5	-1.07
0.6	-0.96
0.7	-0.85
0.8	-0.74
0.9	-0.64
1.0	-0.56
1.1	-0.49
1.2	-0.44
1.3	-0.40
1.4	-0.36
1.5	-0.32
1.6	-0.27
1.7	-0.22
1.8	-0.16
1.9	-0.10
2.0	-0.03

6-9 Determine the depth to the center of each of the three spheres represented by anomaly curves 1, 2, and 3.

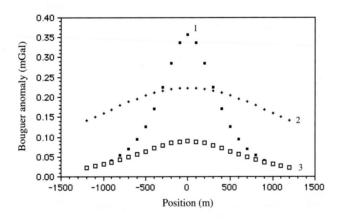

6-10 (a) Derive the half-maximum relationship for a horizontal cylinder (Eq. 6-55) using Eq. 6-37. (b) Determine the depth to the centers of the two horizontal cylinders represented by anomaly curves 1 and 2.

6-11 New boring information obtained along the traverse illustrated in Figure 6-35 fixes bedrock depth at 50 m at 3 km along the traverse and at 100 m at position 8 km. Assuming all other parameters are correct as discussed in the preceding section, "Revisiting Some Bouguer Anomaly Values," discuss how the maximum probable thickness of the Mesozoic sedimentary rocks are affected by this new information. The gravity data is file "Connecticut Valley data" and the model is file "Connecticut Valley model," which are on Disk 2 included with this text.

6-12 The following Bouguer gravity curve was obtained on a traverse oriented perpendicular to a vertical dike. The dike has a density contrast of 0.9 g/cm³. Seismic surveys which penetrated to 0.5 km did not encounter the top of the dike. Exposed dikes in the region tend to be no wider than 0.4 km. Deduce as much as possible about the dimensions and position of the dike. (Graph values are supplied in the table following the graph.)

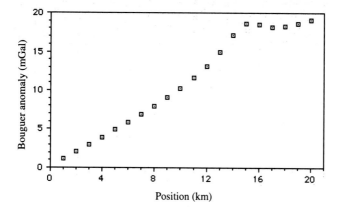

Horizontal Position (km)	Bouguer Anomaly (mGal)
1	1.11
2	2.04
3	2.98
4	3.93
5	4.90
6	5.88
7	6.89
8	7.93
9	9.04
10	10.23
11	11.56
12	13.10
13	14.96
14	17.12
15	18.61
16	18.48
17	18.18
18	18.25
19	18.59
20	19.12

6-13. Pumping tests on a major water supply well create a cone of depression with a top surface that extends more than 1000 m from the well. The area is underlain with totally saturated sand possessing a 40 percent porosity and a density of 2.60 g/cm^3. The specific yield of the pumping is estimated at 30 percent (30 percent of the fluid occupying the 40 percent void space). Can the removal of water be detected by a gravity measurement with an accuracy of 0.01 mGal? For modeling purposes assume that the cone of depression can be represented sufficiently well by a horizontal cylinder with a radius of 300 m and a center at a depth of 300 m.

6-14. The following data represent a traverse across a faulted contact between an uplifted crystalline basement complex and an overlying sedimentary sequence. Use a second-derivative technique to determine the dip of the fault and the depth to the top of the uplifted basement.

Position (km)	Bouguer Anomaly (mGal)
19	0.530
20	0.597
21	0.677
22	0.775
23	0.897
24	1.048
25	1.239
26	1.478
27	1.777
28	2.143
29	2.569

Horizontal Position (km) *(cont.)*	Bouguer Anomaly (mGal) *(cont.)*
30	3.018
31	3.415
32	3.719
33	3.944
34	4.110
35	4.235
36	4.331
37	4.406
38	4.464
39	4.508
40	4.543
41	4.568

REFERENCES CITED

Butler, Dwain K., 1984, Microgravimetric and gravity gradient techniques for detection of subsurface cavities: Geophysics, v. 49, p. 1084–1096.

Chandler, William E., Jr., 1978, Graben mechanics at the junction of the Hartford and Deerfield basins of the Connecticut Valley, Massachusetts: University of Massachusetts, Department of Geology and Geography Contribution No. 33, 151 p.

Grant, F. S., and West, G. F., 1965, Interpretation theory in applied geophysics: New York, McGraw-Hill, 584 p.

Hammer, Sigmund, 1939, Terrain corrections for gravimeter stations: Geophysics, v. 4, p. 184–194.

Hubbert, M. King, 1948, Line-integral method of computing the gravimetric effects of two-dimensional masses: Geophysics, v. 13, p. 215–225.

Kick, John F., 1985, Depth to bedrock using gravimetry: Geophysics: The Leading Edge of Exploration, Society of Exploration Geophysicists, v. 4, p. 38–42.

Maslyn, R. Mark, 1984, Case histories of computers in oil exploration–polynomial trend surface analysis: Computer Oriented Geological Society Computer Contributions, v. 1, p. 12–19.

Nettleton, L. L., 1942, Gravity and magnetics calculations: Geophysics, v. 7, p. 293–310.

Plouff, Donald, 1976, Gravity and magnetic fields of polygonal prisms and applications to magnetic terrain corrections: Geophysics, v. 41, p. 727–739.

Roberts, R. L., Hinze, W. J., and Leap, D. I., 1990, Application of the gravity method to investigation of a landfill in the glaciated midcontinent, U.S.A. *in* Ward, Stanley H., ed., Geotechnical and environmental geophysics; Volume 2: Environmental and groundwater: Society of Exploration Geophysicists Investigations in Geophysics No. 5, p. 253–266.

Robinson, Edwin S., and Coruh, Cahit, 1988, Basic exploration geophysics: New York, John Wiley, 562 p.

Robinson, Joseph E., 1982, Computer applications in petroleum geology: Stroudsburg, PA, Hutchinson Ross, p. 72–81.

Rudman, Albert J., Ziegler, Robert, and Blakely, Robert F., 1977, FORTRAN program for generation of earth tide gravity values: Indiana Geological Survey Occasional Paper 22, 14 p.

Smith, R. A. 1959, Some depth formulae for local magnetic and gravity anomalies: Geophysical Prospecting, v. 7, p. 55–63.

Stanley, John M., 1977, Simplified gravity interpretation by gradients—the geological contact: Geophysics, v. 42, p. 1230–1235.

Talwani, M., Worzel, J. W., and Landsman, M., 1959, Rapid gravity computations for two-dimensional bodies with application to the Mendocino submarine fracture zone: Journal of Geophysical Research, v. 64, p. 49–59.

Telford, W. M., Geldart, L. P., Sheriff, R. E., and Keys, D. A., 1976, Applied geophysics: London, Cambridge University Press, 860 p.

Telford, W. M., Geldart, L. P., and Sheriff, R. E., 1990, Applied geophysics, 2nd ed.: London, Cambridge University Press, 770 p.

Tsuboi, Chuji, 1979, Gravity: London, Allen & Unwin, 254 p.

SUGGESTED READING

Nettleton, L. L., 1976, Gravity and magnetics in oil prospecting: New York, McGraw-Hill, 464 p.

Tsuboi, Chuji, 1979, Gravity: London, Allen & Unwin, 254 p.

7

Exploration Using the Magnetic Method

The magnetic method as used in geophysical exploration is possibly the most versatile of the methods discussed in this text, as it can be applied to both deep and shallow structures, and relative to other methods, measurements can be relatively cheaply obtained for both local and regional studies. On the other hand, quantitative interpretation of magnetic information is much more difficult than interpretation of gravity data even though these two methods share many similarities.

A great deal of the time exploration of the shallow subsurface is directed toward characteristics of the sediments above bedrock or a definition of the bedrock surface, and in the majority of these cases magnetic information is not very useful. This arises in part from the difficulty in obtaining useful quantitative information from magnetic data but also is due to the magnetic characteristics of rocks and sediments. For the most part, magnetic signatures depend on magnetite content, which tends to be low in most sediments and in many rocks constituting basement. In general, therefore, magnetic field work is not commonly employed for many shallow subsurface objectives. Nevertheless, magnetic data can contribute to certain exploration efforts when targets are not deep. Buried ore deposits constitute an obvious example. Others include the mapping of rock units with sufficiently large magnetic signatures to be traceable beneath glacial or other cover and the discovery and delineation of a variety of features important to archaeological investigations.

In this chapter our goal is to provide a basic and practical grounding in magnetic surveys rather than to develop an in-depth theoretical basis. We cover fundamental relationships that control the nature of magnetic anomalies, govern field methods, and affect interpretation efforts. We dwell on a few relatively straightforward examples in detail so you can appreciate the applications of magnetic surveying and gain an understanding of what can and cannot be achieved. Although you are more likely to engage in, for instance, seismic exploration than in magnetic work, the magnetic method could be the most appropriate method to apply during a particular assignment, and we want to prepare you for that eventuality.

FUNDAMENTAL RELATIONSHIPS

Virtually all of you should be familiar with the simple bar magnet and its behavior when placed in the proximity of a second similar magnet. You likely also remember the great delight of sprinkling iron filings on a glass plate underneath which was a strong bar magnet. The iron filings (which became very small magnets) aligned themselves in a pattern about the magnet (Fig. 7-1). This pattern outlined what we refer to as *magnetic lines of force,* which follow curved trajectories and are most closely spaced at the ends of the magnet. If you were lucky, you were also able to map these lines of force by placing a small compass at various positions above the magnet and recording the orientation of the compass needle. The pattern that results from such an investigation confirms the pattern of iron filings and demonstrates that the lines of force converge at the ends of the magnet. These points of convergence are referred to as *magnetic poles.* The lines of force record the orientation of the magnetic field produced by the magnet. At any point the orientation of the field is illustrated by a line drawn tangent to the lines of force. The spacing of the lines of force is an indication of the magnetic force at that position; the force is greatest where the lines of force are closest together.

Each bar magnet has two poles and is referred to as a *dipole.* In fact, a single pole cannot exist—poles always come in pairs. If a bar magnet is broken in half, each half still has two poles. Because the properties of magnets and their associated magnetic fields are closely related to magnetic fields produced by electric current, and by analogy with electric

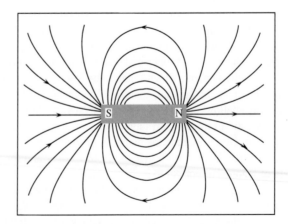

Figure 7-1   Magnetic lines of force produced by a simple bar magnet.

charges, the magnetic poles are referred to as positive (+m) and negative (–m). Poles of like sign repel, and those of opposite sign attract. It is the positive pole (north-seeking pole) of a compass needle that points toward the north magnetic pole of the earth.

It is convenient in visualizing many magnetic effects to conceive of a magnet, or any magnetized body, as composed of a very large number of small dipoles which are similarly aligned and stacked closely together (Fig. 7-2). Within the magnet, effects of the positive and negative poles cancel, but on one end there is an accumulation of positive poles, and on the other end is an accumulation of negative poles which give rise to the dipolar nature of the large magnet. If we cut the magnet across its length, we simply produce two new surfaces, and each has its own accumulation of poles of like sign. We assume, of course, that the dipoles within the magnet are too small to be cut or, if cut, are themselves composed of yet smaller dipoles.

Magnetic Force

We've mentioned attraction and repulsion, but what is the actual force exerted between two magnetic poles? This force is stated in Coulomb's law,

$$F = \frac{1}{\mu} \frac{m_1 m_2}{r^2} \tag{7-1}$$

where r is the distance between poles of strength m_1 and m_2. The remaining factor μ, or *magnetic permeability*, is a property of the medium in which the poles are located. Because the magnetic permeability is very nearly equal to unity in water and air, for simplicity we do not include the term in the remaining equations in this chapter. However, in some instances the permeability of the medium is important, so its effect cannot be neglected in all cases. Consider Figure 7-3, which illustrates an iron bar placed within a wire coil. When current is passed through the wire, a magnetic field is generated which is greater than had the iron bar not been present. Thus, the strength of the field generated depends on some property (magnetic permeability) of the medium (iron bar) in which the field is generated.

Magnetic Field Strength

Although poles cannot exist individually and always occur in pairs, it is convenient to utilize a unit magnetic pole m'. Such a pole repels a similar pole with a force of 1 dyne when

Figure 7-2   Representation of a magnet as an assemblage of small dipoles.

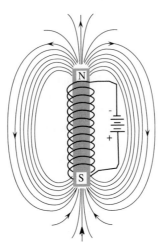

Figure 7-3   Effect of magnetic permeability
on magnetic field strength.

separated by a distance of 1 cm. Such a situation can be visualized by considering two very
long, slender bar magnets with their two positive (or negative) ends close together. In this
arrangement we effectively are dealing with the interaction of two poles, as the remaining
poles are far removed from the region of interaction.

The *magnetic field strength H* is the force a unit magnetic pole would experience if
placed at a point in a magnetic field which is the result of some pole strength m and where r
is distance of the point of measurement from m

$$H = \frac{F}{m'} = \frac{m}{r^2} \tag{7-2}$$

H is a vector quantity having a magnitude given by Eq. 7-2 and with a direction determined
by assuming our unit pole is *positive*. The SI unit of magnetic field strength is the *nanotesla*
or *nT*, whereas the cgs unit of magnetic field strength is the *oersted*. The oersted is 1
dyne/unit pole strength and is equal to 10^5 *gammas*. The gamma is numerically equivalent
to the nanotesla. Until recently the gamma was the common unit employed in geophysics,
but current use favors the nanotesla.

Magnetic Moment

If a bar magnet is placed in a uniform magnetic field H (Fig. 7-4), it will experience a pair
of equal forces acting parallel to each other but in opposite directions (a *couple*). The magnitude of the couple is

$$C = 2(ml)H \sin \theta \tag{7-3}$$

where θ specifies the original orientation of the magnet in the field. The motion produced
by the couple is dependent on the magnitude of H as well as the value of θ (no motion is

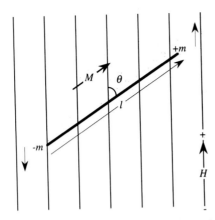

Figure 7-4 Diagram illustrating role of magnetic moment in determining forces acting on a bar magnet placed in a magnetic field.

produced if $\theta = 0$). The other quantity (ml) which also affects the magnitude of the couple is termed the *magnetic moment*, typically represented by M, so that

$$M = ml \tag{7-4}$$

Intensity of Magnetization

A bar magnet possesses a fundamental property per unit volume known as the *intensity of magnetization I*. The magnitude of I is defined as the magnetic moment M per unit volume or

$$I = \frac{M}{\text{volume}} \tag{7-5}$$

and, therefore,

$$I = \frac{ml}{\text{volume}} = \frac{m}{\text{area}} \tag{7-6}$$

In order to develop an appreciation of the relation of intensity of magnetization consider once again the bar magnet in Figure 7-2. If we cut the magnet into unequal pieces perpendicular to its length, the force at a distance is proportional to the length of each piece. We have not altered the fundamental magnetic properties of the material but simply have changed the volume of each. Similarly, if we cut the magnet longitudinally into two sections (as suggested by the shading in Fig. 7-2), the force at a distance is proportional to the cross-sectional area of each section. Of course, the volumes are once again different, but as the lengths are the same, we can visualize the effect in terms of the cross-sectional areas, which are not the same. Recall Figure 7-2 and our visualization of the magnet as a concentration of poles on its end surfaces. As I increases, the concentration of poles increases, and the pole strength per unit area is greater.

Magnetic Susceptibility

If we place a material that can be magnetized in an external magnetic field H, the intensity of magnetization I will be proportional to the strength of the field. When considering geologic materials, the direction of magnetization is parallel to the earth's magnetic field, which functions as the inducing field. We thus can consider the result of this magnetization as magnetic poles induced on surfaces perpendicular to the field direction, analogous to our representation of the bar magnet in Figure 7-2. The magnitude of I is

$$I = kH \tag{7-7}$$

where k is the constant of proportionality generally referred to as *magnetic susceptibility*. We'll examine representative values of magnetic susceptibility when we discuss the magnetic effects of various geologic bodies, but first let's take a moment to examine the process of magnetization in a bit more detail.

Based on their magnetic susceptibilities, minerals are classified as *diamagnetic* or *paramagnetic*. Diamagnetic minerals have negative susceptibilities, but the values are so low that these have no effect with regard to subsurface exploration. Quartz and feldspar are common diamagnetic minerals. Paramagnetic minerals have positive susceptibilities, but in general the values are quite low. Examples include the Fe-Mg silicates such as pyroxene, amphibole, and olivine. However, in a few paramagnetic materials the magnetic moments imparted to atoms by orbital motion and spin of electrons interact strongly. The result is the alignment of magnetic moments in areas with dimensions on the order of 10^{-4} cm. These small areas are referred to as *magnetic domains*. If these domains are parallel (as they are in iron, nickel, and cobalt), the material is termed *ferromagnetic* (Fig. 7-5(a)). Ferromagnetic materials have very high susceptibilities, but they do not occur naturally on Earth (although they are present in meteorites) and so are not of interest within the realm of geophysical exploration. If the domains are parallel and antiparallel in equal numbers (Fig. 7-5(b)), the net magnetic moment is zero, and the material is said to be *antiferromagnetic*. An example of an antiferromagnetic mineral is hematite, which certainly is common, but because susceptibilities of antiferromagnetic materials are low, such minerals need not be considered for exploration purposes.

Of primary importance to the exploration effort are *ferrimagnetic* minerals. In these few minerals (magnetite, titanomagnetite, ilmenite, and pyrrhotite are the common ones) the domains are oriented as in antiferromagnetic materials, but one direction of orientation is preferred (Fig. 7-5(c)). This imparts a net magnetic moment to such materials and results in relatively high susceptibilities. These susceptibilities are dependent on temperature and the strength of the inducing field. It is these minerals, the most common of which by far is magnetite, which impart induced magnetizations to rocks in the earth's crust.

If a ferrimagnetic mineral is placed in an external magnetic field, an induced magnetization is produced, and the external field causes some magnetic domain walls to move. If the external field is weak, as is the earth's magnetic field, only a limited movement of domain walls occurs, and no permanent magnetization occurs when the field is removed. This behavior is represented in Figure 7-6. As the inducing field is increased to H_1, an induced magnetization I_i is generated and will return to zero if H_1 is removed. However, if

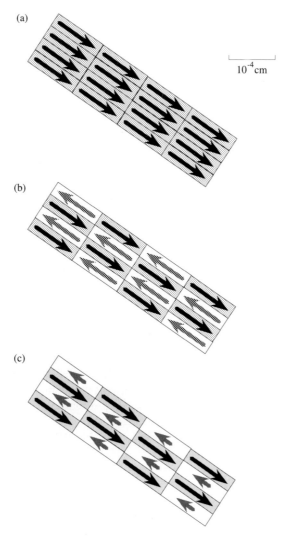

(a)

(b)

(c)

10^{-4} cm

Figure 7-5   Schematic representation of magnetic domains in (a) ferromagnetic, (b) anti-ferromagnetic, and (c) ferrimagnetic materials.

the external field is raised to much larger values, more domain walls move, favorably oriented domains increase at the expense of unfavorably oriented domains, and reordering of unfavorably oriented domains occurs. Eventually, the induced magnetization will increase to the saturation level (point S in Fig. 7-6). Further increases in H produce no further increase in I. If the external field now is reduced to zero, the values of I follow the curve in Figure 7-6 from point S to point I_r. Because a substantial portion of the domain changes are retained during the decrease in the inducing field, the mineral now possesses a net permanent magnetization. This generally is referred to as the *remanent magnetization* I_r. In order

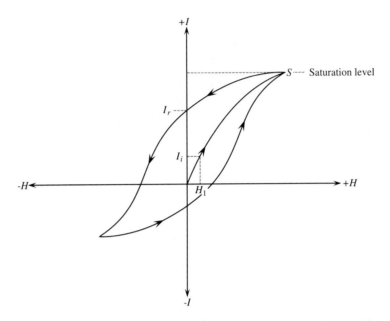

Figure 7-6 Hysteresis curve for a ferrimagnetic material in the presence of a magnetizing field.

to return the magnetization to zero, the inducing field must first be applied with a reversed polarity and then returned to its normal polarity, as indicated by the curve in Figure 7-6. This path defined by the values of induced magnetization and external field values is known as an *hysteresis curve*.

Since the earth's field is weak, need we be concerned with remanent magnetizations? When we measure the magnetic effect of magnetizations induced in geologic bodies, any remanent magnetizations with an order of magnitude similar to the induced magnetization will either increase or decrease the total anomalous field depending on the respective orientations of the two magnetizations. Remanent magnetizations in rocks can be produced in a number of ways, all of which are important in paleomagnetic studies, but one especially strong remanent magnetization is *thermoremanent* magnetization. Thermoremanent magnetization typically is produced when ferrimagnetic minerals crystallize and cool during the solidification of igneous rocks. Since the ferrimagnetic minerals are initially at high temperatures, even a weak field such as the earth's can facilitate preferential alignments of magnetic domains, thus creating a substantial I_r. In general, rock susceptibilities are controlled by magnetite content, and many rocks with sufficient magnetite to produce magnetic anomalies have an igneous origin. In many situations, therefore, thermoremanent magnetization may be present.

Field instruments used in magnetic surveying normally do not determine the magnitude and orientation of remanent magnetizations. Such information more commonly is determined in the laboratory. Typically, however, survey results are interpreted by assuming that the recorded anomalous field is due entirely to induced magnetization, and possible effects of remanent magnetizations are ignored. Such neglect is not as serious as it might at first appear because of the inherent difficulty in obtaining quantitative results from magnetic surveys. In many situations the qualitative information sought by a survey is not sufficiently affected by the presence of remanent magnetizations to warrant the extra time and expense of laboratory analysis for determining remanent magnetizations and then to correct field data for their effect. When discussing magnetic anomalies due to geologic forms of various shapes and compositions, we assume all magnetization is induced only.

Magnetic Potential

Magnetic, gravitational, and electrical fields all are potential fields. A characteristic of such fields is that the work done by moving a pole (or mass or charge) from one point to another is path independent. If a complete circuit is made by returning the pole to its original position, then the work done is zero. It is convenient to define the potential V as the negative of the work done on a unit pole in a magnetic field as the pole is moved against the field from infinity to a point in the field. Utilizing Eq. 7-2 and remembering that we assume $\mu = 1$, we have

$$V = -\int_{\infty}^{r} \frac{m}{r^2} = \frac{m}{r} \tag{7-8}$$

An especially useful feature of the potential is that we can find the magnetic field in a given direction by taking the negative of the derivative of the potential in that direction. We utilize this feature in the following section to derive equations for a dipole.

THE EARTH'S MAGNETIC FIELD

The earth's magnetic field at any point on the earth's surface is a vector quantity that is defined by measuring its total intensity and direction. Intensity can be measured by any number of instruments, some of which are described in a following section. Orientation is readily determined by allowing a compass needle to rotate freely in all directions. Just as in the case of a bar magnet, the needle will rotate into parallelism with the earth's field.

Field Elements

The total-field vector is defined by its intensity F_E; its *inclination i*, which is the angle the vector makes with a horizontal plane; and its *declination d*, which is the angle the vertical plane containing the total-field vector makes with geographic north. These relationships are illustrated in Figure 7-7. F_E can be resolved into a vertical component Z_E and a horizontal component H_E. The vertical plane containing F_E, Z_E, and H_E is a magnetic meridian. H_E can also be resolved into horizontal components directed toward geographic north (X_E) and geographic east (Y_E). These seven geomagnetic elements are interrelated in several ways that are readily apparent from Figure 7-7. Any three elements are sufficient to determine the

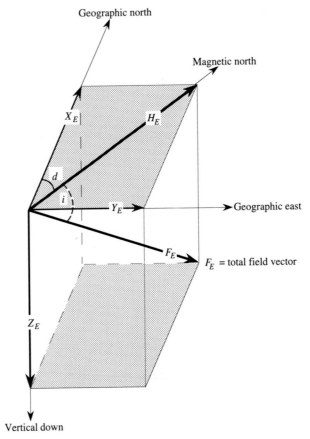

Figure 7-7   The elements of the earth's magnetic field: F_E = total-field vector, H_E = horizontal component, Z_E = vertical component, d = declination, and i = inclination.

remaining four. A thorough grasp of these relationships is essential for later derivations, so be sure you can reproduce each of the following:

$$F_E = \sqrt{H_E^2 + Z_E^2} = \sqrt{X_E^2 + Y_E^2 + Z_E^2}$$

$$Z_E = F_E \sin i, \quad H_E = F_E \cos i, \quad \text{and} \quad \tan i = \frac{Z_E}{H_E} \tag{7-9}$$

$$X_E = H_E \cos d \quad \text{and} \quad Y_E = \sin d$$

The positions on the earth's surface where $i = 90°$ are known as the *magnetic dip poles* (see Fig. 7-11), and the *magnetic equator* is defined by positions of $i = 0°$. At the dip poles $Z_E = F_E$, and the intensity is approximately 70,000 nT (nanotesla). At the magnetic equator, $H_E = F_E$, and the intensity is approximately 30,000 nT. Note that the earth's magnetic field varies in intensity by more than 200 percent, whereas the gravity field varies only by approximately 0.5 percent.

Dipolar Nature of the Earth's Field

Magnetic data collected at many points on the earth's surface can be displayed on a map of the world as contours of F_E, i, and d. Figures 7-8 through 7-10 illustrate such maps for a portion of the western hemisphere. If world maps are examined carefully, the contour patterns appear somewhat similar to those that would be produced by a dipole. In fact, the earth's field can be quite well approximated by placing a small dipole of large moment at the earth's center and tilting the dipole at an angle of 11.5° to the earth's axis of rotation (Fig. 7-11). The points where an extension of the axis of this imaginary dipole intersect the earth's surface are referred to as the *geomagnetic north and south poles*. Note that these do not coincide with the dip poles, which are referred to as the *north and south magnetic poles*.

Although treating the earth's field as dipolar in nature is useful in deriving a number of relationships that we will investigate, the fact that the geomagnetic and dip poles do not

Figure 7-8   Values of total intensity for the geomagnetic field F_E for a portion of the western hemisphere. Geomagnetic epoch 1985.0. Contour interval = 2000 nT. (Based on information in United States Geological Survey Total Intensity Chart GP-987-F).

Figure 7-9 Values of inclination i of the geomagnetic field for a portion of the western hemisphere. Geomagnetic epoch 1985.0. (Based on information in United States Geological Survey Inclination Chart GP-987-I).

coincide, together with the irregularities evident on maps of inclination, declination, and total-field intensity demonstrate that this field cannot be accurately portrayed by one dipole. Actually, about 10 percent of the field cannot be accounted for by a single dipole, and it is necessary to employ numerous other dipoles of smaller magnetic moments at various other positions near the outer boundary of the earth's outer core to model accurately the existing field.

Variations of the Earth's Field

As one acquires magnetic data at various points on the earth's surface, intensity and orientation of the total field vary. In this section we examine the reasons for such variations with position and also consider variations with time.

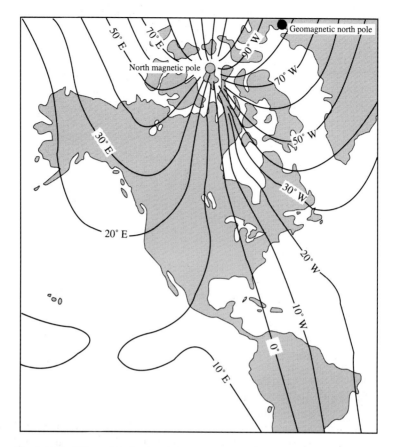

Figure 7-10 Values of declination d of the geomagnetic field for a portion of the western hemisphere. Geomagnetic epoch 1985.0. (Based on information in United States Geological Survey Declination Chart GP-987-D).

By far the major portion of the geomagnetic field arises from currents within the earth's fluid core. As discussed, this *main magnetic field* consists of a primary dipole portion and a secondary nondipole component which produce variations in the total-field vector over the earth's surface. In addition, this main field is not constant but changes slowly in both intensity and direction. This slow or *secular variation* occurs over long periods and can be ignored when one is conducting an exploration exercise that takes place over intervals of days or weeks. However, data must be adjusted for secular variation when one is compiling data from different years or decades. In order to facilitate such adjustments geomagnetic reference maps are produced by various government agencies at 5-year intervals. These maps are constructed from a network of magnetic observatories from which repeat readings are obtained. Such readings used in conjunction with very sophisticated models for

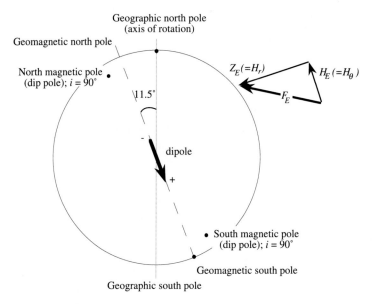

Figure 7-11   Some important features of the earth's magnetic field.

the dipolar and nondipolar components of the geomagnetic field result in the International Geomagnetic Reference Field (IGRF), which is presented in the form of maps and data tables. The direction and intensity of the main field at any position on the surface can be obtained from this information. Figures 7-8 to 7-10 are based on information taken from a 1985 calculation of the IGRF. IGRF information is also is critical for adjusting field values from magnetic surveys that encompass more than a very limited area. Such adjustments are discussed in the sections dealing with field procedures.

Flow of charged particles in the earth's ionosphere produces an *external magnetic field*. The external field is small in contrast to the main field (normally several tens of nanotesla as compared with several ten-thousands of nanotesla). This flow of particles toward the magnetic poles is affected by the gravitational attraction of the sun and moon and varies with time. Such *diurnal variations* must be determined and corrected for during surveys, as modern magnetometers are capable of precisions of 1 nT, and one naturally wants to maintain this level of precision. Occasionally, and unpredictably, solar activity increases substantially, which leads to an abrupt increase in ionized particles arriving in the ionosphere. These *magnetic storms* can produce variations of hundreds of nanotesla. Due to these large and erratic variations in magnetic intensities, field work usually halts during such activity.

Of course, variations in magnetic susceptibilies of rocks lead to local variations in induced magnetization, which affect total-field values. This *anomalous field* is the one we are interested in isolating, just as we strive to isolate gravity anomalies by reducing gravity observations.

Dipole Equations

The magnetic potential constitutes a convenient approach to describe the magnetic field at a point P due to a dipole. Consider the system illustrated in Figure 7-12(a). We assume that r is much larger than l. From Eq. 7-8 the potential at P is

$$V = \frac{m}{r_1} - \frac{m}{r_2} \qquad (7\text{-}10)$$

Using our assumption of $r \gg l$, we have the relations

$$V = \frac{m}{r - \left(\dfrac{l}{2}\cos\theta\right)} - \frac{m}{r + \left(\dfrac{l}{2}\cos\theta\right)}$$

$$V = \frac{ml\cos\theta}{r^2 - \left(\dfrac{l}{2}\right)^2 \cos^2\theta}$$

Figure 7-12   Parameters used in deriving the dipole equations.

and, approximating,

$$V = \frac{ml \cos \theta}{r^2} = \frac{M \cos \theta}{r^2} \tag{7-11}$$

For our purposes at this time, we wish to derive the radial and tangential components of the field at P (see Fig. 7-12(b)). Recalling that we can determine the magnetic field in a given direction by taking the negative of the derivative of the potential in that direction and noting that θ is in radians, we see that

$$H_r = -\frac{dV}{dr} = \frac{2 M \cos \theta}{r^3} \tag{7-12}$$

and

$$H_\theta = -\frac{1}{r}\frac{dV}{d\theta} = \frac{M \sin \theta}{r^3} \tag{7-13}$$

Because slightly more than 90 percent of the earth's field can be represented by a dipole at the earth's center, these equations provide a good approximation of some of the properties of this field. By consulting Figure 7-11 it should be clear that the radial field is equivalent to the vertical field ($H_r = Z_E$) and that the tangential field is equivalent to the horizontal field ($H_\theta = H_E$). Given that we are using cgs-units, the magnetic moment (M) of the imaginary dipole responsible for the major portion of the earth's field is 8×10^{25} emu (electromagnetic units). Noting that r must be expressed in centimeters and that θ is equivalent to colatitude, Eqs. 7-12 and 7-13 will provide a rough approximation of the vertical and/or horizontal components of the earth's field (and also, therefore, the total field) at any point on the surface. Resulting field strength is expressed in oersteds, which must be multiplied by 10^5 to be expressed in nanoteslas.

Determine if Eqs. 7-12 and 7-13 provide a reasonable approximation for the horizontal and vertical components of the earth's field at the magnetic equator and poles.

Equations 7-12 and 7-13 are also useful in determining gradients of magnetic field components at various locations on the earth's surface. Assume we are interested in the vertical gradient of the vertical component of the field. We use Eq. 7-12 to obtain

$$\frac{dZ_E}{dr} = \frac{dH_r}{dr} = -\frac{6 M \cos \theta}{r^4} = -\frac{3}{r} H_r = -\frac{3}{r} Z_E \tag{7-14}$$

A similar approach can be used to determine approximations for other gradients of interest.

> Determine the vertical gradient of the vertical component of the earth's field at a point on the surface where $Z_E = 52,950$ nT and the earth's radius equals 6.3×10^6m.

MEASURING THE MAGNETIC FIELD

The two instruments most commonly used at present for field-based measurements of the magnetic field are the *flux-gate* and *proton-precession* magnetometers. We briefly describe their operation here as well as some of their advantages and disadvantages. However, since most measurements today are total-field measurements, we concentrate on how such measurements relate to the local anomalous field.

Flux-Gate Magnetometer

In the common design for flux-gate magnetometers two bars of metal with high permeability are placed close together and oriented parallel to one another. Each bar is wound with a primary coil but the sense of the windings is reversed, so that the magnetic fields induced when alternating current is applied to the coils will be opposite in orientation. A secondary coil surrounds the primary coils. In the absence of an external field, the voltage induced in the secondary coil will be zero when current is passed through the primary coils due to the opposite magnetizations they induce. If the metal cores are aligned parallel to a component of a weak external field (e.g., the earth's field), one core will produce a field in the same sense as the earth's field and reinforce it, whereas the other will be in opposition to the field. This difference is sufficient to induce a voltage in the secondary core which is proportional to the strength of the external field.

Note that this instrument measures only relative changes in the earth's field but that it can measure the component of the earth's field parallel to which the metal bars are aligned. It is therefore possible to obtain readings for F_E, Z_E, and H_E. With modern versions of this instrument it is routine to obtain readings rapidly for F_E that are accurate to ± 1 nT. However, without elaborate orientation efforts values for Z_E and H_E typically are not accurate to more than a few nanoteslas.

Proton-Precession Magnetometer

The "sensor" component of a proton-precession magnetometer is a cylindrical container filled with a liquid rich in hydrogen atoms and surrounded by a coil. The sensor is connected by a cable to a small unit in which is housed the power supply and other necessary electronics such as an amplifier and frequency measuring device. When power is applied to the coil, a magnetic field is created which is aligned parallel to the coil axis. The hydrogen nuclei (protons), which behave as minute spinning dipoles, become aligned along the direction of this field. Power is then removed, and the hydrogen nuclei precess around the earth's total field. This precession induces a small alternating current to flow in the coil at the precession frequency. Because the frequency of precession is proportional to the strength of the total field and because the constant of proportionality is the well-known gyromagnetic ratio of the proton, the total-field intensity can be determined very accurately.

One of the important advantages of the proton-precession magnetometer is its ease of use and reliability. Sensor orientation need only be at a high angle to the earth's field, which is easily achieved. No other leveling or orientation requirements exist. The lack of moving parts ensures generally trouble-free operation. Readings can be obtained very rapidly. Equally important is that a precision of ±1 nT is routinely achieved. Limitations of this magnetometer are minor but must be kept in mind during field work. A large magnetic field gradient on the order of 600 nT/m will differentially affect regions of the sensor and seriously degrade readings. Also, because the signal generated by precession is small, the instrument cannot be operated close to ac power sources.

Although the flux-gate magnetometer can measure the horizontal and vertical components of the total field as well as F_E itself, the proton-precession magnetometer measures only the absolute strength of the earth's total field. Note that this gives us only the absolute intensity of the total field and not its direction. This is somewhat a disadvantage because total-field measurements tend to be more difficult to interpret than are vertical component measurements. However, its ease of use (both on ground and in aircraft), the speed with which a survey can be completed, and its precision all contribute to the predominance of the proton-precession magnetometer in modern surveys. Thus, we must concentrate on understanding what these total-field measurements encompass and how to interpret them.

Total-Field Anomalies

As a first step in this process, let's decide on our terminology. We have already chosen H_E, Z_E, and F_E as references to the earth's main field. We use H_A, Z_A and F_A in a similar sense for a magnetic field induced in a geologic body by the earth's field. We normally derive the vertical and horizontal anomalous-field components (H_A and Z_A) in an attempt to compute the induced field due to various subsurface geometries. F_A is the resultant of these components and is computed from them just as F_E is determined from H_E and Z_E (Eq. 7-9).

At any point on the surface at which a total-field measurement is taken, the measurement includes the main field plus the anomalous field (plus the very small contribution from the external field, which is discussed in the following section). For instance, in the situation illustrated in Figure 7-13(a) the total-field reading from a magnetometer would equal 55,005 nT, which includes the undisturbed main-field reading F_{E_U}, which is 55,000 nT, plus the contribution of the anomalous field, which is 5 nT. We refer to this total as F_{E_T}. However, since we do not know the orientation of F_{E_T}, how do we proceed with interpretations? First, we assume that $F_{E_U} \ggg F_A$, which is almost always the case, especially in most shallow subsurface surveys, unless the survey is close to large steel objects or encounters rocks with very high susceptibilities. If this assumption holds, then the total-field vector F_{E_T} that includes the contribution from the anomalous field is for all practical purposes parallel

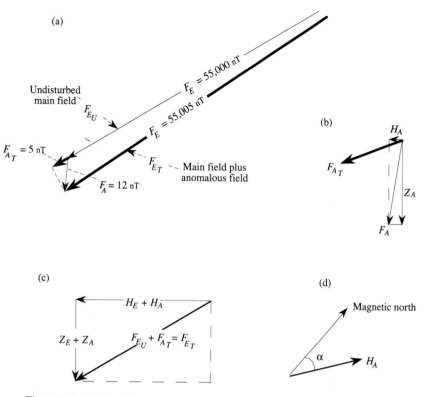

(a)

Undisturbed main field

F_{E_U}

$F_{A_T} = 5$ nT

$F_E = 55,000$ nT

$F_E = 55,005$ nT

F_{E_T} -- Main field plus anomalous field

$F_A = 12$ nT

(b)

H_A

F_{A_T}

Z_A

F_A

(c)

$H_E + H_A$

$Z_E + Z_A$

$F_{E_U} + F_{A_T} = F_{E_T}$

(d)

Magnetic north

α

H_A

Figure 7-13 Relationships surrounding the meaning of the total-field anomaly. (a) Vectors of the main field and anomalous field. (b) Components of the anomalous field. (c) Components of the undisturbed main field and the anomalous field. (d) Correction for horizontal component of the anomalous field when it does not lie along a magnetic meridian.

to the direction of the main field in the absence of the anomaly (the undisturbed main field F_{E_U}). Thus, the effect of the anomaly on the magnetometer reading essentially is the component of the anomaly F_A in the *direction of the undisturbed main field*. We refer to this anomaly component as F_{A_T} (see Fig. 7-13(b)). This relationship allows us to develop equations to determine total-field anomalies F_{A_T} from H_A and Z_A and to reduce total-field magnetometer readings for interpretation.

In general, for simplicity, we will continue to refer to the undisturbed main field as F_E and use the more specific terminology introduced in the previous paragraph only when we need to make such specific distinctions. Study Figure 7-13(b) once again and note our goal. If we derive values for Z_A and H_A, it is straightforward to determine F_A, but this is of no use in interpreting or computing total-field values. We want to determine F_{A_T}. Now consider Figure 7-13(c). If the anomalous field is oriented such that H_A is directed toward magnetic north (i.e., the H_A - Z_A plane is parallel to a magnetic meridian or, equivalently, α in Fig. 7-

13(d) = 0) and given the assumptions in the preceding paragraph, we have the relationships depicted in Figure 7-13(c). Then

$$\left(F_E + F_{A_T} \right)^2 = \left(Z_E + Z_A \right)^2 + \left(H_E + H_A \right)^2$$

and

$$F_E^2 + 2 F_{A_T} F_E + F_{A_T}^2 = Z_E^2 + 2 Z_E Z_A + Z_A^2 + H_E^2 + 2 H_E H_A + H_A^2 \quad (7\text{-}15)$$

Because $F_E >>> F_A$, we ignore $F_{A_T}^2$, H_A^2, and Z_A^2, which gives

$$F_E^2 + 2 F_{A_T} F_E = Z_E^2 + 2 Z_E Z_A + H_E^2 + 2 H_E H_A \quad (7\text{-}16)$$

But $F_E^2 = Z_E^2 + H_E^2$, so Eq. 7-16 reduces to $F_{A_T} F_E = Z_E Z_A = H_E H_A$, and

$$F_{A_T} = Z_A \left(\frac{Z_E}{F_E} \right) + H_A \left(\frac{H_E}{F_E} \right) \quad (7\text{-}17)$$

Finally, we apply the relationships among the geomagnetic elements (Eq. 7-9) to arrive at

$$F_{A_T} = Z_A \sin i + H_A \cos i \quad (7\text{-}18)$$

If H_A does not lie along a magnetic meridian, we use the component of H_A parallel to the meridian, because this is the only effect of H_A on the total anomaly. In such a case

$$F_{A_T} = Z_A \sin i + H_A \cos \alpha \cos i \quad (7\text{-}19)$$

We return to these relationships in our discussion of magnetic anomalies and interpretation of magnetic data. First, however, we take a slight diversion and summarize field techniques relevant to the collection of such field data.

BASIC FIELD PROCEDURES

In general the requirements for ground-based magnetic surveying are not as demanding as those related to gravity exploration. Nevertheless, you must follow a number of procedures in order to compile useful data with an accuracy comparable to the instrument used. At present the majority of magnetic exploration surveys are conducted from aircraft. Such surveys impose a number of considerations in addition to those detailed here, but as this text is concerned chiefly with the shallow subsurface, we do not include aeromagnetic survey procedures. However, aeromagnetic maps constitute an extremely valuable resource for certain types of shallow subsurface investigations, and you always should check the availability of such maps for any area in which you work. If you are interested in this topic, an excellent introductory summary is presented in Robinson and Coruh (1988).

Magnetic Cleanliness

When using the sensitive portable magnetometers so common today, you cannot be too obsessed with magnetic cleanliness. This simply means that a number of seemingly innocent objects (belt buckles, metal eyeglass frames, clipboards, watches, pens, etc.) can disrupt readings if too close to the instrument in use. The best rule is for the person holding the instrument or sensor to divest herself or himself of all possible contaminants.

In addition, nearby objects can be just as problematic or even more so. Cars, fences, metal poles, ac power lines, metal pipes near the surface, buildings with metal beams or doors, and many, many other items should be avoided—period. Although it might be somewhat inconvenient to move 20 m or more from such objects, the final integrity of your data is worth the effort.

When using any magnetometer, the sensor never should be closer to the ground than 1 m and in most cases should be placed at least 3 m above ground level. This is to ensure that local accumulations of magnetite such as heavy-mineral pockets in sands do not affect readings.

Diurnal Corrections

Because of the short-term variations in the external field that often cause variations of several nanoteslas in main-field intensities, it is necessary to correct magnetic field data in much the same way that gravity data were corrected for drift and tidal effects.

A common approach for surveys of limited extent or resources is to reoccupy base stations at regular intervals and follow the same procedure used for gravity drift corrections. Returns to base ideally should occur about every hour in order to be reasonably certain that fluctuations in the external field are not missed. Figure 7-14 illustrates a hypothetical, but not unrealistic, record for diurnal variations. Note that much of the record contains linear changes that would be appropriately managed by hourly returns to base. However, occasional fluctuations of several nanoteslas would be missed by such hourly visits.

A better approach is to place a continuously recording magnetometer at a base station, begin the day's activities at this site, and then return to base at the end of the day. Corrections to magnetometer readings then consist simply of adjusting field readings by the

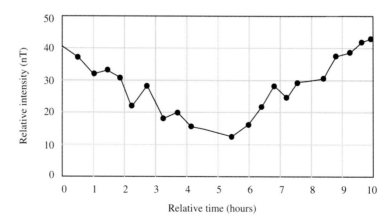

Figure 7-14 Representative variation in magnetic field intensity compiled by returning to a base station at short time intervals during a day's work.

fluctuations observed on the magnetometer at the base station. Even this procedure is not sufficient during magnetic storms, as variations in magnitude as great as several hundred nanoteslas occur and differences on the order of 5 to 10 nT are likely between a base and a field position just a few kilometers away. Field work obviously should be suspended during such periods.

Elevation Corrections

Magnetic field data normally are not corrected for elevation differences among recording sites. If you worked out the answer to the second boxed question in the section entitled "Dipole Equations," you discovered that the vertical gradient of the vertical component of the earth's magnetic field at 42° N latitude is 0.025 nT/m. Although this gradient clearly varies with position, it does give us an idea of how the magnetic field varies with elevation. Recall that we made a free-air correction for gravity data which amounted to a factor of roughly 0.3 mGal/m in a total field of some 983,000 mGal, and yet we do not make a correction for a variation of roughly 0.03 nT/m in a total field of some 70,000 nT. It certainly is legitimate to ask why such a correction is not made.

The answer lies in the variation of the main field and the relative susceptibilities of rocks. The total variation of gravity amounts only to some 5200 mGal in a total field of 983,000 mGal, whereas the magnetic field varies by some 40,000 nT in a field of only 70,000 nT. In addition, the susceptibility of a sedimentary rock rarely would be greater than 0.0003 emu, but an average susceptibility for a basalt dike is likely to be 0.003 emu. The relative contributions of these two rock types to an induced magnetic field are vastly different. Typical densities for these same two rocks, however, are quite similar (2.65 g/cm^3 versus 2.8 g/cm^3). In essence, the elevation effect in magnetic exploration is "lost in the noise," and, quite simply, it is not worth the effort to make the correction.

We previously noted that substantial magnetic gradients can negatively affect magnetometer readings. Such gradients must be on the order of hundreds of nanotesla per meter in order to pose a serious problem. These conditions normally arise only in areas of quite irregular topography when rocks of very high susceptibilities are at or near the surface or in areas of subdued topography when rocks of extremely high susceptibilities are present. In these situations a proton-precession magnetometer would produce erratic readings for only very minor changes in position, which should alert any operator to the possible presence of gradients. If substantial gradients exist due to topography, corrections can be made, but the process is more complex than terrain corrections in gravity work. An approach to magnetic terrain corrections is described in Gupta and Fitzpatrick (1971).

Correcting for Horizontal Position

How can we determine the rate at which the magnetic field varies with position? One approach is to use the same technique as we did for elevation changes. Let's determine the change of the vertical field as horizontal position changes along a magnetic meridian. To do this we once again resort to the dipole equation and the derivative of the vertical field with respect to horizontal position θ. Thus,

$$\frac{dZ_E}{d\theta} = \frac{1}{r}\frac{-2M\sin\theta}{r^3} = \frac{-2M\sin\theta}{r^4} = -2\frac{H_E}{r} \qquad (7\text{-}20)$$

Using appropriate values for 42° N latitude ($r = 6.3 \times 10^6$ m and $H_E = 18{,}200$ nT), we obtain 0.00578 nT/m, or roughly 6 nT/km. This gradient appears sufficiently great to warrant a correction for any but very local surveys, but note that positioning requirements will not be as severe as in gravity work.

The best solution to corrections for horizontal position is to consult IGRF data for the area in which a field survey is planned. Such data are available from several sources (e.g., U.S. Geological Survey, Branch of Global Seismology and Geomagnetism, Box 25046, MS 968, Denver Federal Center, Denver CO 80225). For example, the total intensity values used in Figure 7-15 were obtained over a computer link to the USGS source for the four specified latitude and longitude coordinates. These coordinates were chosen to extend just beyond the planned survey area (Fig. 7-15).

Let's assume we are collecting total-field data. Each field site is noted on a topographic base map, and accurate latitude and longitude values are obtained following the overlay method as described in the gravity correction discussion. We might designate the F_{E_U} control value for the survey to be 55172 nT, as this value is located in the southeastern corner of the rectangle in Figure 7-15. The main field increases to the north and to the west from this position at the rates noted in Figure 7-15. As an example of such a position cor-

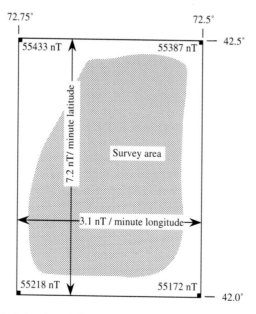

Figure 7-15 Variations in total-field anomaly intensity as a function of longitude and latitude for a given survey area.

rection, assume we recorded a total-field value (F_{E_T}) of 55193 nT at a site located 0.45' north and 0.18' west of the control point. The correction should be

$$F_{E_U} = 55172 \text{ nT} + 0.45' (7.2 \text{ nT/minute}) + 0.18' (3.1 \text{ nT/minute}) = 55176 \text{ nT}$$

If we now subtract this value from the total-field reading, we have a total-field anomaly of 17 nT. Of course, if we are recording vertical-field intensities, we follow the same procedure using IGRF vertical-field values.

At this point we've corrected field data for short-term variations of the external magnetic field and for variations of the undisturbed field with geographic latitude and longitude. In general we can ignore the effect of elevation differences in sampling localities, and as most modern instruments record absolute values of the magnetic field, we do not have to tie our readings into base stations as we did in gravity surveying. What now remains is the anomalous magnetic field. In order to begin to interpret these results, we follow a path similar to that in our discussion of gravity interpretation and begin by considering the magnetic effects of subsurface bodies of simple shape.

MAGNETIC EFFECTS OF SIMPLE GEOMETRIC SHAPES

The derivation and interpretation of magnetic anomalies is considerably more complex than was the case for gravity anomalies. This is because in most cases we must consider the effect of both positive and negative poles in order to derive the anomalous field. We must then compute the total anomalous field so that during the data reduction process we can remove the main field contribution. In addition, remanent magnetization can have a significant effect on the form of the final anomaly. Therefore, in the material that follows we consider only the most simple shapes in order to emphasize the important considerations relevant to anomaly character.

Rock Susceptibilities

Before embarking on an analysis of magnetic anomalies, we pause for a moment to reconsider the topic of magnetic susceptibility and to review several other considerations relevant to computation of magnetic intensities. Recall that we elected to use the cgs system of units in this chapter. Although susceptibility is dimensionless, values normally are expressed as cgs emu units to call attention to the fact that cgs units are in use rather than SI units. The susceptibility values given here can be converted to SI units by multiplying by 4π.

Although several familiar minerals have high susceptibilities (magnetite, ilmenite, and pyrrhotite), magnetite is by far the most common. Rock susceptibility almost always is directly related to the percentage of magnetite present. The true susceptibility of magnetite varies from 0.1 to 1.0 cgs emu depending on grain size, shape, and impurities. Average susceptibility as quoted in several sources ranges from 0.2 to 0.5 cgs emu, so we will adopt a value of $k = 0.35$ cgs emu for our computational purposes. Average susceptibilities for ilmenite and pyrrhotite are 0.15 and 0.125 cgs emu, respectively.

As you might guess, rock susceptibilities vary over quite a range even within a given rock type. The only reliable approach is to measure k in rocks in an exploration area if detailed information is necessary. Representative values for a selection of common rocks and minerals is presented in Telford *et al.* (1990, p.74). For our purposes we believe it is preferable to concentrate on average susceptibilities for common rock groups. This is suffi-cient for the following discussions of anomaly interpretation and will make representative values easier to remember, but keep in mind that these values vary widely. Approximate averages for common rock groups are as follows:

Sedimentary rocks	0.00005 cgs emu
Metamorphic rocks	0.0003 cgs emu
Granites and rhyolites	0.0005 cgs emu
Gabbros and basalts	0.006 cgs emu
Ultrabasic rocks	0.012 cgs emu

Application of magnetic exploration to the shallow subsurface also may encompass archaeological targets or searches for metallic objects. Susceptibilities of many buried objects (e.g., stone walls) will, of course, possess the susceptibility of the rock itself. It is not uncommon for soils to have susceptibilities on the order of 0.0001 cgs emu due to the selective sorting of heavy minerals (including magnetite) and the possible presence of maghemite, which has a relatively high k-value. Many iron and steel objects possess a sus-ceptibility from 1 to 10 cgs emu.

For our purposes the preceding k-values are sufficient. You should know, however, that the apparent susceptibility k_a may be quite different from the true susceptibility k_t when the latter value is greater than 1. In such cases a shape factor η may be present, which serves to reduce the value of induced magnetization that would be predicted on the basis of k_t alone. This is known as the *demagnetization* effect and can be computed if η is known using the relationship

$$k_a = \frac{k_t}{(1 + \eta k_t)} \tag{7-21}$$

As η varies between 0 and 4π, let's compute the demagnetization effect for a thin horizontal sheet (with true susceptibility = 0.0001 cgs emu) magnetized perpendicular to its surface (in which case $\eta = 4\pi$).

$$k_a = \frac{k_t}{(1 + \eta k_t)} = \frac{0.0001}{(1 + 0.0001\eta)} = 0.00009987 \text{ cgs emu}$$

In this case it is obvious that the demagnetization effect is so small that it can be ignored.

Is there a serious demagnetization effect for a sphere (true susceptibility = 0.001 cgs emu and $\eta = 4\pi/3$)?

Finally, we wish to emphasize once again that remanent magnetization I_r is not considered in the material that follows. You are already aware that the effect of I_r may be substantial, especially if it is close in magnitude to the induced magnetization I_i and oriented in a different direction. In such cases the resulting anomaly is the vector sum of the two magnetizations, which clearly can be very different from I_i alone. In basalts I_r / I_i often is 10 or greater, so you should be especially alert to the role of remanent magnetization when working in areas in which basaltic rocks are present. Other approximate values for the ratio I_r / I_i are ~1.0 in granitic rocks, <1.0 in metamorphic rocks, and <0.1 in sedimentary rocks.

Magnetic Effect of an Isolated Pole (Monopole)

Our first effort focuses on the magnetic field above a single pole. Although such an isolated pole cannot exist, the exercise is convenient because of its simplicity. In addition, a very long and thin body oriented vertically and magnetized along its length essentially functions as a monopole, as the top surface has a pole strength of $-m$ and the bottom surface $(+m)$ is sufficiently far removed for its effect to be negligible. In our derivation we utilize Eqs. 7-5, 7-7, and 7-8. Refer to Figure 7-16 for relevant relationships and notation. These considerations produce

$$V = \frac{m}{r}, \quad m = IA = kF_EA, \quad \text{and} \quad r = \left(c^2 + z^2\right)^{1/2} = \left(x^2 + y^2 + z^2\right)^{1/2}$$

Remembering that we determine the magnetic field in a given direction by taking the negative of the derivative in that direction, we can state

$$Z_A = -\frac{dV}{dz} = -\frac{2z(-\frac{1}{2})(kF_EA)}{\left(x^2 + y^2 + z^2\right)^{3/2}} = \frac{z(kF_EA)}{\left(x^2 + y^2 + z^2\right)^{3/2}} \qquad (7\text{-}22)$$

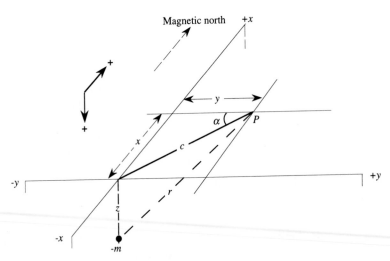

Figure 7-16   Relationships and notation used to derive the magnetic effect of a single pole.

where the inducing field is F_E, and A is cross-sectional area. When determining the horizontal field due to the monopole, it is convenient to orient our coordinate system so that the $+x$-axis is oriented toward magnetic north (Fig. 7-16). This orients the horizontal components of the anomalous field H_{A_x} and H_{A_y} parallel to X and Y of the earth's field. Vertical down is the $+z$-axis. Thus, magnetic field components oriented in the sense of the black arrows in Figure 7-16 are considered positive. Now we can determine H_{A_x} and H_{A_y} using the same approach as before.

$$H_{A_x} = -\frac{dV}{dx} = \frac{x\left(kF_EA\right)}{\left(x^2 + y^2 + z^2\right)^{3/2}} \tag{7-23}$$

and

$$H_{A_y} = -\frac{dV}{dy} = \frac{y\left(kF_EA\right)}{\left(x^2 + y^2 + z^2\right)^{3/2}} \tag{7-24}$$

The total anomalous field is calculated using the form of Eq. 7-18. Since H_{A_x} represents the component of the horizontal anomalous field in the direction of magnetic north, we have

$$F_{A_T} = Z_A \sin i + H_{A_x} \cos i \tag{7-25}$$

Following our practice in previous chapters, let's put these equations into a dynamic table (Table 7-1) and analyze the results. The table as illustrated provides results for a traverse along the x-coordinate axis but provides the capability to obtain results for any position in xy-space.

Figure 7-17 is a graph of the vertical, horizontal, and total anomaly fields for the traverse values in Table 7-1. The vertical-field curve is quite straightforward, because you would expect intensity to be greatest directly over the monopole and to decline equally in both directions away from this position. Usually we are interested only in the vertical and total fields, as these are the fields measured during magnetic exploration. However, we present the horizontal-field component here so you can visualize (with the aid of Eq. 7-18) the contribution it makes to the total field. The horizontal-field components are easy to understand by referring to Figure 7-17(b). Small inserts illustrate orientations of the vertical- and horizontal-field vectors for $-x$ and $+x$ values. Based on the convention we adopted in Figure 7-16, the vertical-field component always is positive, but the horizontal-field component is positive for $-x$ values and negative for $+x$ values. The total-field values and curve form are best understood by noting curves for Z_A and H_A in addition to the value of i and referring once again to Eq. 7-18.

Magnetic Effect of a Dipole

The next step up in complexity is to derive the magnetic field due to a dipole without the constraints of distance and dipole length imposed in our previous derivation of dipole equations. For the present derivation we utilize Eq. 7-2. Figure 7-18(a) and (b) illustrate the gen-

TABLE 7-1 Magnetic Effect of a Monopole

| Depth to negative pole (m) | 5 |
| Horizontal increment (m) | 1 |

(All anomaly values are in nanoteslas.)

Pole cross-sectional area (m^2)	0.5
Susceptibility of pole (cgs emu)	0.003
Earth's field (nT)	55000
Earth's field inclination (°)	70

X-position (m)	Y-position (m)	Z_A	H_{A_x}	H_{A_y}	F_{A_T}
−15	0	0.10	0.31	0.00	0.21
−14	0	0.13	0.35	0.00	0.24
−13	0	0.15	0.40	0.00	0.28
−12	0	0.19	0.45	0.00	0.33
−11	0	0.23	0.51	0.00	0.40
−10	0	0.30	0.59	0.00	0.48
−9	0	0.38	0.68	0.00	0.59
−8	0	0.49	0.79	0.00	0.73
−7	0	0.65	0.91	0.00	0.92
−6	0	0.87	1.04	0.00	1.17
−5	0	1.17	1.17	0.00	1.50
−4	0	1.57	1.26	0.00	1.91
−3	0	2.08	1.25	0.00	2.38
−2	0	2.64	1.06	0.00	2.84
−1	0	3.11	0.62	0.00	3.14
0	0	3.30	0.00	0.00	3.10
1	0	3.11	−0.62	0.00	2.71
2	0	2.64	−1.06	0.00	2.12
3	0	2.08	−1.25	0.00	1.53
4	0	1.57	−1.26	0.00	1.05
5	0	1.17	−1.17	0.00	0.70
6	0	0.87	−1.04	0.00	0.46
7	0	0.65	−0.91	0.00	0.30
8	0	0.49	−0.79	0.00	0.19
9	0	0.38	−0.68	0.00	0.12
10	0	0.30	−0.59	0.00	0.08
11	0	0.23	−0.51	0.00	0.04
12	0	0.19	−0.45	0.00	0.02
13	0	0.15	−0.40	0.00	0.01
14	0	0.13	−0.35	0.00	0.00
15	0	0.10	−0.31	0.00	−0.01

eral geometry and notation for the derivation. In this case we assume that the dipole is magnetized along its axis (parallel to its length). If the magnetization is induced, such a situation requires both dipole orientation and the orientation of the earth's field F_E to be parallel. Of course, in the general case such a correspondence would be most fortuitous, so you might question the wisdom of pursuing this derivation. We do so for two reasons. First, the deriva-

(a)

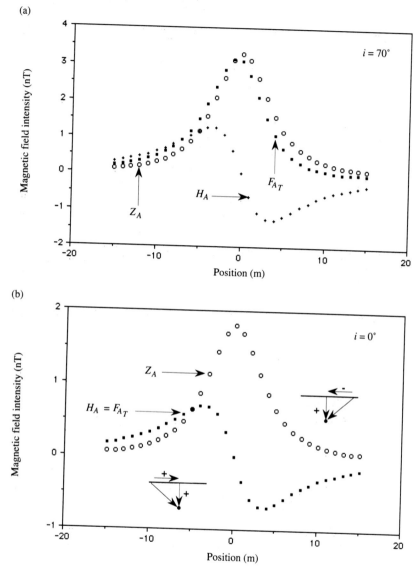

Figure 7-17 Intensities of vertical (Z_A), horizontal (H_{A_x}), and total (F_{A_T}) magnetic field anomalies for a monopole. Traverse directly over the monopole in a direction parallel to a magnetic meridian. (a) Inclination of earth's field = 70°. (b) Inclination of earth's field = 0°.

tion is straightforward, and some of the resulting anomalies provide useful information. Second, although the dipole as pictured in Figure 7-18 and its polarization would be unlikely in the real world, we soon will learn that dipole behavior is contained within bodies of geologic interest. So bear with us.

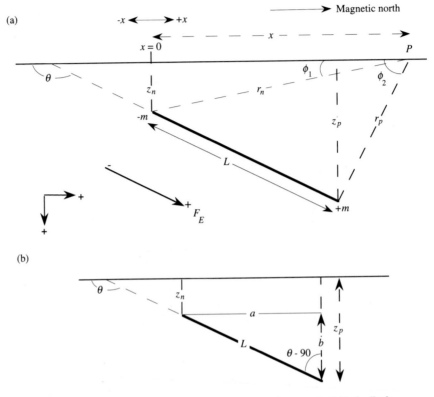

Figure 7-18   Relationships and notation used to derive the magnetic field of a dipole.

Recalling the definition we adopted for a unit pole, the magnetic field intensity at P due to the negative pole of the dipole is

$$R_{A_n} = +\frac{m}{r_n^2} = +\frac{kF_E A}{r_n^2} \tag{7-26}$$

and that due to the positive pole is

$$R_{A_p} = -\frac{m}{r_p^2} = -\frac{kF_E A}{r_p^2} \tag{7-27}$$

Our next step is to determine the horizontal and vertical components of the magnetic field at P due to each of the poles ($-m$ and $+m$). These components are

$$Z_{A_n} = R_{A_n} \sin\phi_1, \quad Z_{A_p} = R_{A_p} \sin\phi_2, \quad H_{A_n} = R_{A_n} \cos\phi_1, \quad \text{and} \quad H_{A_p} = R_{A_p} \cos\phi_2 \tag{7-28}$$

The final expressions for the vertical and horizontal components simply are the sums of each pole's contribution to the component, or

$$Z_A = Z_{A_n} + Z_{A_p} \quad \text{and} \quad H_A = H_{A_n} + H_{A_p} \tag{7-29}$$

The total field is found as in the monopole example except that for the present derivation H_A is oriented parallel to a magnetic meridian, so we can use Eq. 7-18. Now let's place these results in a dynamic table. Once again check Figure 7-18(a) and (b), from which we develop the following relationships:

(1) $\quad a = L\cos(180 - \theta),\qquad b = L\sin(180 - \theta),\qquad$ and $\quad z_p = z_n + b$

(2) $\quad r_n = \left(x^2 + z_n^{\,2}\right)^{1/2},\quad$ and $\quad r_p = \left[(x - a)^2 + z_p^{\,2}\right]^{1/2}$

(3) $\quad \sin\phi_1 = \dfrac{z_n}{r_n},\qquad \cos\phi_1 = \dfrac{x}{r_n},\qquad \sin\phi_2 = \dfrac{z_p}{r_p},\quad$ and $\quad \cos\phi_2 = \dfrac{(x - a)}{r_p}$

Equations 7-26 through 7-29 and the relationships noted above all are used in Table 7-2. In order to save space several cells are not shown in Table 7-2 but are included in the spreadsheet template on disk. These include the inducing field F_E (55,000 nT), the susceptibility k (0.003 cgs emu), and the cross-sectional area A (1.0 m^2). These all may be changed if so desired.

   Following our standard procedure, let's graph the values in Table 7-2 and then investigate some other relationships facilitated by this table. Figure 7-19 illustrates the vertical, horizontal, and total magnetic field anomalies as represented in Table 7-2. The angle of inclination of the dipole (120°) represents an inclination of the earth's field of 60°. Immediately evident is the fact that the vertical-field anomaly has negative as well as positive values (as does the total-field anomaly). You now should begin to realize that magnetic anomalies are going to be more difficult to interpret, because as in most cases both positive and negative poles will affect final curve form. Note also that the total-field curve differs to a fair degree from the vertical-field curve. The reason for this should be evident from the form of the horizontal-field curve and Eq. 7-18. Maxima and minima are not in the same position, nor do they have the same values. We already mentioned that vertical-field curves are simpler to work with, so this disparity likely will increase the difficulty of quantitatively interpreting magnetic field curves.

   In general, total-field curves and vertical-field curves for anomalies are very similar at high magnetic latitudes (approximately 70° and greater), so for normal purposes we can analyze total-field curves as if they were vertical-field curves if our data come from such latitudes. Figure 7-20 illustrates vertical and total anomaly curves for several inclinations of the earth's field. Of course, the total-field and vertical-field curves are identical for an inclination of 90°. For $i = 60°$ the two curves are somewhat close in form, position, and magnitude of maxima and minima (as noted previously). These correspondences decrease progressively as i is decreased to 30° and 0°.

   In general, due to complexity, we do not derive equations nor provide diagrams of magnetic anomalies as they appear on maps. However, in the dipole case, equations for xyz-space are convenient to generate. Figure 7-21 provides the general relationships. Most quantities are the same as in Figure 7-18. The dipole still is contained within the xz-plane. The only adjustment we need to make is to realize that at each point P we compute H_x and H_y components. Similarly to the previous derivation, H_x is generated by determining H_{x_n}

TABLE 7-2　Magnetic Effect of a Dipole

Dipole inclination (°)	120
Depth to negative pole (m)	5
Length of dipole (m)	15
Horizontal increment (m)	2

(All anomaly values are in nanoteslas)

Position (m)	R_{A_n}	R_{A_p}	Z_{A_n}	Z_{A_p}	H_{A_n}	H_{A_p}	Z_A	H_A	F_{A_T}
− 30	0.18	−0.10	0.03	− 0.04	0.18	−0.09	−0.01	0.09	0.03
− 28	0.20	−0.10	0.04	− 0.05	0.20	−0.09	−0.01	0.11	0.04
− 26	0.24	−0.11	0.04	− 0.05	0.23	−0.10	−0.01	0.13	0.06
− 24	0.27	−0.13	0.06	− 0.06	0.27	−0.11	−0.01	0.16	0.07
− 22	0.32	−0.14	0.07	− 0.07	0.32	−0.12	0.00	0.20	0.10
− 20	0.39	−0.15	0.09	− 0.08	0.38	−0.13	0.01	0.25	0.13
− 18	0.47	−0.17	0.13	− 0.10	0.46	−0.14	0.03	0.32	0.18
− 16	0.59	−0.19	0.18	− 0.11	0.56	−0.15	0.06	0.41	0.26
− 14	0.75	−0.21	0.25	− 0.13	0.70	−0.16	0.12	0.54	0.37
− 12	0.98	−0.23	0.38	− 0.16	0.90	−0.17	0.22	0.73	0.55
− 10	1.32	−0.26	0.59	− 0.19	1.18	−0.18	0.40	1.00	0.85
− 8	1.85	−0.29	0.98	− 0.22	1.57	−0.19	0.76	1.38	1.35
− 6	2.70	−0.33	1.73	− 0.26	2.08	−0.20	1.47	1.88	2.21
− 4	4.02	−0.36	3.14	− 0.30	2.51	−0.19	2.84	2.32	3.62
− 2	5.69	−0.40	5.28	− 0.35	2.11	−0.19	4.93	1.93	5.23
0	6.60	−0.43	6.60	− 0.40	0.00	−0.17	6.20	−0.17	5.29
2	5.69	−0.47	5.28	− 0.45	−2.11	−0.14	4.84	−2.25	3.06
4	4.02	−0.49	3.14	− 0.48	−2.51	−0.09	2.66	−2.61	1.00
6	2.70	−0.51	1.73	− 0.50	−2.08	−0.04	1.23	−2.12	0.00
8	1.85	−0.51	0.98	− 0.51	−1.57	0.01	0.47	−1.56	−0.37
10	1.32	−0.50	0.59	− 0.50	−1.18	0.07	0.09	−1.11	−0.47
12	0.98	−0.48	0.38	− 0.47	−0.90	0.12	−0.09	−0.78	−0.47
14	0.75	−0.45	0.25	− 0.42	−0.70	0.15	−0.17	−0.55	−0.42
16	0.59	−0.42	0.18	− 0.38	−0.56	0.18	−0.20	−0.38	−0.37
18	0.47	−0.38	0.13	− 0.33	−0.46	0.19	−0.20	−0.26	−0.31
20	0.39	−0.34	0.09	− 0.28	−0.38	0.20	−0.19	−0.18	−0.25
22	0.32	−0.31	0.07	− 0.24	−0.32	0.19	−0.17	−0.12	−0.21
24	0.27	−0.28	0.06	− 0.20	−0.27	0.19	−0.15	−0.08	−0.17
26	0.24	−0.25	0.04	− 0.17	−0.23	0.18	−0.13	−0.05	−0.14
28	0.20	−0.22	0.04	− 0.15	−0.20	0.17	−0.11	−0.03	−0.11
30	0.18	−0.20	0.03	− 0.12	−0.18	0.16	−0.09	−0.02	−0.09

and H_{x_p} and adding these latter two quantities. H_y is determined in the same way. Final values can be calculated in a spreadsheet similar to Table 7-2, but we do not include it here due to file size and the amount of space available on the enclosed disks. The results of such a derivation are illustrated in Figure 7-22.

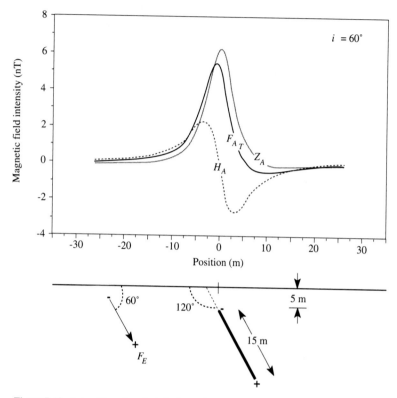

Figure 7-19 Intensities of vertical, horizontal, and total magnetic field anomalies over a dipping dipole polarized along its axis. The dipole is oriented parallel to a magnetic meridian.

Magnetic Effect of a Sphere

The derivation of equations to determine the magnetic effect of a sphere is more complex than our derivation for a dipole. Our purpose here is to present the general steps in this derivation in order to give you a flavor of the procedure. With this in mind, we consider the simplest case—that of a sphere which is vertically polarized. We follow this with the case of a sphere with inclined polarization. Figure 7-23 functions as a reference for both derivations. In the case of the vertically polarized sphere, $i = 90°$. Thus, this derivation is strictly applicable only to cases in which a spherical mass is located where the earth's field has an inclination of $90°$.

The crucial relationship for this derivation is referred to as *Poisson's relation* (Dobrin, 1988, p. 640–641), which states that the magnetic potential V is proportional to the derivative of the gravity potential U in the direction of magnetization, or

$$V = -\frac{I}{\rho G}\frac{dU}{dw} \tag{7-30}$$

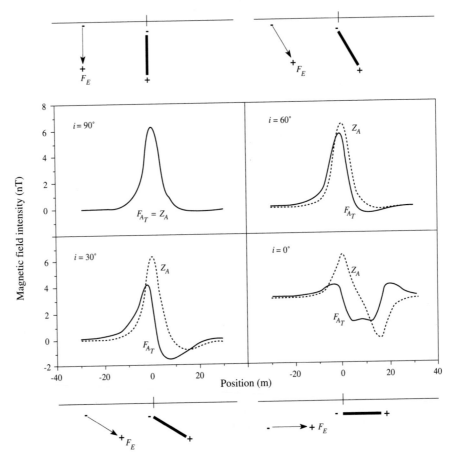

Figure 7-20 Total (solid) and vertical (dashed) magnetic anomalies over a dipole polarized along its axis for various inclinations of the earth's field. In each case the dipole is parallel to a magnetic meridian.

assuming the body has uniform susceptibility and density. Remembering that the direction of magnetization for our derivation is vertical or z, it follows that vertical- and horizontal-field anomalies Z_A and H_A must be defined as

$$Z_A = -\frac{dV}{dz} = \frac{I}{\rho G}\frac{d^2U}{dz^2} \qquad (7\text{-}31)$$

and

$$H_A = -\frac{dV}{dx} = \frac{I}{\rho G}\frac{d}{dx}\left(\frac{dU}{dz}\right) \qquad (7\text{-}32)$$

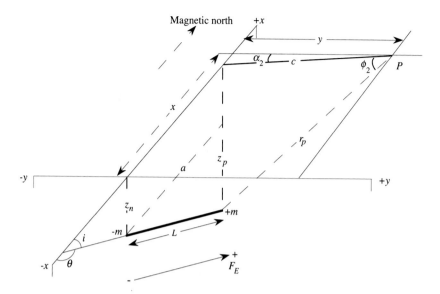

Figure 7-21   Relationships and notation used to derive the magnetic effect of a dipole oriented parallel to a magnetic meridian at any point on an xy-surface.

As the gravitational potential of a sphere is

$$U = \frac{GM}{r} = \frac{G\frac{4}{3}\pi R^3 \rho}{r} = \frac{G\frac{4}{3}\pi R^3 \rho}{\left(x^2 + z^2\right)^{1/2}} \tag{7-33}$$

then

$$\frac{d^2U}{dz^2} = \frac{\left(G\frac{4}{3}\pi R^3 \rho\right)\left(2z^2 - x^2\right)}{\left(x^2 + z^2\right)^{5/2}} \tag{7-34}$$

This gives us the vertical anomaly, which is

$$Z_A = \left(\frac{I}{G\rho}\right)\frac{\left(G\frac{4}{3}\pi R^3 \rho\right)\left(2z^2 - x^2\right)}{\left(x^2 + z^2\right)^{5/2}} = \frac{\left(\frac{4}{3}\pi R^3 I\right)\left(2z^2 - x^2\right)}{\left(x^2 + z^2\right)^{5/2}} \tag{7-35}$$

The horizontal anomaly follows similarly.

If we approach the more general case, that of a sphere uniformly magnetized where the earth's field is inclined, the derivation is more complex, but it can be developed in much the same fashion. Thorough explanations are presented in Telford et al. (1976, p.158–160) and in Parasnis (1979, p. 31). These yield the following equations for the vertical and horizontal anomalies:

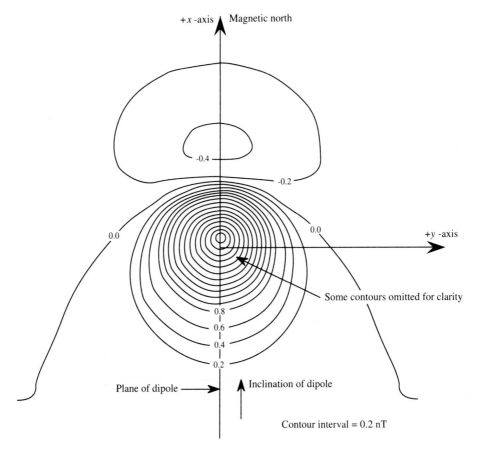

Figure 7-22   Contours of total-field anomaly intensities above a dipole polarized along its axis. Input parameters are similar to those in Table 7-2.

$$Z_A = \frac{\left(\frac{4}{3}\pi R^3 kF_E\right)\sin i}{\left(x^2 + z^2\right)^{3/2}}\left[\frac{3z^2}{\left(x^2 + z^2\right)} - \left(\frac{3xz}{\left(x^2 + z^2\right)}\cot i\right) - 1\right] \quad (7\text{-}36)$$

and

$$H_A = \frac{\left(\frac{4}{3}\pi R^3 kF_E\right)\cos i}{\left(x^2 + z^2\right)^{3/2}}\left[\left(\frac{3x^2}{\left(x^2 + z^2\right)} - 1\right) - \left(\frac{3xz}{\left(x^2 + z^2\right)}\tan i\right)\right] \quad (7\text{-}37)$$

Table 7-3 is based on Eqs. 7-36 and 7-37. Before you work with this table, one comment is in order. No provision is made (nore is one made in Table 7-2) to adjust the earth's field value depending on the inclination value. This does not affect the general shape of the

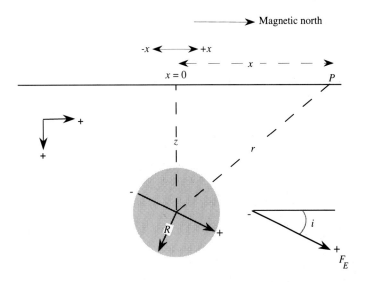

Figure 7-23 Notation used for the derivation of magnetic-field anomalies over a uniformly magnetized sphere.

curves, but anomaly values will be incorrect in magnitude unless realistic values are input both for F_E and i. Also, Eq. 7-36 generates a "divide-by-zero" error when $i = 0$, so if you wish to use such a low inclination value, substitute $i = 0.001$.

Curves based on the values in Table 7-3 (Earth field inclination of 60°) are illustrated in Figure 7-24. Compare Figure 7-24 with Figure 7-19. They are amazingly similar! If we continue this comparison by graphing anomaly curves for the sphere with inducing field inclinations of 0°, 30°, and 90°, we finish with the results in Figure 7-25. Again, this figure is remarkably similar to Figure 7-20. If we were even more careful and ensured that the magnetic moment of the sphere was the same as that of the dipole, these curves would have similar magnitudes as well as form.

It now seems warranted to conclude that if a sphere is uniformly magnetized by an inducing field, it produces anomalies similar to those resulting from a *dipole magnetized along its axis* (see Fig. 7-23). As in gravity exploration, a sphere can represent a somewhat equidimensional mass—which is much more likely in the subsurface than a dipole that happens to be in the same orientation as that of the inducing field. Thus, our use for the dipole is clearly very valuable even though it achieves its status through the intermediary of the sphere. Studying Figure 7-25 once again of course supports our former contention that vertical and total anomalies are very similar in form at high magnetic latitudes. In addition note how very different the total field is from the vertical field at low magnetic latitudes. If we substitute a dipole for the sphere, it's quite easy to understand the vertical anomaly for an inclination of low value, but it takes a bit more effort to work out the reason for the form of the total-field anomaly.

Although we presented the case of the horizontal cylinder in the previous chapter, we do not do so in this chapter. As long as the strike length is long relative to cross-sectional

TABLE 7.3 Magnetic Effect of a Sphere

Position (m)	Z_A	H_A	F_{A_T}
−30	−0.02	0.05	0.01
−28	−0.02	0.06	0.01
−26	−0.03	0.08	0.02
−24	−0.03	0.10	0.02
−22	−0.04	0.14	0.04
−20	−0.04	0.18	0.05
−18	−0.05	0.25	0.08
−16	−0.05	0.36	0.14
−14	−0.03	0.53	0.24
−12	0.02	0.81	0.43
−10	0.19	1.30	0.82
−8	0.67	2.18	1.67
−6	2.06	3.66	3.61
−4	5.78	5.40	7.71
−2	12.68	4.04	13.00
0	14.51	−4.19	10.47
2	5.74	−7.97	0.99
4	−0.05	−4.71	−2.40
6	−1.18	−1.96	−2.01
8	−1.01	−0.73	−1.24
10	−0.71	−0.25	−0.74
12	−0.48	−0.07	−0.45
14	−0.13	0.00	−0.29
16	−0.24	0.03	−0.19
18	−0.17	0.04	−0.13
20	−0.13	0.04	−0.09
22	−0.10	0.03	−0.07
24	−0.08	0.03	−0.05
26	−0.06	0.03	−0.04
28	−0.05	0.02	−0.03
30	−0.04	0.02	−0.02

Depth to sphere center (m)	**5**
Sphere radius (m)	**1**
Susceptibility (cgs emu)	**0.005**
Horizontal increment (m)	**2**

Earth's field (F_E) (nT)	**50000**
Earth's field inclination (°)	**60**

(All anomaly values are in nanoteslas)

radius, we can treat the cylinder as a two-dimensional case. Curves are qualitatively similar to those of the sphere. At this point we change gears just a bit and revert to a previous derivation in order to understand magnetic anomalies due to thin sheets.

Magnetic Effect of a Thin Horizontal Sheet

Several informative observations stem from a consideration of the magnetic effect of thin sheets, so we provide a brief presentation of how such effects are calculated. In this derivation we follow almost precisely the same path as used for the Bouguer derivation and the

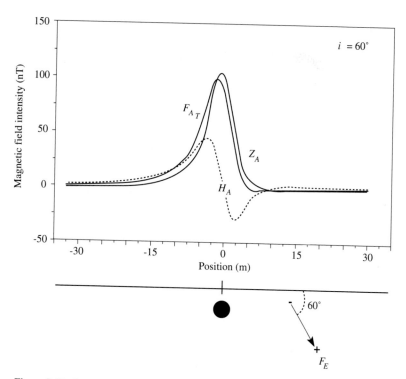

Figure 7-24 Intensities of vertical, horizontal, and total magnetic field anomalies over a uniformly magnetized sphere.

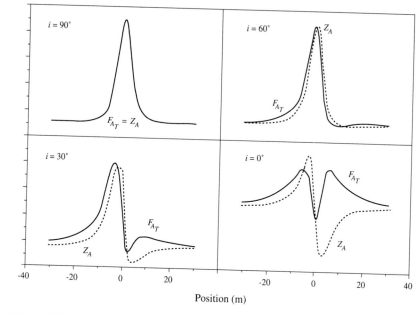

Figure 7-25 Total (solid) and vertical (dashed) magnetic anomalies over a uniformly magnetized sphere for various inclinations of the earth's field.

derivation of the gravity effect of a horizontal sheet. As the steps are so similar, the presentation here is abbreviated, so be sure to refer to Chapter 6 for details.

In all considerations of horizontal sheets presented here, we assume the inducing field is vertical. If we consider a uniform strip of negative poles (Fig. 7-26(a)), then, from Eq. 7-2, the magnetic field intensity at point P due to a small area of the strip is

$$R_A = \frac{\left(m/_{area}\right) dx\, dy}{r^2} \qquad (7\text{-}38)$$

Following the Bouguer effect derivation, we can say that the vertical field at P due to the strip of poles is

$$Z_A = \int_{x=-\infty}^{x=+\infty} \frac{\left(m/_{area}\right) dx\, dy}{r^2} \cos\theta \qquad (7\text{-}39)$$

(a)

(b)

(c)

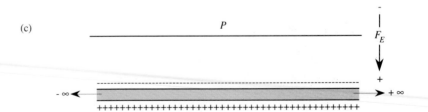

Figure 7-26 Generalized relationships used to derive the expression for the magnetic effect of a thin horizontal plate of infinite extent.

$$Z_A = \int_{-\pi/2}^{\pi/2} \frac{\left(m/_{area}\right)dy}{d} \cos\theta d\theta \tag{7-40}$$

and

$$Z_A = \frac{2\left(m/_{area}\right) dy}{d} \tag{7-41}$$

We next sweep out a sheet of poles using the strip. Recall that in the Bouguer derivation we rotated our view with respect to the coordinate axes for the second step. We do the same here and then equate the distance d in Figure 7-26(a) with the distance r in Figure 7-26(b). Therefore,

$$R_{A_{sheet}} = \frac{2\left(m/_{area}\right) dy}{r} \tag{7-42}$$

$$Z_{A_{sheet}} = \int_{y=-\infty}^{y=+\infty} \frac{2\left(m/_{area}\right)dy}{r} \cos\theta \tag{7-43}$$

$$Z_{A_{sheet}} = \int_{-\pi/2}^{\pi/2} 2\left(m/_{area}\right)d\theta \tag{7-44}$$

and

$$Z_{A_{sheet}} = 2\left(m/_{area}\right)\pi = 2\pi I \tag{7-45}$$

We can apply Eq. 7-45 in order to estimate the magnetic anomaly due to such a thin horizontal plate of infinite extent. For such a case and with a vertical inducing field, negative poles are induced on the upper surface of the plate and positive poles on the bottom, as illustrated in Figure 7-26(c). The magnetic anomaly due to such a feature is

$$Z_A = Z_{A_{top}} - Z_{A_{bottom}} = 2\pi I - 2\pi I = 0$$

Our next step is to truncate the slab as illustrated in Figure 7-27. As in our gravity example, we can demonstrate that the total included angle can be substituted for π in Eq. 7-45, so that the vertical-field anomaly for the truncated thin plate is

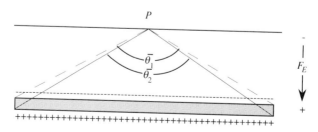

Figure 7-27 Notation used to derive the expression for a thin plate of limited extent.

$$Z_A = Z_{A_{top}} - Z_{A_{bottom}} = 2I\bar{\theta}_1 - 2I\bar{\theta}_2 = 2I(\bar{\theta}_1 - \bar{\theta}_2) \qquad (7\text{-}46)$$

One of the most interesting applications of these relationships is to determine magnetic anomalies due to basement that is overlain by nonmagnetic material. One example of such a situation is a thick sequence of Paleozoic sedimentary rocks overlying Precambrian metamorphic and igneous rocks in the midcontinent region of the United States. Another example consists of a thin cover of Pleistocene sediments overlying Mesozoic sedimentary rocks cut by basaltic volcanics, which is similar to the geology of the Connecticut Valley of Massachusetts and Connecticut.

Let's take a look at a representation of this latter geology and assume a vertical inducing field. Consider Figure 7-28(a). If the basement units are very thick, the effective result is a sheet of negative poles induced on the basement surface (the positive poles effectively are removed from consideration). The susceptibility of the Pleistocene sediments is so low that we can ignore their magnetic effect. The *effective susceptibility* k' of the basalt is 0.0045. Using Eq. 7-46 and remembering that the included angle is in radians, we obtain the vertical-field anomaly at P:

$$Z_A = Z_{A_{top}} - Z_{A_{bottom}} = [2(0.0045 \text{ cgs emu})(60,000 \text{ nT})(2.38)] - 0 = 1286 \text{ nT}$$

Figure 7-28(b) diagrams an instance of structural relief on basement which, although only 4 m, is significant considering the thickness of the overlying sediments. Due to this relief we also have to contend with a strip of positive poles at the level of the deeper basement surface. Therefore, the anomaly at P due to this structural relief is

$$Z_A = Z_{A_{top}} - Z_{A_{bottom}} = [2(0.005 \text{ cgs emu})(60,000 \text{ nT})(2.52)]$$

$$- [2(0.005 \text{ cgs emu})(60,000 \text{ nT})(2.38)]$$

$$= 1513 \text{ nT} - 1429 \text{ nT} = 84 \text{ nT}$$

Such results demonstrate that for situations similar to that diagramed in Figure 7-28 magnetic exploration is much more effective in mapping lithologic contrasts (assuming sufficient k-values are present) than in determining structural relief. Of course, relief on the order of that in Figure 7-28 in rocks with such a high k-value should be quite evident.

As a final exercise in deriving equations for magnetic effects, we once again apply Eq. 7-46 to the problem of a thin, semi-infinite plate or sheet. The geometry is presented in Figure 7-29 and is almost identical to a similar derivation we presented for the gravity effect of such a model. The only difference in the magnetic case is that we are dealing with two included angles due to negative and positive poles. Once again refer to the derivation in Chapter 6 for details. In this case

$$\bar{\theta}_1 = \frac{\pi}{2} + \tan^{-1}\left(\frac{x}{z}\right), \quad \bar{\theta}_2 = \frac{\pi}{2} + \tan^{-1}\left(\frac{x}{(z+t)}\right)$$

and so

$$Z_A = 2I(\bar{\theta}_1 - \bar{\theta}_2) = 2kF_E\left[\frac{\pi}{2} + \tan^{-1}\left(\frac{x}{z}\right) - \frac{\pi}{2} - \tan^{-1}\left(\frac{x}{z+t}\right)\right]$$

or

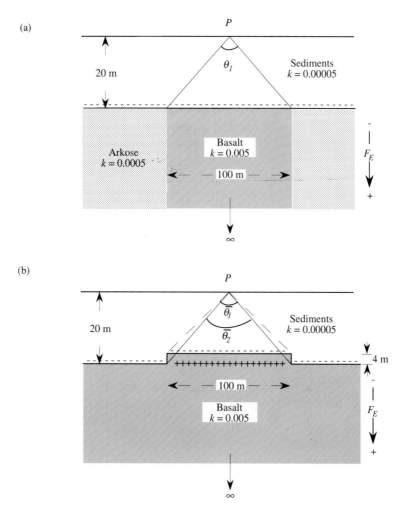

Figure 7-28 Comparison of the magnetic effects of a lithologic contrast and structural relief.

Figure 7-29 Relationships and notation for derivation of semi-infinite sheet equation.

$$Z_A = 2kF_E \left[\tan^{-1}\left(\frac{x}{z}\right) - \tan^{-1}\left(\frac{x}{z+t}\right) \right] \qquad (7\text{-}47)$$

Finally, we place this equation in a dynamic table, Table 7-4. This table is organized the same way as Table 6-9, so we not only can produce curves for a semiinfinite sheet as in Figure 7-29 but also can simulate faulted sheets (Fig. 7-30) as we did in the gravity case. For the representation of Table 7-4 the magnitude of the vertical inducing field is set at 60,000 nT.

The results from Table 7-4 are illustrated in Figure 7-30. Let's focus on the curve for sheet A for a moment. Note the pronounced asymmetry (remember that this is for the verti-

TABLE 7-4 Magnetic Effect of Thin Semi-infinite Sheets

	Sheet A			Sheet B
Depth to sheet (m)	5	Depth to sheet (m)		10
Sheet thickness (m)	1	Sheet thickness (m)		1
Susceptibility (cgs emu)	0.001	Susceptibility (cgs emu)		0.001
Horizontal increment (m)	2	Horizontal increment (m)		2

Horizontal Position (m)	Z_A(nT)	Horizontal Position (m)	Z_A(nT)
24	4.75	−24	−4.20
22	5.13	−22	−4.44
20	5.58	−20	−4.70
18	6.10	−18	−4.97
16	6.71	−16	−5.24
14	7.42	−14	−5.49
12	8.26	−12	−5.67
10	9.21	−10	−5.71
8	10.19	−8	−5.51
6	10.88	−6	−4.93
4	10.41	−4	−3.81
2	7.05	−2	−2.11
0	0.00	0	0.00
−2	−7.05	2	2.11
−4	−10.41	4	3.81
−6	−10.88	6	4.93
−8	−10.19	8	5.51
−10	−9.21	10	5.71
−12	−8.26	12	5.67
−14	−7.42	14	5.49
−16	−6.71	16	5.24
−18	−6.10	18	4.97
−20	−5.58	20	4.70
−22	−5.13	22	4.44
−24	−4.75	24	4.20

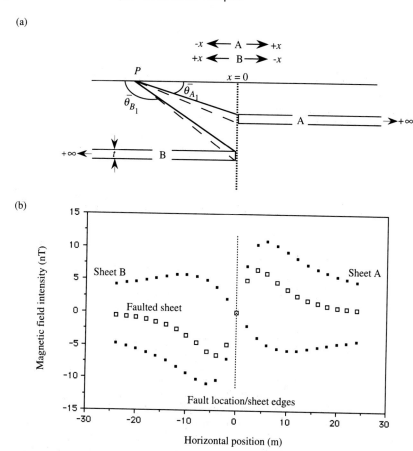

Figure 7-30 Using semi-infinite sheets to simulate a faulted horizontal bed. (a) Notation used in Table 7-4 to derive magnetic anomaly values. (b) Graph illustrating effects of sheet A, sheet B, and sum of A and B (the faulted relationship shown in (a)).

cal-field anomaly Z_A) and the steep slope directly over the edge of the sheet. The form of the curve for Z_A is sufficiently different from that of the monopole or dipole that we might suspect some sheetlike subsurface form. However, the curve for the faulted sheet is so similar in form that it seems unlikely we could demonstrate the existence of a fault without additional evidence. Matters become even more complex if we do not restrict the inducing field to vertical. Representative curves for the general solution to faulted sheets can be consulted in Telford et al. (1990, p. 100–102).

Magnetic Effects of Polygons with Infinite Strike Length (Using MagModel)

At this point we have investigated anomaly curves for a monopole, dipole, sphere, and thin horizontal sheets. After some study, curves for the first three forms should make reasonable sense. You probably also realize that the anomaly patterns on contour maps for these shapes are roughly equidimensional, as Figure 7-22 demonstrated. The contour pattern over the edge of a semiinfinite sheet is marked by a steep gradient and has a pronounced linear trend which is parallel to the strike of the sheet's edge. We now turn to an analysis of forms that have infinite strike length and a cross-sectional shape of a polygon. Such shapes provide a good approximation to real forms that have strike lengths on the order of 10 times their width. To investigate anomalies related to such shapes we turn to a computer program, MagModel (see Appendix E), which is extremely similar in operation and design to GravModel.

The detailed mathematics used in MagModel is in Talwani and Heirtzler (1964, p. 464–480) and is not presented here. The approach, however, is not all that different than those presented in this and the previous chapter. The goal is to derive a formula for a semi-infinite two-dimensional prism. The first step is to define the magnetic potential of a small rod with dimensions $dx \, dy \, dz$. The vertical and horizontal magnetic field strengths of this small rod are then determined by taking the appropriate derivatives. After these equations are in hand, the small rod is used to sweep out a lamina through integration, and the lamina is used to create a prism. Effects of a polygon can then be determined by obtaining anomalies for a number of prisms and summing both positive and negative contributions to the polygon anomaly.

In our preceding discussion of magnetic anomalies we investigated the differences in curve form for vertical-, horizontal-, and total-field anomalies and learned how these curves vary with inclination of the earth's field. We also made a number of simplifying arrangements such as orienting our dipole parallel to a magnetic meridian. Of course, in the real world we won't be so fortunate, so our main emphasis in this section is to study the effect of model orientation on anomaly magnitude and form. We begin with a vertical prism and are primarily concerned with the relationship between prism strike and the direction of magnetic north. In order to keep everything relatively simple, we present curves only for Z_A. After you have a good understanding of the following material, create total-field curves for the same input parameters and see if you can explain resulting curve forms. For the material that follows we utilize several files included on disk: *zero magnetics data—2 km*, *shallow vertical prism*, and *inclined shallow prism*.

First, let's orient the vertical prism so its strike is parallel to a magnetic meridian. We choose a total-field value of 52,000 nT and an inclination of 65°. The resulting curve for the vertical-field anomaly is presented in Figure 7-31(a). The curve is quite symmetrical about the midpoint of the prism top and resembles that for a vertical dipole. If we now rotate the prism through 90° so that its strike is west-east, curve shape changes and is clearly more asymmetric. Note that the only parameter changed is prism strike. Why does this produce a curve shape change and noticeable differences in field intensities? The most direct answer involves a consideration of the horizontal and vertical components of the inducing field (Fig. 7-32(a)). Z_E acts normal to the top and bottom of the prism and therefore induces a magnetization that can be presented as an accumulation of negative poles on the upper surface of the prism and positive poles on the bottom surface. H_A, however, acts normal to the ends of the prism that are infinitely far removed from the position of measurement. These

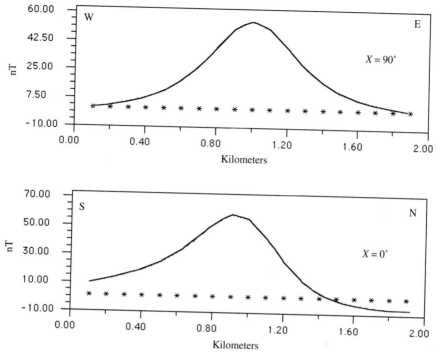

Figure 7-31 Anomaly curves (Z_A) for a vertical prism produced by MagModel. (a) Prism strike is parallel to magnetic north (X-angle = 0°; traverse direction is west-east); (b) Prism strike is perpendicular to magnetic north (X-angle = 90°; traverse direction is south-north).

pole accumulations therefore do not affect local field intensities, which are due to the effects produced by the top and bottom of the prism—hence, the symmetric curve form.

In the case of a west-east striking prism, the horizontal component of the earth's field does have an effect. Pole accumulations are induced on the top and bottom surfaces of the prism as before. Now, however, the broad, vertical side of the prism is perpendicular to H_E, and negative and positive pole accumulations will be induced on these sides. Negative poles are indicated on the side closest to an observer (shaded surface in Fig. 7-32(b)), but the positive poles are not shown on the far side so as not to clutter the diagram. These pole accumulations are not as dense so those produced by Z_A because H_A is significantly less than Z_A when the inclination is 65°. For a north-south traverse over this prism, the pole accumulations on the vertical sides add to the effect of those on the top and bottom and produced the observed asymmetry.

Next let's take the prism with north-south strike (and symmetrical curve) and incline it to the west (Fig. 7-33(b)). Curve form now becomes markedly asymmetric with a long "tail" in the direction of dip. What is different? We can analyze this in the same qualitative way that we used for prism strike. H_A still has no effect for this prism orientation. However,

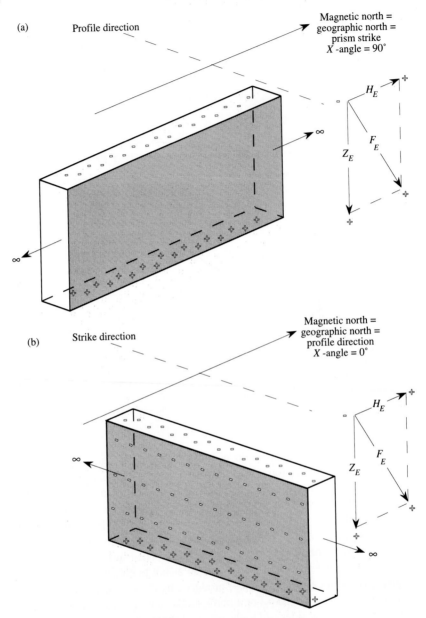

Figure 7-32 Prism orientations responsible for curves in Figure 7-31. (a) Corresponds to Figure 7-31(a), and (b) corresponds to Figure 7-32(b).

Z_A now induces pole accumulations on the inclined sides of the prism as well as on its top and bottom. Although the accumulations on the inclined sides can be viewed as much more sparse than those on the top and bottom, they create a small but detectable effect as we pro-

(a)

(b)

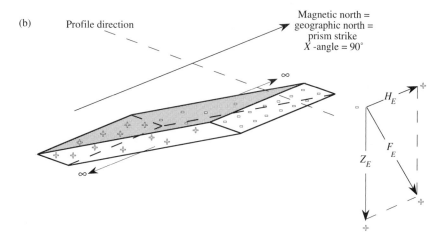

Figure 7-33   Effects of an inclined prism. (a) Anomaly curve (Z_A) for an inclined prism produced by MagModel. (b) Prism strike is parallel to magnetic north (*X*-angle = 0°; traverse direction is west-east).

ceed from west to east along the traverse. This effect is felt a considerable distance from prism top and bottom and explains the pronounced asymmetry of the anomaly curve.

You likely now are overwhelmed with the myriad possibilities that affect anomaly curve form and anomaly magnitudes. Gravity anomalies may appear as a panacea! In magnetic exploration we not only contend with shape and depth of subsurface bodies but also must consider the effect of the inclination of the earth's field, the orientation of the subsurface body, and its dip if it has a tabular or sheetlike form. Add to these variables the fact that we deal mostly with total-field anomalies, and you should appreciate why a detailed interpretation of profiles and maps of magnetic data is viewed as a demanding exercise.

INTERPRETATION OF MAGNETIC DATA

Lest our previous statement be construed as overly negative in regard to the likely utility of magnetic prospecting, let's take a bit more detailed look at some of the pluses and minuses (other than pole concentrations!) of this exploration method.

Disadvantages and Advantages

We face the same difficulty as we did in gravity interpretation when attempting to match model curves with field curves in that there are no unique solutions. The need for independent geologic and geophysical controls is just as important, if not more so, when interpreting magnetic maps and profiles. We also face a difficulty not present in gravity surveys—that of remanent magnetization. If sufficient remanence is present, the anomaly we view is a combination of this remanence and the induced magnetization and may be very different in form from what we would see if no remanence existed. Because data on remanent magnetization characteristics most often are absent, a degree of uncertainty commonly is added to many interpretations.

Not only is the range of magnetic susceptibility quite variable even for one rock type, but there is no guarantee that the minerals responsible for the susceptibility are uniformly distributed throughout the rock mass. In one basalt flow with which we worked susceptibility is quite high due to abundant magnetite. However, the magnetite is concentrated in pockets distributed throughout the flow, which would give somewhat discontinuous anomalies if mapped in great detail on a large scale. On the other hand, these pockets are distributed throughout the extent of the flow, so that the overall magnetic signature of this unit is quite clear.

The common practice of working with total-field measurements does nothing to reduce the difficulties of interpreting magnetic data. Since the direction of the magnetic vector is not known when these data are collected, a number of data-processing techniques that are utilized in gravity work either cannot be applied or are much more difficult to apply. In concluding this list of disadvantages, we emphasize once again the dependence of anomaly characteristics on direction of magnetization.

Possibly the greatest advantage of the magnetic method is cost relative to area surveyed. This is especially true when aeromagnetic surveying is considered but is also true for ground surveys. This low relative cost coupled with its high precision makes it a very attractive method for several purposes even if quantitative results are difficult to achieve.

Although the local orientation of the earth's field affects anomaly form, this orientation is, of course, essentially constant for all surveys in a given area even when the area is quite large. A geophysicist is therefore able to become thoroughly familiar with characteristic curve shapes for the range of geologic model forms typically employed in interpretation methods.

Because magnetic susceptibilities are low in a wide variety of rock types, it usually is easier to focus on the likely source material when large anomalies are encountered, as high susceptibilities belong to such a restricted range of rocks. Of course, some external geologic information is necessary, but a reduced set of such information often is sufficient as compared to the task of correlating gravity anomalies with specific rock types. In addition, the simple result of finding a large anomaly (e.g., in an area of suspected mineralization) often is sufficient, and more elaborate interpretation procedures are not necessary. Good examples of utilizing anomalies to focus other exploration techniques without analysis of the anomalies themselves are presented during the discussion of case studies in a following section. Marked susceptibility contrasts also lend themselves to geologic mapping, which relies heavily on tracing of high magnetic gradients but not necessarily on detailed analysis of anomaly form.

You are already aware of the utility of *Poisson's relation* from our discussion of a sphere's magnetic effect. This ability to relate the magnetic and gravitational potentials makes it possible to take magnetic data and transform them into a *pseudogravitational* field. Not only should this field be less complex to analyze than the magnetic field from which it was derived, but it can be compared with gravity data directly if such are available. A good correlation between anomalies in the pseudogravitational and gravitational fields suggests a similar source for both and more narrowly defines the interpretation process.

Many data-processing techniques that are used in gravity work can be adapted for magnetic work, often with little additional effort. The separation of regional trends from local anomalies continues to be an important step in the interpretation process. Second-derivative methods emphasize local anomalies and remove more regional trends. Upward continuation removes anomalies from shallow sources and is particularly useful when survey objectives include basement-depth determinations. Downward continuation enhances shallow anomalies and can be used to separate overlapping anomalies from nearby shallow sources. Now might be a good time to review the descriptions of these approaches in Chapter 6.

Quantitative Interpretation Techniques

Half-Maximum Techniques.   As in the case of gravity, we can use half-maximum techniques to provide a reasonably good value with which to begin other modeling endeavors provided we have some information about the body responsible for the observed anomaly. The procedure for developing such half-maximum relationships is identical to those followed for gravity anomalies, but owing to the complexity of anomaly shape from even simple subsurface shapes, the process provides even more of an approximation than in the gravity case.

As a straightforward example consider the case of the slender vertical rod with bottom far removed from the observer. Recall that this shape can be represented by a monopole. If we assume our traverse is directly over the rod, Eq. 7-22 simplifies to

$$Z_A = \frac{z(kF_E A)}{\left(x^2 + z^2\right)^{3/2}} \tag{7-48}$$

Because the maximum is directly over the center of the rod, one-half this maximum value is

$$Z_{A_{(1/2)\,max}} = \frac{(kF_E A)}{2z^2} \tag{7-49}$$

At a position where the maximum is one-half its value, we have

$$Z_{A_{(1/2)\,max}} = \frac{\left(kF_E A\right)\left(z^2 + x^2_{(1/2)\,max}\right)^{1/2}\cos\theta}{\left(z^2 + x^2_{(1/2)\,max}\right)^{3/2}} \tag{7-50}$$

since $\cos\theta = \dfrac{z}{r}$ and $r = \left(z^2 + x^2_{(1/2)\,max}\right)^{1/2}$ (see Fig. 7–16).

As before, we equate Eqs. 7-49 and 7-50 such that

$$\frac{\left(z^2 + x^2_{(1/2)\,max}\right)^{1/2}\cos\theta}{\left(z^2 + x^2_{(1/2)\,max}\right)^{3/2}} = \frac{1}{2z^2} \tag{7-51}$$

This reduces to

$$2z^3 = \left(z^2 + x^2_{(1/2)\,max}\right)^{3/2} \tag{7-52}$$

and, finally,

$$0.766\,z = x_{(1/2)\,max} \quad \text{or} \quad z = \frac{x_{(1/2)\,max}}{0.766} \tag{7-53}$$

Relationships for several other shapes include (Telford et al., 1990, p. 85–102):

> *Sphere*—total width of anomaly curve at $Z_{A_{max}}/2$ roughly equals depth to sphere center.
> *Cylinder*—total width of anomaly curve at $Z_{A_{max}}/2$ roughly equals depth to cylinder center.
> *Semi-infinite sheet*—depth roughly equals one-half the distance from $Z_{A_{max}}$ to $Z_{A_{min}}$.

Slope Methods. Figure 7-34(a) pictures Z_A curves for three vertical prisms with tops at different depths. As you no doubt would expect, the prism closest to the surface produces the curve with the greatest amplitude and greatest slopes. We've already observed similar effects in our study of gravity anomalies, so perhaps you are wondering if some precise relation exists between curve shape characteristics and source depth. Many geophysicists explored this possibility and developed a number of useful techniques for analysis of anomaly curves. As you also might expect, due to the complexities inherent in magnetic anomaly analysis, most of these methods provide only approximations, work only in somewhat restricted circumstances, and are not universally applicable. Nevertheless, slope methods are widely applied in efforts to determine source depths and do give useful first approximations. These approximations may be sufficient or may be used as a starting point for more sophisticated analysis. Two of the most frequently cited approaches are those of Peters (1949) and Vacquier et al. (1951). In order to provide some sense of how these methods work, we present a summary of Peters' solution to the depth problem.

Peters' slope method utilizes Z_A curves related to vertical prisms by first identifying the maximum slope on the anomaly curve (Fig. 7-35). The next step is to create a line that has a slope equal to one-half the maximum slope value. This "one-half-maximum slope" line is moved about on the anomaly curve until its two points of tangency with the curve are identi-

(a)

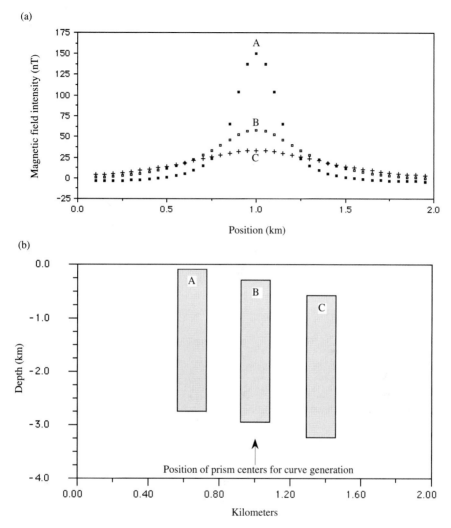

Figure 7-34 (a) Magnetic anomaly curves (Z_A) for vertical prisms with tops at various depths. (b) Depths of prisms used to generate curves in (a).

fied (Fig. 7-35). The distance on the horizontal axis between these two points of tangency is denoted as d. The normal relationship between depth to the top of the prism z and d is

$$z = \frac{d}{1.6} \qquad (7\text{-}54)$$

In Figure 7-35 $d = 0.43$ km, which results in a value for z of 0.3 km. This corresponds to the depth of the top of prism B in Figure 7-34(b), which was used to generate the anomaly curve.

Figure 7-35 Illustration of the Peters slope method as used to analyze curve B in Figure 7-34.

This approach assumes that prism length is great compared to width, that the depth to prism top is the same order of magnitude as prism width, and that the prism has infinite strike length and is oriented parallel to a magnetic meridian. Naturally, all these constraints rarely are realized, so that in most cases Eq. 7-54 provides at best an approximation. Peters does note various adjustment that can be made to this equation, such as changing the value 1.6 to 1.2 if prism width is very narrow.

The Peters slope method has been widely applied in spite of its stated constraints and limitations. For instance, total-field anomalies can be analyzed at high magnetic latitudes, as they are so similar to vertical-field anomalies in these regions. In this as in other similar methods a great deal of experience is necessary in order to apply slope approaches with an expectation of useful estimates to anomaly depths.

Computer Modeling. As in the interpretation of gravity anomalies, modeling programs such as MagModel or ones that accept three-dimensional models often are used for magnetic interpretation. Anomaly values calculated by the model are compared with measured values in order to evaluate the model which presumably is based as much as possible on geologic and/or geophysical controls.

When a satisfactory fit is achieved, the model represents one possible interpretation of the subsurface configuration responsible for the observed magnetic measurements. We emphasize once again in conclusion that many different models can produce similar magnetic anomalies. Unique solutions do not exist. As geologic and geophysical controls diminish, the number of solutions increase rapidly. Thus, interpretative techniques such as those outlined above are extremely useful in supplying appropriate boundary conditions. However, the most useful and reliable analyses will utilize every bit of information available from well logs and surface mapping as well as from other geophysical studies. This is why it is so important to be a good geologist as well as a good geophysicist and why no geophysical approaches should be neglected in the quest to penetrate beyond the surface and to shed light on the complexities that await beneath.

APPLICATIONS OF THE MAGNETIC METHOD

The past examples of geophysical applications featured reasonably routine cases of using geophysics to shed light on geological problems. In this concluding chapter we thought it might be useful as well as interesting to document somewhat different approaches as well as the more routine.

Archaeological Surveys

A number of geophysical techniques are useful in archaeological studies, but resistivity and magnetics have a special utility in locating buried archaeological sites and objects. Magnetic techniques are especially useful due to the high susceptibility of iron objects often associated with ancient sites and the strong thermoremanent magnetism produced during manufacture of bricks, tiles, pottery, and kilns and hearth use.

Standard surveys typically employ proton-precession magnetometers, which are capable of revealing significant details under carefully controlled field conditions. Readings normally are collected along a grid with station spacings on the order of 1 m. This fairly dense station network is necessary to separate anomalies due to "trash" from those with archaeological significance. The sensing element of the magnetometer usually is kept quite close to the ground—approximately 0.5 m. Because of the station density and number of readings collected (due to the ease of field work with the proton-precession magnetometer) computer data storage often is mandatory. Figure 7-36 illustrates the result of such a survey. The major anomaly, that of a burned and filled pit house, has an average anomaly of 27 nT (Weymouth, 1986, p. 546).

An example of an especially detailed survey is highlighted in Figure 7-37. This survey employed an adaptation of the flux-gate magnetometer we described previously. Two detectors were rigidly mounted and aligned in order to measure the *gradient* of the vertical-field component. Readings were taken continuously along a specified traverse direction and plotted on an *xy*-recorder. If numerous traverses are aligned side-by-side as pictured in Figure 7-37, an extremely detailed site map is produced. This amount of detail makes it much easier to differentiate anomalies of true archaeological significance from peaks due to recent rubbish, small accumulations of magnetic minerals in soils, and other disturbing

0   10 m

Figure 7-36 Total-field magnetic map from the Dolores Archaeological Program, Colorado. Contour interval is 1 nT. The main anomaly is from a burned and filled pit house. (From Weymouth, John W., 1986; Archaeological site surveying program at the University of Nebraska: Geophysics, v. 51, p. 538–552.)

influences. In Figure 7-37 the two large circles are ditches of Bronze Age round barrows. The upper circle contains an inner ditch and a central grave pit (Clark, 1986, p. 1407). Such wonderful detail illustrates what magnificent dividends can result from a little ingenuity and careful attention to the basic fundamentals of geophysical exploration techniques.

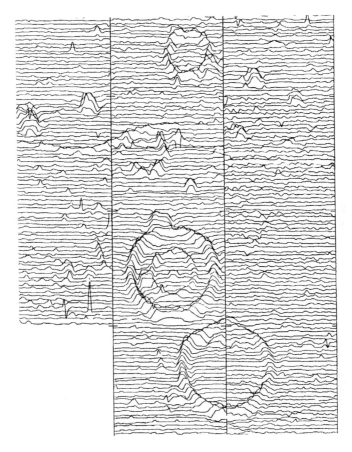

Figure 7-37 Results of a flux-gate magnetometer gradiometer survey at Radly, Oxfordshire, England. The major circular features are ditches of Bronze Age burial mounds. (From Clark, Anthony J., 1986, Archaeological geophysics in Britain: Geophysics, v. 51, p. 1404–1413.)

Detection of Voids and Well Casings

At first it might not seem logical to search for caves or voids using the magnetic method. However, if rocks have sufficient magnetic susceptibility, then a void within these rocks produces a negative susceptibility contrast. An excellent example of a study located in an area with such a contrast is presented in Arzate et al. (1990). This study describes a magnetic survey at Teotihuacan, Mexico, conducted to map tunnels in the Oztoyohualco area. Standard magnetic survey procedures included the use of a proton-precession magnetometer, measurements taken 2 m above ground level, and removal of a regional field. Residual magnetic lows such as those illustrated in Figure 7-38 (positions 2 and 3) correlate well with known tunnel locations. Another pronounced low (position 1) is believed to be due to a previously unknown tunnel location. The susceptibility of the pyroclastic rocks in which the tunnels are located is approximately 0.006 cgs emu. Using a susceptibility contrast of -0.006 cgs emu and an average magnitude of 300 nT for the anomaly at position 1, the authors of the study arrived at a tunnel size of 4.5 m radius and depth to center of 7.7 m (Arzate et al., 1990, p. 159).

Such interpretations appear straightforward, but note that they are possible only because previous independent knowledge of the subsurface was available which enabled correlation of anomalies with known sources. Even so, the identification of the anomaly at position 1 in Figure 7-38 would be considerably strengthened by conducting an additional survey with a geophysical method known to be capable of detecting voids (such as the studies described involving electrical resistivity and gravity).

Magnetic methods are, as you might expect, very efficient at detecting iron and steel objects buried at depths of only a few meters. Often, buried barrels and old well casings are of great concern, since these frequently pose a significant contamination hazard.

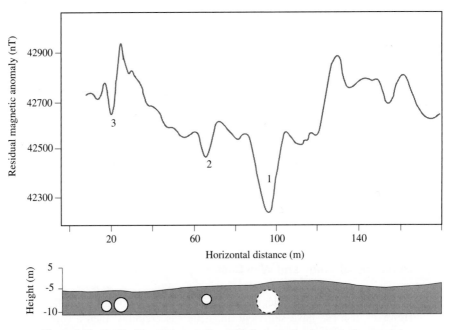

Figure 7-38   Residual magnetic anomaly profile in nanoteslas demonstrating correlation of magnetic lows with known tunnels (solid circles at positions 3 and 2) and revealing an undiscovered tunnel (dashed circle at position 1). (Modified from Arzate, J. A., Flores, L., Chávez, R. E., Barba, L. and Manzanilla, L., 1990, Magnetic prospecting for tunnels and caves in Teotihuacan, México *in* Ward, Stanley H., ed., Geotechnical and environmental geophysics; Volume 3: Geotechnical: Society of Exploration Geophysicists Investigations in Geophysics No. 5, p. 1–30.)

Frischknecht and Raab (1984) demonstrated that abandoned well positions can be identified by the magnetic method. Figure 7-39 is taken from this study and confirms that well casings provide an identifiable magnetic signature even at distances of several tens of meters above the well.

Defining Landfill Geometry

In the preceding chapter we described a detailed investigation of the Thomas Farm landfill designed to assess the effectiveness of the gravity method in studying landfills. As part of this same study, magnetic data were collected using a proton-precession magnetometer

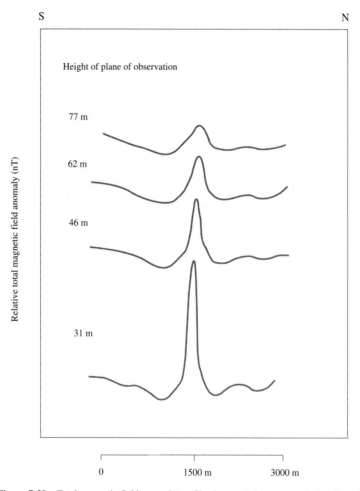

Figure 7-39 Total magnetic field anomaly profiles in nanoteslas measured at various elevations above a cased well at Piney Creek, Colorado. (Modified from Frischknecht, F. C., and Raab, P. V., 1984, Location of abandoned wells with geophysical methods: Environmental Protection Agency 600/4-84-085.)

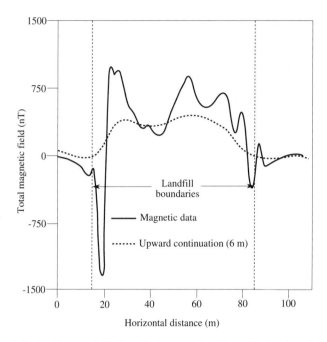

Figure 7-40   Total magnetic field profile in nanoteslas. The profile is oriented along an east-west line and crosses the southern third of the Thomas Farm landfill. (Modified from Roberts, R. L., Hinze, W. J., and Leap, D. I., 1990, Data enhancement procedures on magnetic data from landfill investigations *in* Ward, Stanley H., ed., Geotechnical and environmental geophysics, Volume 2: Environmental and groundwater: Society of Exploration Geophysicists Investigations in Geophysics No. 5, p. 261–266.)

(Roberts et al., 1990). Stations were located on a 2- m by 2- m grid, observations were at 1 m above ground level, and data were corrected for diurnal variations. The magnetic pattern is characterized by intense magnetic variations. Figure 7-40 provides a sample of the amplitudes of these variations, which are due to buried iron and steel objects. Some variations also are due to surface objects beyond landfill boundaries. A 6-m upward continuation of the magnetic data is plotted in Figure 7-40. As this upward continuation subdues the effect of shallow sources, the 6-m curve provides a more generalized view of landfill character. This should be apparent when you compare the form of the 6-m curve and the location of landfill boundaries (see Fig. 7-40).

    This point marks the end of the last chapter in this text. Before closing, however, we urge you to reread Chapter 1 and to pay particular attention to the discussions about fundamental considerations, defining objectives, limitations of geophysical methods, and the advantage of using multiple methods. These words perhaps will be more significant now than when you first read them. Consider the admonitions carefully and, above all, remember the importance of linking geophysical interpretations to geologic information.

PROBLEMS

7-1 What is the horizontal gradient in nT/m of the earth's vertical field (Z_E) in an area where the horizontal field (H_E) equals 20,000 nT and the earth's radius is 6.3×10^8 cm?

7-2 Use Eq. 7-2 as a basis for deriving an equation for the magnetic effect of a thin, vertical, infinitely long rod magnetized along its axis (monopole).

7-3 A buried stone wall constructed from volcanic rocks has a susceptibility contrast of 0.001 cgs emu with its enclosing sediments. The wall is very long compared to its other dimensions. Approximate the wall with a cylinder and assume the anomaly from a sphere is sufficiently close to that of a cylinder to allow a reasonable estimate of field magnitudes. Determine the wall's detectability with a typical proton-precession magnetometer when $i = 50°$.

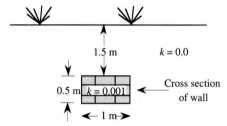

7-4 How large would an iron artifact ($k = 1.0$) have to be for detection with a proton-precession magnetometer?

7-5 Could a total-field magnetic survey detect the illustrated burial chamber (spherical void) in a region where $F_E = 55,000$ nT and $i = 70°$?

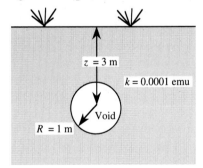

7-6 Determine the vertical-field anomaly (Z_A) for the intrusive diagrammed. The intrusive has a very long strike length. F_E is vertical and equal to 55,000 nT.

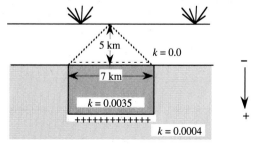

7-7   The following magnetic data represent vertical-field measurements (Z_A) in an area where shallow crystalline basement is overlain by nonmagnetic sediments. The basement gneisses are intruded by numerous thin kimberlite pipes. Both gneisses and kimberlite are eroded to a common, level surface. Determine the likely depth to basement. $F_E = 58{,}000$ nT and $i = 80°$.

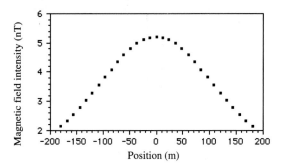

7-8   The curve illustrated is from a magnetic traverse in an area of numerous basalt flows. A 25-m-deep drillhole at position "-15 m" encountered no basalt. Comment as fully as possible on the near-surface geology.

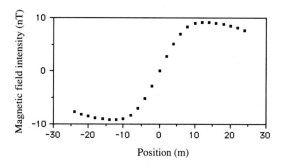

7-9   Discuss the wisdom of using the Peters slope method to determine depth to the top of a vertical prism in the following situation. The vertical prism is 2 km wide and has a large strike length and large vertical length. The prism strikes N70W and has a susceptibility contrast of 0.0013 cgs emu. Data for the earth's field at the prism location is $F_E = 50{,}000$ nT, $d = 0°$, and $i = 70°$. Only a proton-precession magnetometer is available for survey use.

7-10   The following data represent total-field anomaly intensities in a region where tilted basalt flows are known to exist. The data is also on disk as file *7-10 data*. Important magnetic field information includes: $F_E = 55{,}000$ nT, $i = 65°$, $d = 20°$, and $k = .001$ cgs emu. The basalt flows generally strike N70E, so the traverse direction (X-angle) for the total-field measurements is 340°. Comment on the direction of dip and likely dip amount of the flow over which the data was gathered.

Traverse Position (km)	Total-Field Anomaly Intensity (nT)
0.1	-0.5
0.2	0.2
0.3	1.4

Traverse Position (km) *(cont.)*	Total-Field Anomaly Intensity (nT) *(cont.)*
0.4	3.6
0.5	7.5
0.6	14.9
0.7	28.4
0.8	47.9
0.9	60.0
1.0	51.7
1.1	32.9
1.2	17.9
1.3	8.7
1.4	2.9
1.5	-0.9
1.6	-3.6
1.7	-5.4
1.8	-6.8
1.9	-7.8

REFERENCES CITED

Arzate, J. A., Flores, L., Chávez, R. E., Barba, L. and Manzanilla, L., 1990, Magnetic prospecting for tunnels and caves in Teotihuacan, México *in* Ward, Stanley H., ed., Geotechnical and environmental geophysics; Volume 3: Geotechnical: Society of Exploration Geophysicists Investigations in Geophysics No. 5, p. 1–30.

Clark, Anthony J., 1986, Archaeological geophysics in Britain: Geophysics, v. 51, p. 1404–1413.

Dobrin, Milton B., and Savit, Carl H., 1988, Introduction to geophysical prospecting, 4th ed.: New York, McGraw-Hill Book, 896 p.

Frischknecht, F. C., and Raab, P. V., 1984, Location of abandoned wells with geophysical methods: Environmental Protection Agency 600/4-84-085.

Gupta, V. K., and Fitzpatrick, M. M., 1971, Evaluation of terrain effects in ground magnetic surveys: Geophysics, v. 36, p. 582–589.

Parasnis, D. S., 1979, Principles of applied geophysics, 3rd ed., London, Chapman and Hall, 275 p.

Peters, Leo J., 1949, A direct approach to magnetic interpretation and its practical application: Geophysics, v. 14, p. 290–320.

Roberts, R. L., Hinze, W. J., and Leap, D. I., 1990, Data enhancement procedures on magnetic data from landfill investigations *in* Ward, Stanley H., ed., Geotechnical and environmental geophysics, Volume 2: Environmental and groundwater: Society of Exploration Geophysicists Investigations in Geophysics No. 5, p. 261–266.

Robinson, Edwin S., and Coruh, Cahit, 1988, Basic exploration geophysics: New York, John Wiley, 562 p.

Talwani, M., and Heirtzler, J. R., 1964, Computation of magnetic anomalies caused by two-dimen sional structures of arbitrary shape: Computers in the mineral industries, Part 1, Stanford University Publication, p. 464–480.

Telford, W. M., Geldart, L. P., Sheriff, R. E., and Keys, D. A., 1976, Applied geophysics: Cambridge, Cambridge University Press, 860 p.

Telford, W. M., Geldart, L. P., and Sheriff, R. E., 1990, Applied geophysics, 2nd ed.: Cambridge, Cambridge University Press, 770 p.

Vacquier, Victor, Steenland, Nelson C., Henderson, Roland G., and Zietz, Isadore, 1951, Interpretation of aeromagnetic maps: Geological Society of America Memoir 47, 151 p.

Weymouth, John W., 1986, Archaeological site surveying program at the University of Nebraska: Geophysics, v. 51, p. 538–552.

SUGGESTED READING

Breiner, S., 1973, Applications manual for portable magnetometers: Palo Alto, Geometrics, 58 p.

Nagata, Takesi, 1961, Rock magnetism: Tokyo, Maruzen Publishing Company Ltd., 350 p.

Nettleton, L. L., 1971, Elementary gravity and magnetics for geologists and seismologists: Society of Exploration Geophysicists Monograph Series, No. 1, 121 p.

APPENDICES

APPENDIX A

RefractModel

RefractModel produces time-distance data for a subsurface model as defined by the user. The data are presented both in a travel-time curve and in a table of values. The data are examined "as is" or you can save the data, transfer to RefractSolve, and analyze the data.

The instructions in the following appendices assume that you possess basic knowledge of how to use the Macintosh.

It is possible to run RefractModel either by double-clicking on a preexisting data file produced by the program or by double-clicking on the program icon. Assuming you enter the program directly, select **New** from the **File** menu. Figure A-1 illustrates all menu items. Bold arrows point at particularly important items. After **New** is selected, the first data-input dialog box will appear (see Figure A-2). The information requested by the dialogs is self-evident. You can move through the dialog sequence at any time by clicking on the arrow buttons, on **Cancel**, or on **OK** depending where you are in the sequence. It is easiest to move through the data-entry boxes in each dialog by pressing the **Tab** key, but you also may point and click as in any Macintosh word-processing operation. After you are satisfied with your data input, you should click on the **Done** button.

Nothing will occur until you select **Plot Graph** from the **File** menu. This selection creates the data graph (travel-time curve) and the data table (Fig. A-3). After examining the results of your input parameters, you can select **Revise Data** from the **Edit** menu, transfer

(a) These all are standard Macintosh commands with the exception of Plot Graph. No calculations will take place at any time until Plot Graph is selected.

(b) These are standard Macintosh menu items with the exception of Revise Data. Revise Data can be selected at any time. After data is revised, Plot Graph must be selected before any computations will take place.

(c) These items determine which window is active.

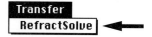

(d) Units may be changed at any time. No change will take effect until Plot Graph is selected. Once this selection is made, all displays will use the new units, and all input data will be converted.

(e) Selecting this menu item transfers you to RefractSolve. You are given an option to save the time-distance data for the current model and a subsequent option to cancel the transfer.

Figure A-1   Menu items from RefractModel.

to RefractSolve by selecting **RefractSolve** from the **Transfer** menu, or select **Quit** from the **File** menu.

A major point to remember is that no computations will occur after any data revision or file opening until you select **Plot Graph**. Also, the **Save** command saves the model parameters, not the time-distance data. To save the time-distance data select **RefractSolve** from the **Transfer** menu. You then will be asked if you want to save the time-distance data for the current model. After saving the data, you then can elect to transfer or can elect to stay in RefractModel and try another model.

When using RefractModel for the first time, you may wish to utilize the following data. Accept all values in the initial data-input dialog box, except change the number of geophones to 24. Change the number of layers to 3 in the second dialog box. Type in the following values for layer parameters:

Thickness	Velocity	Dip
10	500	4
15	1500	−5
—	5000	—

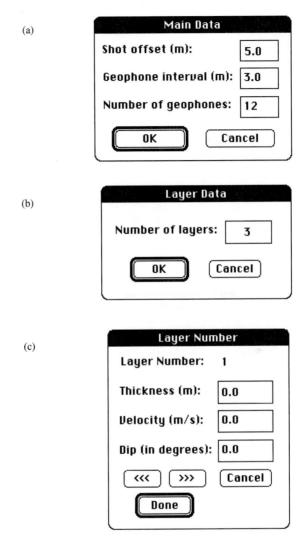

(a)

(b)

(c)

Figure A-2   Data input dialog boxes for RefractModel illustrated in the order they appear in the program when selecting **New** from the **File** menu or **Revise Data** from the **Edit** menu. The arrow buttons in (c) move through the data for each layer.

RefractSolve

RefractSolve allows you to fit lines to time-distance data in order to determine a subsurface structure based on standard refraction seismology principles. Many menu items function in the same manner as in RefractModel. You either can open a preexisting data file or create a

(a)

(b)

Site 2a-90 Data

Number of geophones : 12
Shot offset (m):   5.0
Geophone spacing (m):   5.0

Layer:	Thickness (m):	Velocity (m/s):	Angle (°):
1	20.0	500.0	0.0
2	Infinite	4000.0	----

Geophone:	Forward Time (ms):	Reverse Time (ms):
1	10.0	10.0
2	20.0	20.0
3	30.0	30.0
4	40.0	40.0
5	50.0	50.0
6	60.0	60.0
7	70.0	70.0
8	80.0	80.0
9	90.0	90.0
10	91.9	91.9
11	93.1	93.1
12	94.4	94.4

Figure A-3 Window displays from RefractModel. (a) Graph: time-distance data for a given model. (b) Table: model parameters and resultant times for each geophone.

new file by selecting the appropriate menu item (Fig. A-4). When creating a new file or revising an existing file, you encounter two dialog boxes.

The first, illustrated in Figure A-5(a), is straightforward. The shot point is assumed to be in line with the geophone array. Each geophone position is calculated from this information. The second dialog seeks first-arrival time data for each geophone. Reverse times are requested if you indicated that these should be included when answering the appropriate dialog query. When entering geophone time data, be sure you understand the geophone numbering system used in this program. Geophones are numbered relative to the shot point for both the forward and reverse traverses. Therefore, geophone 1 for the forward shot occupies the same position as geophone 12 for the reverse shot if the geophone string consists of 12 geophones. In the case of Figure A-5(b) the forward time is for the eighth geophone from the forward shot, and the reverse time is for the eighth geophone from the reverse shot. If this system still is not clear, generate some time-distance data using RefractModel, save it, enter RefractSolve, and examine it by selecting **Revise Data**.

(a) These menu items are the same as in RefractModel. Draw Structure functions similarly to Plot Graph.

(b) These menu items are identical to those in RefractModel.

(c) This governs which travel-time curve data the user can fit lines to.

(d) These items determine which window is active.

(e) These menu items function similarly to those in RefractModel.

(f) These menu items govern operations on lines in the travel-time curve window.

Figure A-4   Menu items from RefractSolve.

(a)

(b)

Figure A-5 Data-input dialog boxes for RefractSolve illustrated in the order they appear in the program when selecting **New** from the **File** menu or **Revise Data** from the **Edit** menu. The arrow buttons in (b) move through the geophone data.

Once you open an existing file or select **Done** when creating a new file, the data are plotted on a travel-time curve, and the cursor changes to a cross-hair symbol. Note that the forward data are highlighted, and the reverse data are dimmed. You now fit lines to the forward data by drawing lines with the mouse just as you draw lines in any other Macintosh program. You must draw your lines by proceeding from the shallowest to deepest layer (by proceeding from left to right across the time-distance plot when working on forward-traverse data). After fitting lines to the forward-traverse data, select **Reverse** from the **Direction** menu. The forward-traverse data now are dimmed, and the reverse-traverse data are highlighted. Fit lines as before, moving from right to left.

Figure A-6(a) illustrates a ravel-time curve with four lines fit to the data. You may access the **Line** menu at any time in order to adjust the lines you have drawn. Select the option you desire and click on the line you wish to edit. **Move** allows you to move the whole line, which keeps its original orientation. **Edit** allows you to change the orientation or length of a line by dragging on one of the handles (black square) present at each end of the line. Remember that you can edit only the part of the travel-time curve selected in the **Direction** menu.

(a)

(b)

Figure A-6 Window displays from RefractSolve. (a) Graph: interpretation of time-distance data illustrating lines drawn by user. (b) Structure: generalized geologic structure computed from lines drawn on travel-time curve. (c) Table: listing of apparent velocities, intercept times, and computed subsurface structure.

(c)

Site 4b-89 Table				
Input Data		**Spread Length = 100 m**		
Layer	Apparent Forward Velocities (m/s)	Apparent Reverse Velocities (m/s)	Forward Intercept Times (msec)	Reverse Intercept Times (msec)
1	1518.28	1526.12	0.0	0.0
2	4346.67	6095.12	12.8	19.6

Computed Structure						
Layer	Velocity	For. Thickness	Rev. Thickness	Dip	For. Depth	Rev. Depth
1	1522.2	10.2	15.7		10.2	15.7
2	5067.5			-3.02		

Figure A-6 *(continued)*

When everything is to your liking, select **Draw Structure** in the **File** menu. After a pause for computation, the screen display will consist of three windows containing the travel-time curve, the computed structure, and a data table. These may be manipulated in the same fashion as any Macintosh window.

As was the case when working with RefractModel, whenever you make a change in your line arrangements on the travel-time curve, no computation will occur until you select **Draw Structure**. **Save** saves time-distance data, not the travel-time curve with lines you drew. The only way to save a record of your line fits is to obtain a printed copy.

Both RefractModel and RefractSolve use Adachi's formula (Adachi, 1954). The output format and conventions used in RefractSolve are similar to those published by Mooney (1977, p. 11-1–11-8), since these are in common use. A good exposition of an approach similar to RefractModel is presented in Sverdrup (1986).

APPENDIX B

ReflectModel

ReflectModel produces time-distance data for a subsurface model as defined by the user. The results are presented as a simulated seismogram and in a table of values. The data can be examined as is, or you can save the data, transfer them to ReflectSolve, and analyze them.

It is possible to run ReflectModel either by double-clicking on a preexisting data file produced by the program or by double-clicking on the program icon. Assuming you enter the program directly, select **New** from the **File** menu. Figure B-1 illustrates all menu items. Bold arrows point at particularly important items. After **New** is selected, the first data-input

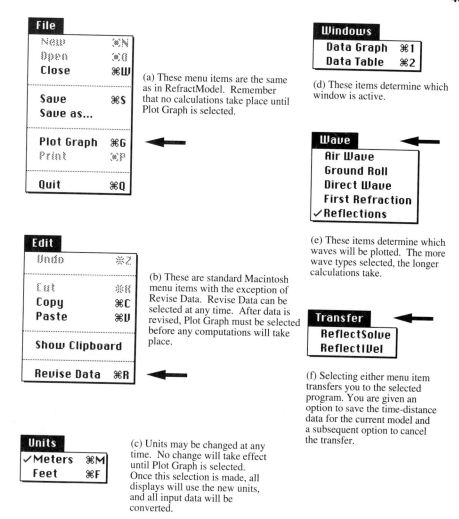

Figure B-1   Menu items from ReflectModel.

dialog box will appear (see Figure B-2). The information requested by the dialogs is almost identical to that required when running RefractModel.

Once again, no computations take place until you select **Plot Graph** from the **File** menu. This selection creates the seismogram and the data table (Fig. B-3). After examining the results of your input parameters, you can select **Revise Data** from the **Edit** menu, transfer to ReflectSolve by selecting **ReflectSolve** from the **Transfer** menu, or select **Quit** from the **File** menu. You also have the option of transferring to ReflectIVel.

As in RefractModel, the **Save** command saves the model parameters, not the time-distance data. To save the time-distance data select **ReflectSolve** from the **Transfer** menu.

(a)

(b)

(c)

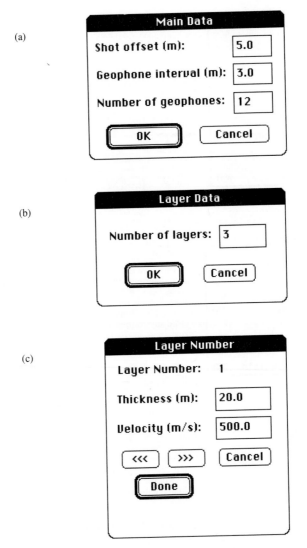

Figure B-2 Data-input dialog boxes for ReflectModel illustrated in the order they appear in the program. These function in precisely the same way as those in the refraction programs described in Appendix A.

You then will be asked if you want to save the time-distance data for the current model. After saving the data, you then can elect to transfer or can elect to stay in ReflectModel and try another model.

The major difference in the menu items in ReflectModel as compared to RefractModel is the **Wave** menu. Time-distance data will be computed for all waves selected in the menu, and all waves will appear on the seismogram and in the data table. Because

Figure B-3 Window displays from ReflectModel. (a) Graph: simulated seismogram for a given model; (b) Table: model parameters and resultant wave arrival times for each geophone.

(a)

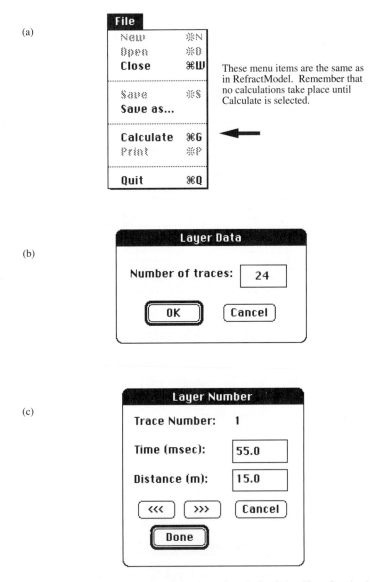

These menu items are the same as in RefractModel. Remember that no calculations take place until Calculate is selected.

(b)

(c)

Figure B-4   File menu and data-input dialog boxes from ReflectSolve. These function just as in the refraction programs and in ReflectModel.

of the calculations involved, computation time can be noticeable if several waves and more than 12 geophones are chosen.

When using ReflectModel for the first time, you may wish to utilize the following data. Accept all default values in the menu items and in the initial data-input dialog box.

(a)

(b)

Figure B-5   Window displays from ReflectSolve. (a) Graph: line fit to x^2 - t^2 data. (b) Table: time-distance values, squared values, and solution.

Change the number of layers to 3 in the second dialog box. Type in the following values for layer parameters:

Thickness	Velocity
5	500
25	1500
—	5000

(a)

(b)

Figure B-6 (a) Data-input dialog from ReflectIVel. The "lower" reflector is deeper than the "upper" reflector, and its time is therefore greater. (b) The table containing the results based on input data.

If the total seismogram record will exceed 250 ms, a dialog requests the length of the record desired. Because details are most readily observed in a 250-ms record, this usually is the one of choice. Arrivals exceeding the selected time simply are not plotted. This usually is not a concern, as the arrival times of ground roll is the most common reason for a long record when dealing with typical shallow horizons.

ReflectSolve

ReflectSolve is an automated x^2 - t^2 method and fits a line to squared time-distance data in order to determine depth and velocity to a reflecting interface. Menu items and data-input dialogs function in the same manner as in the several programs already described (Fig. B-4).

Program output consists of an x^2 - t^2 graph and a table containing time-distance data, these data squared, equation of the line fit to the data points, and velocity and depth information (Fig. B-5).

As is the case when working with all reflection and refraction programs, whenever you make a change in your input data, no computation will occur until you select **Calculate** in the **File** menu. **Save** saves time-distance data.

ReflectIVel

ReflectIVel computes interval velocities and thicknesses based on the Dix equation. Input and output are quite simple (Fig. B-6). All operations function the same as in other programs included with this text. Note that the "upper" reflector is the shallower reflector and that "lower" connotes deeper.

The idea for this particular suite of reflection programs is based on similar programs written for Apple II computers by Lankston (1986).

APPENDIX C

ERModel

ERModel calculates apparent resistivities at preselected electrode spacings for an input model consisting of layer thicknesses and resistivities. Results are posted on a graph and in a table (Fig. C-2). The data can be saved and transferred to ERSolve for practice analysis much in the way RefractModel data can be transferred to RefractSolve.

ERModel functions in much the same way as the applications described in appendices A and B. The only difference in the menu bar is the presence of the **Array** menu (Fig. C-1(a)). ERModel will calculate apparent resistivities for either the Wenner or Schlumberger electrode configurations. The default choice is the Wenner array.

Data-input dialog boxes also function similarly to those in other programs. The first box to appear when **New** is selected from the **File** menu, or **Revise Data** is selected from the **Edit** menu is illustrated in Figure C-2(b). This *Spacing Start* dialog allows you to choose the initial electrode spacing (the smallest spacing). All spacings are preselected so as to occupy equal intervals on a logarithmic axis. This is why the horizontal-axis tick marks on the graph plotted by the program (Fig. C-2(a)) are spaced at even intervals even though the axis is a logarithmic one.

The program will handle up to 10 layers, but, realistically, the maximum useful number is 4 or 5 layers. Calculation times are insignificant up to 4 layers, so one should experiment freely with this program in order to derive a feel for apparent resistivity curves. As before, no calculations occur until **Plot Graph** is selected from the **File** menu.

Calculation of apparent resistivities follows the general scheme used by program "RESIST" (Davis, 1979).

(a)

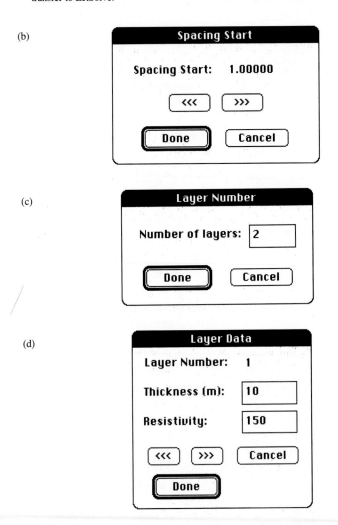

Menus in ERModel are virtually identical to those in programs described in Appendices A and B. The only exceptions are the Array menu, which is self-explanatory, and the Transfer menu, which facilitates transfer to ERSolve.

(b)

(c)

(d)

Figure C-1   Selected menus and data-input dialog boxes from ERModel. (a) Menus which are slightly different than those in other programs. (b) Dialog box to select initial electrode spacing. (c) and (d) These function exactly as programs described in Appendices A and B.

(a)

(b)

Figure C-2   Window displays from ERModel. (a) Plot of apparent resistivity versus electrode spacing. (b) Table listing apparent resistivity values for model.

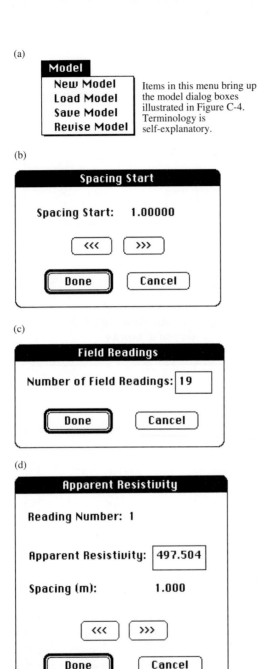

(a)

Model
- **New Model**
- **Load Model**
- **Save Model**
- **Revise Model**

Items in this menu bring up the model dialog boxes illustrated in Figure C-4. Terminology is self-explanatory.

Iterations
- 3
- ✓6
- 12
- 18
- 24
- 30

Normally, 12 or 18 iterations should be selected for best results. For a quick look at model results, select 3 or 6 iterations.

(b)

Spacing Start

Spacing Start:　1.00000

《《《　　》》》

Done　　Cancel

(c)

Field Readings

Number of Field Readings: 19

Done　　Cancel

(d)

Apparent Resistivity

Reading Number: 1

Apparent Resistivity:　497.504

Spacing (m):　　1.000

《《《　》》》

Done　　Cancel

Figure C-3 Menu items and field-data dialog boxes from ERSolve. (a) Menu for operations on model. (b), (c), and (d) Dialog boxes to enter apparent resistivity data. Electrode spacing selected in (b), number of observations entered in (c), and apparent resistivities entered in (d). These function identically to others already described in Appendices A and B.

(a)

(b)

Figure C-4   Dialog box for entering model parameters in program ERSolve. Both (a) and (b) function like other dialogs covered in prior pages. The only exception is in (b), where model parameters can be fixed by clicking on the appropriate radio button.

ERSolve

Program ERSolve accepts apparent resistivities acquired with either the Wenner or Schlumberger electrode configuration and a subsurface model believed to be responsible for the observed data. Apparent resistivities are calculated as in program ERModel. These are then compared with the field data. If the root-mean-square (rms) error is greater than the value set in the first model dialog box (Fig. C-4(a)), the model is adjusted and another trial ensues. This process continues until the number of selected iterations is exceeded or the rms value falls below the designated value.

In general, menu items and data input for ERSolve function similarly to applications discussed in Appendices A and B. **Save** or **Save As** place field data in a file for future access. Field data is entered using the dialog boxes illustrated in Figure C-3(b), (c), and (d).

(a)

(b)

Figure C-5 Results from ERSolve. (a) Graph (b) Data table.

Spacings are assumed to be equally spaced on a logarithmic scale. These cannot be altered. The first dialog sets the electrode spacing for the first apparent resistivity value. Remaining input procedures are straightforward.

One addition to the menu bar is the **Model** item. You can type in a new model, load a model created previously and saved while working either in ERSolve or ERModel, or revise a model. The dialog boxes should be familiar. The first requests the number of layers in the model and the rms cutoff value (Fig. C-4(a)). Figure C-4(b) illustrates the second dialog in the input sequence. ERSolve provides an option to fix layer parameters so that these are not changed as the program attempts to find a model that fits the observed field data. The thickness and/or resistivity of each layer is fixed simply by clicking on the appropriate radio button.

Note that although ERSolve can load models created by itself or by ERModel, models are available only by using the **Model** menu item. Double-clicking on a model icon always launches ERModel. Of course, models created or revised by ERSolve and saved can later be opened by ERModel.

The other addition to the menu bar is the **Iterations** menu. Computations will continue until the rms error is less than specified or the number of iterations is exceeded. Because computation normally is time-consuming, selecting a small number of iterations provides a way to check on model curves without waiting for extended periods of time. Once it appears that a reasonable model is entered, it is a good idea to set the iterations to at least 12.

No computation can occur until both field data and a model are available, so the menu item **Plot Graph** in the **File** menu remains dimmed until both items are loaded or typed in. As in previous programs, **Plot Graph** must always be selected before any calculations take place.

Program output consists of a graph and a data table (Fig. C-5). Field data points on the graph are designated by an asterisk symbol. Values computed by ERSolve for the input model are represented by a solid line. The data table lists the initial input model, the last model computed by ERSolve, and a comparison of field values and model values.

Calculation of apparent resistivities follows the general scheme used by program "INVERSE" (Davis, 1979).

APPENDIX D

GravModel

GravModel calculates Bouguer anomaly values for solids of infinite strike length and polygonal cross section. Model form is created by selecting polygon corners for a section perpendicular to the long dimension of the solid. Gravity values then are computed at points above the polygon that are defined in a data file. Results are presented in a table and on a graph where curves are drawn both for computed gravity and Bouguer values measured in the field.

Many of the features and much of the behavior of GravModel resemble those in the programs described in Appendices A, B, and C. Most of this material is omitted here because it is assumed that you have already used the programs described in these appendices. The more important menu items are illustrated in Figure D-1(a).

The first step in using GravModel is to create a data file (or to open an existing data file) that contains observed Bouguer gravity values and their horizontal (and vertical) positions. Polygons cannot be drawn until a data file exists. Dialog boxes for creating or revising data are shown in Figure D-2(a). They function similarly to dialogs in programs you have already used. Model length and depth control the dimensions contained in the Graph and Polygon Windows, so these values should be selected carefully. Both these windows contain the same horizontal scale. Bouguer gravity values for your model are calculated only at the horizontal positions contained in the data file.

If you simply want to observe model results, you can create a data file for which all measured Bouguer values equal zero. Two data files are contained on the same disk as GravModel that you can use for this purpose: *zero data—2 km* and *zero data—20 km*. These

(a)

(b)

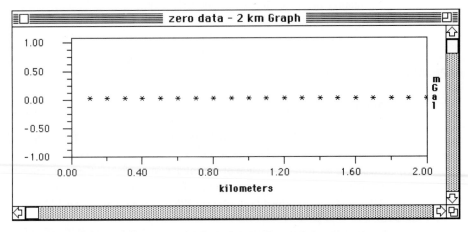

Figure D-1 (a) Important menu items in GravModel; (b) graph display using data file *zero data—2 km*.

(a)

Figure D-2 Screen displays from GravModel. (a) Data-input dialog boxes. (b) Polygon
window containing five polygons.

contain 20 points spaced evenly within the range specified in the file name. A plot of *zero
data—2 km* is illustrated in Figure D-1(b).

Of course, the other option is to supply actual Bouguer anomaly values. The positions
of real field values must be relative to the 0.0 km position, which always is at the far left in

(a)

(b)

Figure D-3 GravModel computation results. (a) Graph window illustrating Bouguer gravity values for model in Figure D-2(b). (b) Table window containing data used to plot curves in (a).

all windows that illustrate horizontal coordinates. Data files can be revised at any time by selecting **Revise Data** in the **Edit** menu.

After a data file is present, models can be created by drawing polygons in the Polygon Window. This window is activated by clicking on it if a portion is visible or by selecting **Polygon** from the **Window** menu. To begin drawing simply press the mouse button at a desired starting position, release the button, and move the mouse to the next position. Continue this process until the polygon is completed. Moving the cross-cursor close to the starting position and clicking *once* will complete and close the polygon. It then will be filled with a pattern to signify it is complete.

Dotted lines on the vertical and horizontal axes which move as the mouse moves help show the general position of the mouse. A coordinate readout is visible in the lower left of the window after the first coordinate position is chosen. The most efficient procedure in drawing polygons is to place corners in their approximate final positions and then to select **Edit** from the **Polygon** menu to fine-tune corner positions. The next step is to click on the polygon of interest. This provides an outline of the polygon with small black squares at the corners. Place the cursor on a square and drag to place it at the desired position. Polygons sometimes balk at being selected for editing. If this occurs, first click in an unoccupied portion of the screen before clicking on the polygon of interest.

Because polygons cannot intersect, it is important to plan polygon positions before drawing on the screen. Polygons can be moved by selecting **Move** from the **Polygon** menu, but in general cannot be moved past other polygons that block their path. Individual polygons can be erased by selecting **Erase** and clicking on a polygon. Of course, at any time **New Model** can be selected from the **Edit** menu. Polygon corners cannot be placed beyond the outline in the Polygon Window. Since you can't draw off-screen, if you want to create, for example, a semi-infinite sheet, select a reasonably large model length in the first data-input dialog box (Fig. D-2(a)) and then confine data points within a more restricted range. There is no special command to revise a model, as the current model is in the Polygon Window and can always be changed by using the options in the Polygon menu.

After your model is complete, select **Plot Graph** in the **File** menu. You then will be prompted for a density contrast for each polygon in your model. Polygon density contrasts are requested in the order polygons were created. The fill pattern of each polygon is used as the identifying feature. The Graph Window contains field-data points as asterisks and a line which connects the Bouguer anomaly values calculated for the model in the Polygon Window. Remember that values are calculated only at field-data points.

The **Save** command in the **File** menu refers to the field-data file. Models also can be saved and reopened. They are not tied to the field-data file open at the time the model was created. However, model files can be opened only from within GravModel and must be opened after a data file is active. Double-clicking on model files will not open GravModel, but double-clicking on a field-data file will open GravModel.

The general scheme of GravModel follows the gravity computation approach of Talwani et al. (1959) as implemented by Salem et al. (1973).

APPENDIX E

MagModel

MagModel calculates vertical-, horizontal-, and total-field anomaly values for solids of infinite strike length and polygonal cross section. Model form is created by selecting polygon corners for a section perpendicular to the long dimension of the solid. Magnetic anomaly values then are computed at points above the polygon that are defined in a data file. Results are presented in a table and on a graph where curves are drawn for both computed magnetic values and values measured in the field.

Most of the features and almost all the behavior of MagModel resemble those in pre-

(a)

(b)

Figure E-1 (a) Important menu items in MagModel. (b) Data-input dialog boxes.

viously described programs and are especially similar to GravModel. Most of this material is omitted here, as it is assumed that you already have used the other programs described in these appendices. New menu items are illustrated in Figure E-1(a). One or more anomaly components can be selected for plotting by clicking on the appropriate item(s) listed in the **Field** menu. In addition this program computes anomaly curves for induced magnetizations or remanent magnetizations. The choice is indicated in the **Magnetization** menu.

Polygon drawing and editing functions similarly to GravModel except that only one polygon can be drawn in the Polygon Window for a given computation (Fig. E-2). After constructing a polygon and choosing **Plot Graph**, data-input dialog boxes appear (Fig. E-1(b)). All requests for values should be quite clear except, possibly, the first entry in the second dialog, which specifies "Profile direction (normal to polygon strike)." The orientation of the strike of the polygon is controlled by the profile direction which is perpendicular to polygon strike. The profile-orientation parameter (referred to as the X-angle because it

Figure E-2 Polygon window containing a dipping prism. If *d* = 0 and *X* = 0, then magnetic north equals geographic north, and the profile direction is north-south with north to the right.

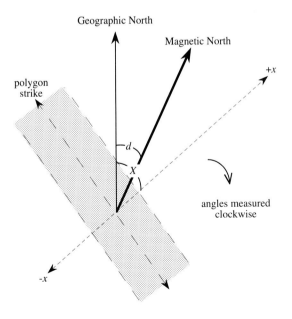

Figure E-3 Diagram illustrating orientation information for data-input dialog boxes in MagModel.

(a)

(b)

Figure E-4 MagModel computation results. (a) Graph window illustrating magnetic anomaly values for model in Figure E-2. (b) Table window containing data used to plot curves in (a).

controls the direction of the +*x* axis) is clearly indicated in Figure E-3. All angles are measured clockwise from geographic north. Thus, the *X*-angle in the diagram is 45°. Default values for declination *d* and *X*-angle both are zero. This orients both magnetic north and the positive end of the *x* axis parallel to geographic north.

If remanent magnetization is specified, an additional dialog box appears (not illustrated). This requests information for the orientation and magnitude of the magnetization vector, all of which should be evident.

After all dialogs are completed, computation takes place, and the Graph Window and Table Window are available for study (Figs. E-3 and E-4).

The general scheme of MagModel follows the magnetics computation approach of Talwani and Heirtzler (1964).

REFERENCES CITED

Adachi, R., 1954, On a proof of fundamental formula concerning refraction method of geophysical prospecting and some remarks: Kumamoto Journal of Science, Series A, v. 2, p. 18–23.

Davis, Philip A., 1979, Interpretation of resistivity data: Computer programs for solutions to the forward and inverse problems: Minnesota Geological Survey Information, Circular 17, 23 p.

Lankston, Robert M., 1986, SEIS: SYNTHESEIS, X2T2, INTERVEL—Three microcomputer programs for exercises with the seismic reflection method: *in* Burger, H. Robert, Ed., Personal computer software for geological education: National Association of Geology Teachers Special Publication No. 1, p. 64–81.

Mooney, Harold M., 1977, Handbook of engineering geophysics: Minneapolis, Minnesota, Bison Instruments.

Salem, Bruce, B., Harbaugh, John W., Abry, Claude G., and Crichlow, Henry, 1973, Interactive CRT display of Bouguer gravity models in computer-assisted instruction in geology: Report to the National Science Foundation, 179 p.

Sverdrup, Keith Allen, 1986, A program for calculating refraction times for layered models with arbitrarily dipping discontinuities: Journal of Geological Education, v. 34, p. 262–264.

Talwani, M., and Heirtzler, J.R., 1964, Computation of magnetic anomalies caused by two-dimensional structures of arbitrary shape: Computers in the mineral industries, Part 1, Stanford University Publication, p. 464–480.

Talwani, M., Worzel, J. W., and Landsman, M., 1959, Rapid gravity computations for two-dimensional bodies with application to the Mendocino submarine fracture zone: Journal of Geophysical Research, v. 64, p. 49–59.

Index